ENVIRONMENTAL RESEARCH ADVANCES

BIOFUELS AND THE ENVIRONMENT

IMPACT ASSESSMENTS AND MITIGATION OPPORTUNITIES

ENVIRONMENTAL RESEARCH ADVANCES

Additional books in this series can be found on Nova's website under the Series tab.

Additional E-books in this series can be found on Nova's website under the E-book tab.

ENERGY SCIENCE, ENGINEERING AND TECHNOLOGY

Additional books in this series can be found on Nova's website under the Series tab.

Additional E-books in this series can be found on Nova's website under the E-book tab.

ENVIRONMENTAL RESEARCH ADVANCES

BIOFUELS AND THE ENVIRONMENT

IMPACT ASSESSMENTS AND MITIGATION OPPORTUNITIES

TOBIAS M. FISCHER
EDITOR

nova publishers
New York

For permission to use material from this book please contact us:
Telephone 631-231-7269; Fax 631-231-8175
Web Site: http://www.novapublishers.com

Library of Congress Cataloging-in-Publication Data

ISBN: 978-1-62948-092-3

Published by Nova Science Publishers, Inc. † New York

CONTENTS

PREFACE

This book examines the first of the Environmental Protection Agency's (EPA's) triennial reports on the current and potential future environmental impacts associated with the requirements of the Clean Air Act. It reviews environmental and resource conservation impacts, as well as opportunities to mitigate these impacts, at each stage of the biofuel supply chain: feedstock production, feedstock logistics, biofuel production, biofuel distribution, and biofuel use. The information included here is considered foundational for future efforts to quantitatively compare the environmental impacts of alternative scenarios for meeting the goals of the RFS2 program, representing the best available information through July 2010 and reflecting the current understanding about biofuel production and use, including input from the U.S. Departments of Agriculture and Energy.

Chapter 1 – This is the first triennial Report to Congress required under Section 204 of the 2007 Energy Independence and Security Act (EISA). EISA increases the renewable fuel standards (RFS) to 36 billion gallons per year by 2022. Section 204 requires an assessment of environmental and resource conservation impacts of the RFS program. Air and water quality, soil quality and conservation, water availability, ecosystem health and biodiversity, invasive species, and international impacts are assessed, as well as opportunities to mitigate these impacts. Feedstocks compared include corn starch, soybeans, corn stover, perennial grasses, woody biomass, algae, and waste. Biofuels compared include conventional and cellulosic ethanol and biodiesel. This report is a qualitative assessment of peer-reviewed literature.

This report concludes that (1) the extent of negative impacts to date are limited in magnitude and are primarily associated with the intensification of corn production; (2) whether future impacts are positive or negative will be determined by the choice of feedstock, land use change, cultivation and conservation practices; and (3) realizing potential benefits will require implementation and monitoring of conservation and best management practices, improvements in production efficiency, and implementation of innovative technologies at commercial scales. This report provides a foundation for comprehensive environmental assessments of biofuel production.

Chapter 2 - Federal policy has played a key role in the emergence of the U.S. biofuels industry. Policy measures include minimum renewable fuel usage requirements, blending and production tax credits, an import tariff, loans and loan guarantees, and research grants. One of the more prominent forms of federal policy support is the Renewable Fuel Standard (RFS)—whereby a minimum volume of biofuels is to be used in the national transportation fuel supply each year. This report describes the general nature of the RFS mandate and its implementation, and outlines some emerging issues related to the continued growth of U.S.

biofuels production needed to fulfill the expanding RFS mandate, as well as the emergence of potential unintended consequences of this rapid expansion.

Congress first established the RFS with the enactment of the Energy Policy Act of 2005 (EPAct, P.L. 109-58). This initial RFS (referred to as RFS1) mandated that a minimum of 4 billion gallons be used in 2006, rising to 7.5 billion gallons by 2012. Two years later, the Energy Independence and Security Act of 2007 (EISA, P.L. 110-140) greatly expanded the biofuel mandate volumes and extended the date through 2022. The expanded RFS (referred to as RFS2) required the annual use of 9 billion gallons of biofuels in 2008, rising to 36 billion gallons in 2022, with at least 16 billion gallons from cellulosic biofuels, and a cap of 15 billion gallons for corn-starch ethanol.

In addition to the expanded volumes and extended date, RFS2 has three important distinctions from RFS1. First, the total renewable fuel requirement is divided into four separate, but nested categories—total renewable fuels, advanced biofuels, biomass-based diesel, and cellulosic biofuels—each with its own volume requirement. Second, biofuels qualifying under each category must achieve certain minimum thresholds of lifecycle greenhouse gas (GHG) emission reductions, with certain exceptions applicable to existing facilities. Third, all renewable fuel must be made from feedstocks that meet an amended definition of renewable biomass, including certain land use restrictions.

The Environmental Protection Agency (EPA) is responsible for establishing and implementing regulations to ensure that the nation's transportation fuel supply contains the mandated biofuels volumes. EPA's initial regulations for administering RFS1 (issued in April 2007) established detailed compliance standards for fuel suppliers, a tracking system based on renewable identification numbers (RINs) with credit verification and trading, special treatment of small refineries, and general waiver provisions. EPA rules for administering RFS2 (issued in February 2010) built upon the earlier RFS1 regulations and include specific deadlines for announcing annual standards, as well as greater specificity on potential waiver requests and RIN oversight.

Over the long term, the RFS is likely to play a dominant role in the development of the U.S. biofuels sector, but with considerable uncertainty regarding potential spillover effects in other markets and on other important policy goals. Emerging resource constraints related to the rapid expansion of U.S. corn ethanol production have provoked questions about its long-run sustainability and the possibility of unintended consequences in other markets as well as on the environment. Questions also exist about the ability of the U.S. biofuels industry to meet the expanding mandate for biofuels from non-corn sources such as cellulosic biomass materials, whose production capacity has been slow to develop, or biomass-based biodiesel, which remains expensive to produce owing to the relatively high prices of its feedstocks. Finally, considerable uncertainty remains regarding the development of the infrastructure capacity (e.g., trucks, pipelines, pumps, etc.) needed to deliver the expanding biofuels mandate to consumers.

In: Biofuels and the Environment
Editor: Tobias M. Fischer
ISBN: 978-1-62948-092-3

Chapter 1

BIOFUELS AND THE ENVIRONMENT: FIRST TRIENNIAL REPORT TO CONGRESS*

Britta Bierwagen, Alice Chen, Christopher M. Clark, Rebecca S. Dodder, Robert J. Frederick, Anne Grambsch, Timothy L. Johnson, Fran Kremer, Stephen D. LeDuc, Brenda Lin, Adrea T. Mehl, Philip E. Morefield, Donna Perla, Caroline E. Ridley and Denice Shaw

ABSTRACT

This is the first triennial Report to Congress required under Section 204 of the 2007 Energy Independence and Security Act (EISA). EISA increases the renewable fuel standards (RFS) to 36 billion gallons per year by 2022. Section 204 requires an assessment of environmental and resource conservation impacts of the RFS program. Air and water quality, soil quality and conservation, water availability, ecosystem health and biodiversity, invasive species, and international impacts are assessed, as well as opportunities to mitigate these impacts. Feedstocks compared include corn starch, soybeans, corn stover, perennial grasses, woody biomass, algae, and waste. Biofuels compared include conventional and cellulosic ethanol and biodiesel. This report is a qualitative assessment of peer-reviewed literature.

This report concludes that (1) the extent of negative impacts to date are limited in magnitude and are primarily associated with the intensification of corn production; (2) whether future impacts are positive or negative will be determined by the choice of feedstock, land use change, cultivation and conservation practices; and (3) realizing potential benefits will require implementation and monitoring of conservation and best management practices, improvements in production efficiency, and implementation of innovative technologies at commercial scales. This report provides a foundation for comprehensive environmental assessments of biofuel production.

---

* This is an edited, reformatted and augmented version of the United States Environmental Protection Agency, EPA/600/R-10/183F, dated December 2011.

EXECUTIVE SUMMARY

This report is the first of the U.S. Environmental Protection Agency's (EPA's) triennial reports to Congress required under the 2007 Energy Independence and Security Act (EISA). EISA requires EPA to revise the Renewable Fuel Standard (RFS) program to increase the volume of renewable fuel blended into transportation fuel from 9 billion gallons per year in 2008 to 36 billion gallons per year by 2022. The revised standards (RFS2), finalized in 2010, establish new specific annual volume requirements for cellulosic biofuel, biomass-based diesel, advanced biofuel, and total renewable fuel in transportation fuel. Increasing the amounts of domestically produced renewable fuels addresses two goals of EISA: decreasing our dependence on foreign sources of energy and decreasing greenhouse gas emissions.

EISA Section 204 calls for EPA to report to Congress on the environmental and resource conservation impacts of the RFS program, including air and water quality, soil quality and conservation, water availability, ecosystem health and biodiversity, invasive species, and international impacts. EPA interpreted the requirements of Section 204 to be those environmental impacts beyond the noteworthy reductions in greenhouse gas emissions associated with the RFS program. For this report, EPA relies upon the existing peer-reviewed scientific literature to review the impacts and mitigation opportunities across the entire biofuel supply chain, including feedstock production and logistics and biofuel production, distribution, and use. The information included here is considered foundational for future efforts to quantitatively compare the environmental impacts of alternative scenarios for meeting the goals of the RFS2 program. Specifically, the report describes the current and potential future environmental impacts from:

- **Seven feedstocks**—The report summarizes information for the two most predominantly used, first-generation feedstocks (*corn starch* and *soybeans*) and five other second-generation feedstocks (*corn stover, perennial grasses, woody biomass, algae*, and *waste*), representing a range currently under development. Because the RFS2 puts a limit of 15 billion gallons on the amount of corn starch– derived biofuel that counts toward the volume requirement in 2022, an increased reliance on other feedstocks is predicted.
- **Two biofuels**—The report summarizes information for *ethanol* (both conventional and cellulosic) and *biomass-based diesel,* because they were the most commercially viable in 2010 and/or projected to be the most commercially available by 2022.

Overall Conclusions

Evidence to date from the scientific literature suggests that current environmental impacts from increased biofuels production and use associated with EISA 2007 are negative but limited in magnitude.

- **Environmental impacts along the supply chain are greatest at the feedstock production stage.** Most activities, processes, and products, particularly those occurring after feedstock production, are regulated and subject to limitations.

- **Current environmental impacts are largely the result of corn production.** Corn starch–derived ethanol constituted 95 percent of the biofuel produced in 2009. In general, feedstock demand has been met by diverting existing corn production or by replacing other row crops with corn, resulting in limited additional environmental impacts.

Published scientific literature suggests a potential for both positive and negative environmental effects in the future.

- **Technological advances and market conditions will determine what feedstocks are feasible, and where and how they will be cultivated.**
- **The magnitude of effects will be largely determined by the feedstock(s) selected, land use changes, and cultivation practices.**
- **Overall impacts given most plausible land use changes and production practices will likely be neutral or slightly negative.** More adverse or beneficial environmental outcomes are possible.
- **Second-generation feedstocks have a greater potential for positive environmental outcomes relative to first-generation feedstocks.** However, current production levels of second-generation biofuels are negligible and limited by economic and technological barriers.

EISA goals can be achieved with minimal environmental impacts if existing conservation and best management practices (BMPs) are widely employed, concurrent with advances in technologies that facilitate the use of second-generation feedstocks.

- **The feedstocks considered in this report all have the potential to support sustainable domestic energy production.** Realizing this potential will require implementation and monitoring of conservation and BMPs, improvements in production efficiency, and implementation of innovative technologies at the commercial scale.
- **International partnerships and federal coordination are needed to accelerate progress toward sustainable and secure energy production.**

Specific Environmental and Resource Conservation Conclusions

- **Land use.** Many potential impacts of biofuel production are the result of land use conversion. An expansion of cropland in response to demand for biofuels is projected, though not yet observed. Production of corn and soybean on land currently enrolled in the Conservation Reserve Program (CRP) will result in the most negative environmental impacts. In comparison, other land use conversions, for example CRP to perennial grasses, would have more moderate environmental impacts.
- **Water quality.** Impacts on water quality from biofuels in the United States are, and likely will be, primarily driven by fertilizer and other chemical inputs at the feedstock production stage. Impacts to date from EISA are considered moderately

negative, resulting primarily from an intensification of corn production contributing to eutrophication, coastal hypoxia, and other areas of concern. In comparison, second-generation feedstocks offer substantial opportunities for improvement regarding water quality impacts.

- **Water quantity.** Most current feedstock production does not require irrigation, but water use will increase if future production expands into drier areas. Per unit volume, water use for feedstock irrigation can be 100 to 1,000 times higher than for feedstock-to-biofuel conversion processes. Adverse water availability impacts will most likely arise in already stressed aquifers and surface watersheds.
- **Soil quality.** Biofuel feedstock production can impact soil quality through erosion, organic matter, and nutrient content. Perennial feedstocks are generally better for soil quality than annual row crops; however, feedstock impacts will be largely determined by which land use changes occur, if any. High corn stover removal rates are of particular concern due to likely increases in soil erosion and decreases in organic matter.
- **Air quality.** While there are some localized impacts, the biofuel volumes required by RFS2 have relatively little impact on national average ambient concentrations of air toxics. Further increases in the use of biofuels will impact emissions and ambient concentrations of "criteria" pollutants (pollutants for which EPA sets ambient air quality standards) and a variety of air toxic compounds. Emissions occur at all stages of the biofuel supply chain and effects will likely vary across the country. Ozone concentrations are expected to rise in many areas, although a few highly populated areas will experience reductions.
- **Ecosystem health.** Feedstock cultivation can significantly affect biodiversity through habitat conversion, especially on CRP lands, from exposure of flora and fauna to pesticides; through sedimentation and eutrophication in water bodies resulting from soil erosion and nutrient runoff, respectively; or from water withdrawals resulting in decreased streamflows.
    - **Forests, grasslands, and wetlands.** Shorter harvest intervals for short- rotation woody crops; residue harvesting; and conversion of pasture, CRP lands, or small, unregulated wetlands can decrease habitat availability and biodiversity. Moderate thinning and best management or conservation practices can increase some habitats, species diversity, and abundance.
    - **Invasive species.** Weed risk assessments predict that switchgrass and some woody crop species or varieties could become invasive in some regions, but that corn, soybean and perennial grasses such as Giant Miscanthus, pose little risk.
- **International.** Increases in U.S. biofuel production and consumption volumes may affect many different countries as trade patterns and prices adjust in response to global supply and demand. This could result in land use change and affect air quality, water quality, and biodiversity, but the location and magnitude of impacts are uncertain.

This first report summarizes and synthesizes peer-reviewed literature through July 2010. The report does not include, nor do its overall findings encompass, a life cycle analysis of greenhouse gas emissions. Such an analysis was previously done for the Regulatory Impact

Analysis (RIA) of the RFS2. EPA's RFS2 RIA found that the EISA-mandated revisions to the RFS2 program are expected to achieve an annual 138 million metric ton reduction in carbon dioxide–equivalent emissions by 2022 compared to continued reliance on petroleum-based fuels.

This first report also does not present environmental impacts relative to petroleum-based transportation fuels; such a comparison is recommended for the next report. Quantitative assessments are presented where there is sufficient scientific literature to support such an assessment; however, in most cases only qualitative assessments are feasible due to methodological and informational limitations.

Recommendations

To promote sustainable approaches, EPA recommends:

- Incorporating the environmental impacts of biofuel production and use described in this report into comprehensive life cycle assessments, including comparisons to fossil fuels and other energy sources.
- Ensuring the success of current and future environmental biofuel research through improved cooperation and sustained support.
- Improving the ability of federal agencies within their respective authorities to develop, implement, and monitor best management and conservation practices and policies that will avoid or mitigate negative environmental effects from biofuel production and use. This will involve coordination among diverse stakeholders, including state agencies, research scientists, and landowners.
- Engaging the international scientific community in cooperative efforts to identify and implement sustainable biofuel and land use practices that minimize environmental impact.

Because biofuel impacts cross many topics and Agency responsibilities, EPA will likely address these recommendations through continued and strengthened cooperation with state and federal agencies, including the U.S. Departments of Agriculture and Energy, and international partners.

1. INTRODUCTION

In December 2007, Congress enacted Public Law 110-140, the Energy Independence and Security Act (EISA), to reduce U.S. energy consumption and dependence on foreign oil, and to address climate change through research and implementation of strategies to reduce greenhouse gases (GHGs). In accordance with these goals, EISA requires the U.S. Environmental Protection Agency (EPA) to revise the Renewable Fuel Standard (RFS) program, created under the 2005 Energy Policy Act,[1] to increase the volume of renewable fuelrequired to be blended into transportation fuel from 9 billion gallons per year in 2008 to 36 billion gallons per year by 2022.

EPA finalized revisions to the RFS program in February 2010. The revised statutory requirements (commonly known as the RFS2) establish new specific annual volume standards for cellulosic biofuel, biomass-based diesel, advanced biofuel, and total renewable fuel that must be used in transportation fuel (see Section 2). The purpose of this report is to examine the environmental and resource conservation impacts of this change, as required under EISA Section 204.

EISA Section 204 calls for EPA to report to Congress every three years on the environmental and resource conservation impacts of increased biofuel production and use as follows:

In General. Not later than 3 years after the enactment of this section and every 3 years thereafter, the Administrator of the Environmental Protection Agency, in consultation with the Secretary of Agriculture and the Secretary of Energy, shall assess and report to Congress on the impacts to date and likely future impacts of the requirements of Section 211(o) of the Clean Air Acton the following:

1. *Environmental issues, including air quality, effects on hypoxia, pesticides,*[4] *sediment, nutrient and pathogen levels in waters, acreage and function of waters, and soil environmental quality.*
2. *Resource conservation issues, including soil conservation, water availability, and ecosystem health and biodiversity, including impacts on forests, grasslands, and wetlands.*
3. *The growth and use of cultivated invasive or noxious plants and their impacts on the environment and agriculture.*
4. *The report shall include the annual volume of imported renewable fuels and feedstocks for renewable fuels, and the environmental impacts outside the United States of producing such fuels and feedstocks. The report required by this subsection shall include recommendations for actions to address any adverse impacts found.*

A key feature of EISA is the establishment of mandatory life cycle GHG reduction thresholds. EPA used state-of-the-art models, data, and other information to assess the GHG impacts of biofuels, as described in the Final Regulatory Impact Analysis (RIA) (U.S. EPA, 2010a). To ensure that it used the best science available, EPA conducted a formal, independent peer review of key components of the analysis. The modeling of GHG emissions in the RIA (U.S. EPA, 2010a) provides a reasonable and scientifically sound basis for making threshold determinations and estimating GHG impacts. Accordingly, this report does not attempt an extensive evaluation of carbon dioxide or other GHGs, nor does it attempt to encompass GHG impacts in its conclusions. Instead, it provides complementary information to the GHG impacts described in the RIA (U.S. EPA, 2010a), which should be consulted for more information on this topic.

This is the first of EPA's triennial reports on the current and potential future environmental impacts associated with the requirements of Section 211(o) of the Clean Air Act. This report reviews environmental and resource conservation impacts, as well as opportunities to mitigate these impacts, at each stage of the biofuel supply chain: feedstock production, feedstock logistics, biofuel production, biofuel distribution, and biofuel use. This foundation supports efforts to quantitatively compare the environmental impacts of alternative scenarios for meeting the goals of the RFS2 program.

This report emphasizes domestic impacts; however, the substantial market created for biofuels by the United States, Brazil, and other countries has important global implications. For example, countries that produce (or will produce) feedstocks that are converted to biofuels that qualify for use in the United States will experience direct impacts; other countries (including the United States) will have to adapt to changing agricultural commodity distributions that result from diversion of food exports to biofuel production. As required under EISA Section 204, this report describes the impacts of increased feedstock and biofuel production in other countries as a result of U.S. policy.

This first triennial Report to Congress represents the best available information through July 2010 and reflects the current understanding about biofuel production and use, including input gained through consultation with the U.S. Departments of Agriculture and Energy. Quantitative assessments are presented, where possible, using the most recently available data through July 2010; however, in most cases only qualitative assessments were feasible due to uncertainties and lack of data and analyses in the peer-reviewed literature. The information included here is considered foundational for future efforts to quantitatively compare the environmental impacts of alternative scenarios for meeting the goals of the RFS2 program. This initial report serves as a starting point for future assessments and for taking action to achieve the goals of EISA. Future reports will reflect the evolving understanding of biofuel impacts in light of new research results and data as they become available.

1.1. Organization of This Report

Section 2 provides background on EISA, including the volume, feedstocks, and GHG thresholds required under the Act; it also describes the approach and coverage of this report. Section 3 focuses on the first stage of the biofuel production process, feedstock production, which includes cultivation and harvest. For each feedstock, this section assesses impacts on water, air, soil, and ecosystems. A summary of impacts on specific habitats is also provided. Section 4 covers the remaining stages of biofuel production: feedstock logistics and biofuel production, distribution, and use. Environmental impacts are evaluated for two main biofuels: ethanol and biodiesel. Section 5 discusses the potential impacts associated with imported biofuels. Currently, imported ethanol and biodiesel supply a highly variable, but relatively small percentage of U.S. biofuel consumption (U.S. EIA, 2009; U.S. ITC, 2010; ERS, 2010a). If this percentage increases, expanded analysis of international impacts associated with imported biofuels may be necessary in future versions of this report. Section 6 provides a synthesis and conclusions based on an assessment of the literature. Since many feedstock technologies are in the early stages of research and development, empirical and monitoring data relevant to environmental impacts are limited, and projections of their potential future use are highly speculative. Section 6 also describes recommendations for improving scientific understanding, as well as practices for minimizing environmental impacts. This report is the first step toward the capability to conduct a biofuels environmental life cycle assessment (LCA), which can provide the basis for future reports to Congress. Section 7 describes a vision for those future reports.

2. Background and Approach

2.1. EISA and RFS2 Requirements for Biofuel Production and Use

The Renewable Fuel Standard as amended by the Energy Independence and Security Act (EISA) (RFS2) establishes new specific annual volume standards for four categories of renewable fuels that must be used in transportation fuel:[5] cellulosic biofuel, biomass-based diesel, advanced biofuel, and total renewable fuel (see the glossary in Appendix A for fuel definitions). Under RFS2, conventional biofuel (i.e., ethanol derived from corn starch) with a maximum volume target and "additional renewable fuels"[6] are included as eligible fuels to meet the total renewable fuel standard. The revised statutory requirements also include new definitions and criteria for both renewable fuels and the feedstocks used to produce them,[7] including new greenhouse gas emission (GHG) reduction thresholds as determined by the life cycle assessment (LCA) that EPA conducted as part of its Regulatory Impact Analysis (RIA) during the final RFS2 rulemaking (U.S. EPA, 2010a).

Table 2-1. RFS2 Renewable Fuel Requirements (Billion Gallons per Year)[a,b]

Year	Renewable Fuel				
	Conventional Biofuel	Advanced Biofuel			Total Renewable Fuel
		Cellulosic Biofuel	Biomass-Based Diesel	Advanced Biofuel[c]	
2008	9.0	n/a	n/a	n/a	9.0
2009	10.5	n/a	0.5	0.6	11.1
2010	12.0	0.1[d]	0.65	0.95	12.95
2011	12.6	0.25	0.80	1.35	13.95
2012	13.2	0.5	1.0	2.0	15.2
2013	13.8	1.0	TBD[e]	2.75	16.55
2014	14.4	1.75	TBD[e]	3.75	18.15
2015	15.0	3.0	TBD[e]	5.5	20.5
2016	15.0	4.25	TBD[e]	7.25	22.25
2017	15.0	5.5	TBD[e]	9.0	24.0
2018	15.0	7.0	TBD[e]	11.0	26.0
2019	15.0	8.5	TBD[e]	13.0	28.0
2020	15.0	10.5	TBD[e]	15.0	30.0
2021	15.0	13.5	TBD[e]	18.0	33.0
2022	15.0	16.0	TBD[e]	21.0	36.0

[a] The requirements for cellulosic biofuel, biomass-based diesel, advanced biofuel, and total renewable fuel are minimum required volumes that must be achieved and may be exceeded. The conventional biofuel requirement is a cap that cannot be exceeded.

[b] Note that the RFS2 volume requirements are nested: cellulosic biofuel and biomass-based diesel are forms of advanced biofuel; and advanced biofuel and conventional biofuel are forms of total renewable fuel.

[c] Note that the sum of the required amounts of cellulosic biofuel and biomass-based diesel is less than the required volume of advanced biofuel. The additional volume to meet the advanced fuel requirement may be achieved by the additional cellulosic biofuel and biomass-based diesel (i.e., beyond the required minimum) and/or by other fuels that meet the definition of advanced biofuel (e.g., sugarcane ethanol).

[d] As described above, and as allowed under EISA, the EPA Administrator determined that the original RFS2 standard of 0.1 bgy for cellulosic biofuel was not achievable for 2010 and therefore decreased this standard to 5 million gallons for 2010.

[e] To be determined by EPA through a future rulemaking, but no less than 1.0 billion gallons. This requirement was designated under EISA as "to be determined" with a minimum requirement because of the uncertainty about future capacity to produce fuel that meets the biomass-based diesel definition.

Source: U.S. EPA, 2010a.

Table 2-1 shows the RFS2 annual renewable fuel standards through 2022. Total renewable fuel under the standard will increase to 36 billion gallons per year (bgy) by 2022 (of which corn starch ethanol is not to exceed 15 bgy).

While EISA establishes the renewable fuel volumes shown in Table 2-1, it also requires the EPA Administrator to set the volume standards each November for the following year based in part on information provided by the U.S. Energy Information Administration (U.S. EIA) and other data indicating the commercial capacity for producing cellulosic biofuels. EISA therefore requires the EPA Administrator to adjust the cellulosic standard, and potentially the total advanced biofuel and total renewable fuel standards, each year based on this assessment. For 2010, the Administrator adjusted the cellulosic standard from 0.1 bgy (100 million gallons per year) in RFS2 to 5.0 million gallons, but did not adjust the total advanced or total renewable fuel standard.[8] Therefore, the final 2010 standard for total renewable fuel is set at 12.95 bgy, with specific targets for cellulosic biofuel (5.0 million gallons per year), biomass-based diesel (1.15 bgy, combining the 2009 and 2010 standards as proscribed in RFS2), and total advanced biofuel (0.95 bgy).

2.1.1. Life Cycle Greenhouse Gas Thresholds

The Act established specific life cycle GHG emission thresholds for each of four types of renewable fuels, requiring a percentage improvement compared to life cycle GHG emissions for gasoline or diesel (whichever is being replaced by the renewable fuel) sold or distributed as transportation fuel in 2005. GHG LCA evaluates emissions resulting from all stages of a product's development—from growth of a feedstock to end use. These life cycle performance improvement thresholds are listed in Table 2-2.

Table 2-2. Life Cycle GHG Thresholds Specified in EISA (Percent Reduction from 2005 Baseline)

Renewable fuel[a]	20%
Advanced biofuel	50%
Biomass-based diesel	50%
Cellulosic biofuel	60%

[a] The 20 percent criterion generally applies to renewable fuel from new facilities that commenced construction after December 19, 2007.

EPA's methodology for conducting the GHG LCA included use of agriculture sector economic models to determine domestic agriculture-sector-wide impacts and international changes in crop production and total crop. Based on these modeling results, EPA estimated GHG emissions using the U.S. Department of Energy's Greenhouse Gases, Regulated Emissions, and Energy Use in Transportation (GREET) model defaults and Intergovernmental Panel on Climate Change (IPCC) emission factors. The GHGs considered in the analysis were carbon dioxide (CO_2), methane (CH_4), and nitrous oxide (N_2O). Biofuel process energy use and associated GHG emissions were based on process models for the different pathways considered. For ethanol and biodiesel, EPA's RFS2 RIA (U.S. EPA, 2010a) projected that:

- Ethanol produced from corn starch at a new, natural-gas-fired facility (or expanded capacity from an existing facility) using advanced efficient technologies will comply with the 20 percent GHG emission reduction threshold.
- Ethanol produced from sugarcane will comply with the 50 percent GHG reduction threshold for the advanced fuel category.
- Biodiesel from soybean oil and renewable diesel from waste oils, fats, and greases will comply with the 50 percent GHG threshold for the biomass-based diesel category.
- Diesel produced from algal oils will comply with the 50 percent GHG threshold for the biomass-based diesel category.
- Cellulosic ethanol and cellulosic diesel (based on the modeled pathways) will comply with the 60 percent GHG reduction threshold applicable to cellulosic biofuels.

Based on the assessment described above, EPA projected a reduction of 138 million metric tons of CO_2-equivalent emissions annually by 2022 compared to projected 2022 emissions without the EISA-mandated changes (see the RFS2 RIA [U.S. EPA, 2010a] for details).

2.1.2. Life Cycle Assessment and Environmental Impacts

LCAs evaluate environmental impacts resulting from all stages of a product's development—from feedstock production through biofuel use and disposal (see the box on the next page). As described above, EPA previously evaluated the aggregate quantity of GHG (including direct emissions and significant indirect emissions such as significant emissions from land use changes) related to the full life cycle, including all stages of fuel and feedstock production, distribution, and use by the ultimate consumer (U.S. EPA, 2010a). Currently, LCA does not include all of the environmental and resource conservation impacts required by Section 204. Extending this methodology to include these effects poses significant challenges, and could not be done for this report. However, it may be possible to draw from the considerable work that has already been done to develop LCA and other methodologies, including ecological and human health risk assessment, to assess impacts of specific biofuel products and processes for future reports.

LCA AND NET ENERGY BALANCE

There has been considerable work to develop LCA frameworks and apply them to specific products or processes, including efforts to standardize the approaches and scope of an LCA to enable comparison across similar products (ISO, 2006). Recent reviews have examined the limitations and biases that have framed a variety of LCA reports on biofuels (Von Blottnitz and Curran, 2007; Gnansounou et al., 2009; Davis et al., 2009). These highlight the importance of understanding that the results of any LCA will depend on how the boundaries for a particular analysis are set. Given the highly interconnected economic, energy, and agricultural systems involved in the production and use of biofuels, it is important to recognize that biofuel production and use will have impacts well beyond the farm-to-vehicle supply chain. Although the impacts will become incrementally smaller as distance from that specific supply chain increases (e.g., the change in steel production to meet demand for construction of biofuel conversion processes), choosing finite boundaries will necessarily exclude some impacts. The National Research Council (NRC, 2010b) used an LCA framework to do a partial analysis intended to provide detailed quantitative assessments of the comparative health and environmental benefits, risks, and cost of existing fossil fuels as well as future mixes of transportation technologies and fuels. While the analysis provided comparative bottom lines for a variety of transportation fuel production processes and uses, the authors acknowledged that they were "constrained—by the limitations of the GREET model and the scarcity of available national databases on many ecosystem impacts and other impacts—to quantify only those impacts from energy use and the air quality emissions produced during these operations." They also limited their assumptions to "reasonable speculations" and selected to report only on direct land use effects.

Net Energy Balances (NEBs) can be used to compare the energy gain or loss associated with different biofuel feedstocks or to compare biofuels and fossil fuels. Analysis of NEB does not directly address environmental impacts, but it is a relevant metric in a full evaluation of biofuels. Putting NEB in the context of other environmental impacts (e.g., environmental impacts per net energy gain/loss) can provide a metric upon which environmental impact comparisons can be made. Hill et al. (2006) estimated an NEB of 1.25 for corn ethanol, only slightly below the lowest energy gain (1.29) estimated by Hammerschlag et al. (2006). On the other hand, two studies reviewed by Farrell et al. (2006) showed a net energy loss when they included the energy associated with manufacturing of farm machinery needed for biofuel feedstock production and the construction of biofuel conversion facilities. Additional studies from U.S. Department of Agriculture (USDA) (Shapouri et al., 2002, 2010) show a continuing improvement in the NEB for corn ethanol production. The picture also improves when the energy embedded in co-products of the fuel conversion process are taken into consideration (Shapouri et al., 2010).

Despite the limitations of LCA and NEB for evaluating many of the environmental impacts specified by Section 204 of EISA 2007, they show great promise. With further development, they can become the basis for estimating and comparing the potential natural resource and environmental impacts of biofuel production and use under a range of scenarios.

2.1.3. Projected Fuel and Feedstock Use to Meet Required RFS2 Targets through 2022

Figure 2-1 summarizes the fuel types and volumes required to meet the targets through 2022, described in the RFS2 (see Table 2-1).

In 2009, corn ethanol constituted 95 percent of total U.S.-produced renewable fuel, with biodiesel made from soybean oil, other virgin vegetable oils, rendered fats, greases, and corn oil from ethanol production accounting for almost all the remaining biofuel consumed (FAPRI, 2010a; U.S. EIA, 2010). However, as technologies improve, EPA expects more advanced cellulosic feedstocks, such as agricultural residues (e.g., corn stover, sugarcane bagasse, wheat residue, sweet sorghum pulp), forestry biomass, urban biomass waste, and dedicated energy crops (e.g., switchgrass) to produce biofuels (U.S. EPA, 2010a). Present research is focused on improving technologies to convert different feedstocks to biofuels in an economically viable manner, and on determining sustainable biofuel production methods.

With respect to biodiesel, EPA expects continued use of soybean oil, which made up 54 percent of feedstock used for biodiesel in 2009 (U.S. EIA, 2010), as well as a varying percentage of other vegetable oils, rendered fats, greases, and corn oil from ethanol production through 2022 (see Table 3-2 for a more detailed breakdown) (U.S. EPA, 2010a). Algae could provide large volumes of oil for the production of biomass-based diesel. However, several hurdles, including technical issues, will likely limit production volumes between now and 2022 (U.S. EPA, 2010a).

Imported sugarcane ethanol, also represents a significant potential supply of biofuel by 2022 (U.S. EPA, 2010a). In 2009, the United States imported 198 million gallons of ethanol (U.S. EIA, n.d.[c]). Import volumes are expected to grow as U.S. demand increases to meet the biofuel targets, although this will depend on the relative costs of U.S. biofuel production and imported ethanol.

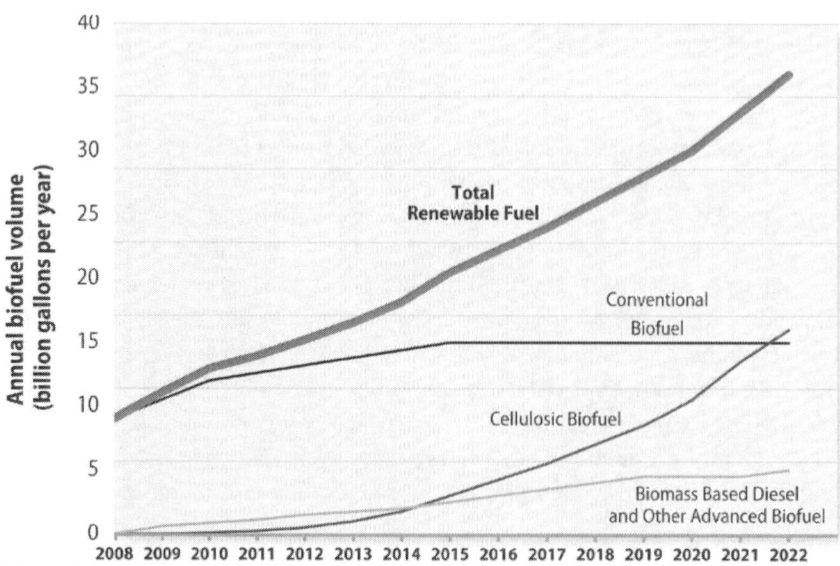

Source: U.S. EPA, 2010a.

Figure 2-1. Renewable Fuel Volumes to Meet RFS2 Targets.

2.2. Regulatory Authority Relevant to Biofuel Environmental Impacts

The EPA, in conjunction with states, tribes, and local environmental agencies, has statutory responsibility to regulate air emissions, water discharges, use of toxic substances, microbial and pesticide use, and waste disposal. Many existing environmental regulations and programs are applicable to the biofuel supply chain, including feedstock production and logistics, biofuel production and distribution, and biofuel use. Thus, the direct point source discharges and emissions associated with the biofuel supply chain are expected to be effectively controlled by existing environmental statutes. It is the impacts associated with non-point pollution and shifts in land-use patterns, however, that pose the greatest concern from an environmental perspective.

EPA's primary federal regulatory authority is derived from the Clean Air Act (CAA), the Clean Water Act (CWA), the Federal Insecticide Fungicide and Rodenticide Act (FIFRA), the Resource Conservation and Recovery Act (RCRA), and the Toxic Substances Control Act (TSCA). Under the CAA, EPA has broad direct statutory authority to regulate fuel quality and emissions from refining and production facilities for all fuels, including biofuels. The CAA also establishes limits for mobile source (vehicular) emissions. The CWA requires permits for point source discharges to waters of the United States, development of water quality standards for receiving waters, and Total Maximum Daily Loads (TMDLs) for water bodies where water quality standards have not been met. FIFRA establishes standards for storage and use of pesticides in a manner that does not harm human health or the environment. RCRA governs the generation, storage, treatment, transport, and disposal of hazardous waste. TSCA requires manufacturers and importers of new chemicals to submit "pre-manufacture" notices for EPA review prior to manufacture and commercial use of new chemicals, including new fuels, new biological materials, and new genetically engineered microorganisms used to produce biofuels or co-products. Through the CWA's Spill Prevention, Control and Countermeasure rule, EPA has enforceable regulations to control water quality impacts from spills or leaks of biofuel products and byproducts. In addition, the Safe Drinking Water Act establishes maximum contaminant levels (MCLs) for more than 90 drinking water contaminants to ensure public health. These statutes provide opportunities within the existing regulatory framework to regulate and mitigate some of the potential adverse health and environmental effects of biofuels. Selected environmental laws relevant to the production and use of biofuels are summarized in Appendix B. A detailed analysis of how each environmental statute might mitigate the direct impacts of biofuels was outside of the scope of this first report.

Generally, EPA program offices develop policies and regulations for these federal statutes, while regional EPA offices, in partnership with the states and tribes, implement these programs, ensure compliance, and enforce regulations. EPA and its regional offices work closely with states and tribes to review permit applications for new facilities and to monitor environmental impacts to ensure compliance with all permit conditions. EPA's Regional Office in Kansas City, representing the major corn-growing states, has prepared two documents to help biofuel facilities understand the full range of regulatory requirements (U.S. EPA 2007, 2008a) that can mitigate a range of direct environmental impacts when appropriately implemented.

2.3. Approach to the Section 204 Report

This report reviews the environmental implications of current and future biofuel production and use, as discussed in the published peer-reviewed literature. An extensive review was conducted on scientific literature published through July 2010, to identify impacts across the biofuel supply chain, including current and anticipated future impacts from feedstock production, feedstock logistics, and biofuel production, distribution, and use. This report summarizes much of the available information and identifies research needed to evaluate potential environmental impacts from a life cycle perspective and quantify them using more substantive and systematic assessment tools.

Any discussion of the environmental impacts of an energy source begs the question, "compared to what?" In the case of biofuels, some studies have provided comparisons to fossil fuels as a reference point for focusing on particular outputs such as GHG emissions or particulate matter emissions. In the case of other environmental impacts (e.g., water quality), most comparisons are between different biomass feedstocks (e.g., corn versus perennial grasses), rather than biomass feedstocks in aggregate compared to fossil fuels. While references to such analyses are made throughout the report, EPA decided a comprehensive, quantitative comparative analysis was beyond the scope of the current effort. Instead, EPA explored the information that will be needed to prepare for such an analysis, and allow for the successful use of methods such as LCA for evaluating and comparing full environmental impacts in future reports.

EISA 2007 mandates the use of increasing volumes of renewable fuel. Ideally, the Section 204 assessment would be based on a comparison between two projections: a baseline (or reference scenario) and an EISA 2007 scenario. There are a number of candidates for the baseline scenario. For example, the RIA (U.S. EPA, 2010a) for the RFS2 included (1) a projection of renewable fuel volumes without the enactment of EISA (e.g., the U.S. EIA's Annual Energy Outlook 2007 reference case), (2) a projection assuming the mandated renewable fuel volumes under the RFS Program from legislation preceding EISA 2007, the Energy Protection Act of 2005, and (3) a specific year that represents conditions prior to the rapid increase in corn acres planted after 2007. Each of these scenarios has important insights to contribute, as each answers slightly different questions. Many other studies assume different reference conditions, baselines, and scenarios, all of which provide useful information on potential environmental impacts. In order to incorporate these insights and information, this report does not restrict analysis to a specific, quantitative baseline against which environmental impacts can be measured. Instead, it uses a broad, qualitative assessment based on the peer- reviewed literature, which provides a variety of baselines.

2.3.1. Qualitative Synthesis of the Literature Reviewed for This Report

Section 6 presents a qualitative synthesis and the underlying assumptions to estimate the range and magnitude of the environmental impacts of producing corn, soybeans, corn stover, perennial grasses, woody biomass, and algae; and corn ethanol, soybean biodiesel, and cellulosic ethanol for transportation biofuel (Tables 6-1 and 6-2, Figure 6-1). This synthesis is based on the information reviewed in preceding sections of this report, and covers environmental impacts attributable to activities across the entire biofuels supply chain, from cultivation of feedstocks through the use of fuel.

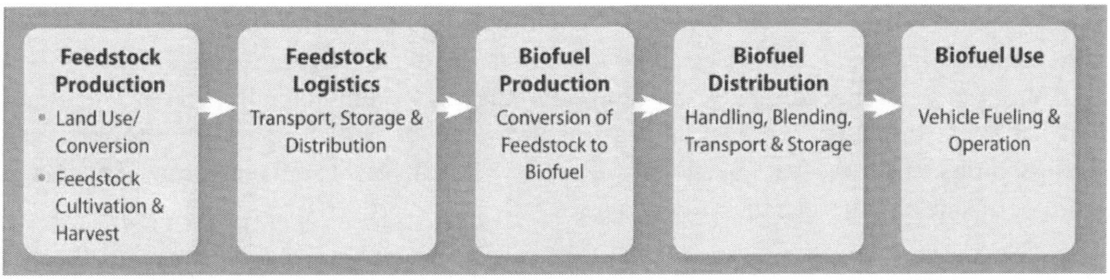

Figure 2-2. Five Stages of the Biofuel Supply Chain.

Tables 6-1 and 6-2 present assumptions underlying the maximum potential range of domestic environmental impacts associated with the production of biofuels under RFS2, as well as conditions for negligible and the most plausible impacts within that maximum potential range. The relative, qualitative values across feedstocks and impact categories presented in these figures are based on the consensus of the authors. Full details of the analysis, including the methodology and conventions used to develop and present the information are described in Sections 6.2 and 6.3.

This synthesis addresses many complexities covered in this report, including the dependence of impacts on the type of feedstock or fuel used, and on management practices, land use change, and conversion technologies, but does not incorporate other environmental (e.g., GHG emission reductions), economic, and social issues relevant to decision making. Further complexities will be explored in subsequent reports.

Although EPA recognizes the limitations of a qualitative literature review (e.g., inconsistent baselines, assumptions, and endpoints across studies), modeling efforts are not sufficiently developed to allow comprehensive quantitative analysis of all environmental impacts required of Section 204 of EISA 2007. Quantitative analyses on subsets of topics addressed in this report have been summarized (e.g., Malcolm et al., 2009), and progress on quantitative and integrated assessments is an important goal for future reports.

2.3.2. Biofuel Production Stages Discussed in This Report

There are five main stages in the biofuel supply chain: feedstock production, feedstock logistics (transport, storage, and distribution), fuel production, fuel distribution, and fuel use (Figure 2-2). Environmental impacts can be generated at all stages of biofuel feedstock production and processing. The specific impacts associated with a particular feedstock or biofuel will vary depending on many factors, including the type, source, and method of feedstock production; the technology used to convert the feedstock to fuel; methods used and distances traveled to transport biofuels; the types and quantities of biofuels used; and controls in place to avoid or mitigate any impacts. This report covers all five of the production stages.

2.3.3. Feedstocks and Fuels Discussed in This Report

There is uncertainty regarding which feedstocks will be used to meet the RFS2 targets in the mid- to long term. A few feedstocks are already in use: corn, soybean, and others in smaller quantities. Other feedstocks are in the early stages of research and development or their potential future commercial viability is still unknown. This report focuses on seven feedstocks: the most predominantly used (corn and soybeans) and five others (corn stover, perennial grasses, woody biomass, algae, and waste materials) that represent a range of feedstocks currently under development. The biofuels highlighted in this report are ethanol (both conventional and cellulosic) and biomass-based diesel. Ethanol and biomass-based diesel are the focus because they are currently the most commercially viable and/or are projected to be the most commercially available by 2022, and they are the primary fuels currently projected to meet RFS2. Future reports will analyze other feedstocks and fuels as technologies and commercial viability change.

2.3.4. Impacts Discussed in This Report

This report focuses on environmental and resource conservation impacts specified in EISA Section 204, as shown in Figure 2-3. It does not extensively discuss CO_2 or other GHGs, nor do its findings encompass environmental benefits gained by GHG emissions reductions established by the RFS2; interested readers are referred to the EPA's RFS2 RIA (U.S. EPA, 2010a). This report is complementary to the RIA.

The environmental and resource conservation impacts discussed in this report reflect a complex set of interactions and feedbacks between land, soil, air, and water; future versions of this report will explore these important complexities as enhanced data and analysis tools become available. The state of knowledge does not permit a fully quantitative analysis of the impacts associated with increased production of biofuels, especially those that are not currently deployed at commercial scales (e.g., cellulosic, algae). Instead, this report compiles available information and analyses on the nature and extent of impacts that might be expected to occur. Thus, it does not use the EISA baselines and volumes, per se, against which impacts can be measured. A number of important findings on the potential impacts of increased biofuel production and use were found in the literature. These findings are based on different baselines and volumes from the RFS2 RIA, but are important in ascertaining the nature and extent of potential impacts.

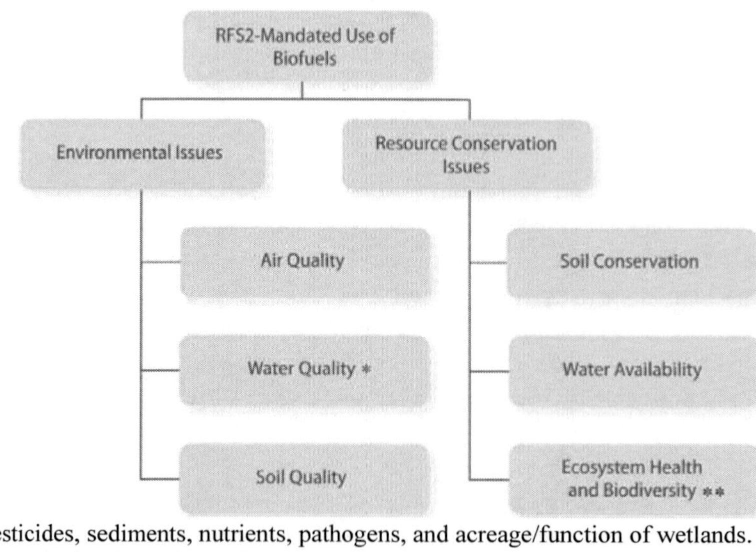

* Includes pesticides, sediments, nutrients, pathogens, and acreage/function of wetlands.
** Includes invasive/noxious plants, forests, grasslands, wetlands, and other aquatic ecosystems.

Figure 2-3. Environmental and Resource Conservation Issues Addressed in This Report.

3. Environmental Impacts of Specific Feedstocks

3.1. Introduction

The Energy Independence and Security Act (EISA) requires that all renewable fuel be made from feedstocks that meet the definition of renewable biomass (see text box). Many different feedstocks meet these requirements and can be used to produce ethanol, other biofuels, or biofuel components.

In 2009, 95 percent—or 10.9 billion gallons—of total renewable fuel produced in the United States was produced from corn and refined almost entirely in the form of conventional corn starch ethanol (FAPRI, 2010a; U.S. EIA, 2010). Soybean oil–based biodiesel accounted for most of the remainder—505 million gallons. EPA expects that corn and soybean feedstocks will continue to account for a large share of U.S. biofuel production in the near future (U.S. EPA, 2010a). As of July 2010, there was neither significant commercial-scale production of ethanol from cellulosic feedstocks, nor significant biodiesel production from oil seed feedstocks other than soybean in the United States.

Requirements for Renewable Fuels

Under EISA, all renewable fuel must be made from feedstocks that meet the Act's definition of renewable biomass:

- Planted crops and crop residue from agricultural lands that were cleared prior to December 19, 2007, and were actively managed or fallow on that date.
- Planted trees and tree residue from tree plantations that were cleared prior to December 19, 2007, and were actively managed on that date.
- Animal waste material and byproducts.
- Slash and pre-commercial thinnings from non- federal forestlands that are neither old-growth nor listed as critically imperiled or rare by a State Natural Heritage program.
- Biomass cleared from the vicinity of buildings and other areas at risk of wildfire.
- Algae.
- Separated yard waste and food waste.

As well, these feedstocks must meet life cycle GHG thresholds (Table 2-2).

As the science and technology of cellulosic biofuel production improve, EPA expects an increase in the use of cellulosic feedstocks to produce advanced biofuel. Such feedstocks include agricultural residues (e.g., corn stover, sugarcane bagasse, and sweet sorghum pulp), forestry biomass, urban waste, and dedicated energy crops (e.g., switchgrass) (U.S. EPA, 2010a). Technologies for producing biodiesel from vegetable oils, recycled oils, rendered fats, greases, and algal oils have been developed and tested at various scales from the laboratory to demonstration plants to commercial facilities. EPA expects biodiesel from these feedstocks to increase its market share as their production becomes more economically and technologically feasible (U.S. EPA, 2010a).

The feedstocks discussed in this section include corn and soybeans, as well as four others currently under development: corn stover, perennial grasses, woody biomass, and algae (see Table 3-1). These feedstocks represent different cultivation and production practices. A brief discussion of waste materials as a feedstock is also included.

The Renewable Fuel Standard as amended by the EISA (RFS2) prescribes allowable conversions of land uses to specific renewable fuel feedstocks. Corn, corn stover, soybeans, and perennial grasses may only be grown on lands that were in agricultural production prior to December 19, 2007 (U.S. EPA, 2010a). These lands comprise cropland, pasture, and currently fallow lands, including land enrolled in the Conservation Reserve Program (CRP). Woody biomass can only be grown on lands that were active forest plantations prior to December 19, 2007, while residue harvesting and thinning can occur on non-federal forestlands (U.S. EPA, 2010a). The scientific literature that precedes finalization of the RFS2 generally uses different assumptions about land use conversions and subsequent environmental effects (e.g., Walsh et al., 2003; Volk et al., 2006). For consistency across feedstocks, impacts are frequently compared with those of row crops, even when such conversions are either not currently economically likely (e.g., conversion of row crops to grasses) or not allowed under the RFS2 (e.g., conversion of forests to row crops). The literature to date repeatedly compares environmental impacts between feedstock types, however, and this report accordingly includes such information.

This section reviews the actual (where known) and potential environmental impacts of producing these feedstocks. Actual environmental impacts will vary, depending on the number of acres in production, cropping techniques, implementation of conservation and best management practices (BMPs), location of the crop acreage, hydrology, soils, species composition, and other geographic factors. Feedstock production impacts are considered during the cultivation and harvest processes (see Figure 2-2). Potential impacts associated with the subsequent four stages of the biofuel supply chain are presented in Section 4. Row crop feedstocks (corn, corn stover, and soybean), which share many common traits, are discussed in Section 3.2. Sections 3.3 to 3.6 present potential effects associated with switchgrass, woody biomass, algae, and waste, respectively. In addition to general ecosystem impacts, Section 3.7 reviews impacts on specific ecosystems (forests, grasslands, and wetlands) as required under EISA Section 204. Section 3.8 reviews environmental concerns associated with genetic engineering of feedstocks.

3.2. Row Crops (Corn, Corn Stover, Soybeans)

3.2.1. Introduction

Overall, U.S. corn and soybean production has increased over the past decade. Increased demand for biofuel provides additional incentive to continue research and development for increasing crop production and yields. As shown in Figure 3-1, plantings of corn in the United States have increased by almost 10 million acres since 2006, an increase of nearly 13 percent. Soybean acres have increased by a more modest 1.9 million acres, or about 2.5 percent, since the previous high in 2006. Yields for both corn and soybean have improved during the past two decades (see Figure 3-2) (NASS, 2010a), moderating the need for increases in acreage.

Table 3-1. Primary Fuels and Feedstocks Discussed in This Report

EISA Biofuel Type	Biofuel	Feedstock
Conventional biofuel	Ethanol	Corn starch
Cellulosic biofuel	Ethanol	Corn stover
		Perennial grasses
		Woody biomass
Biomass-based diesel	Biomass-based diesel	Soybeans
		Algae

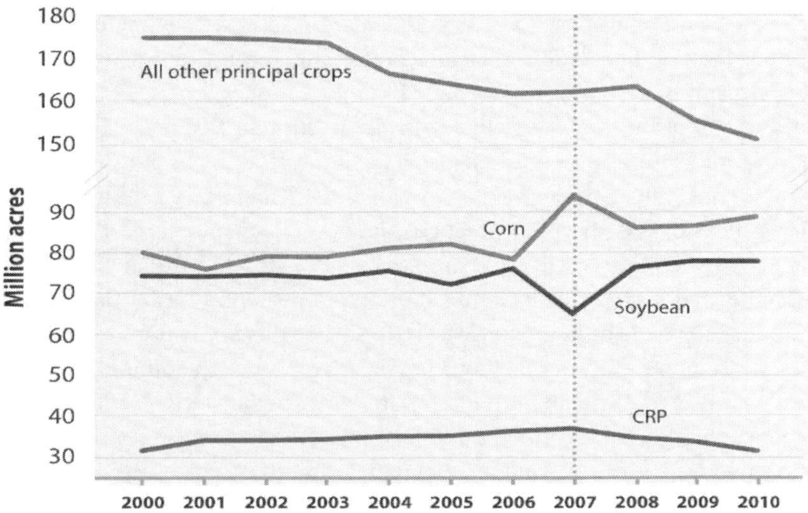

Data source: NASS, 2010a.

Figure 3-1. U.S. Acres of Crops Planted and U.S. Acres Enrolled in the CRP.

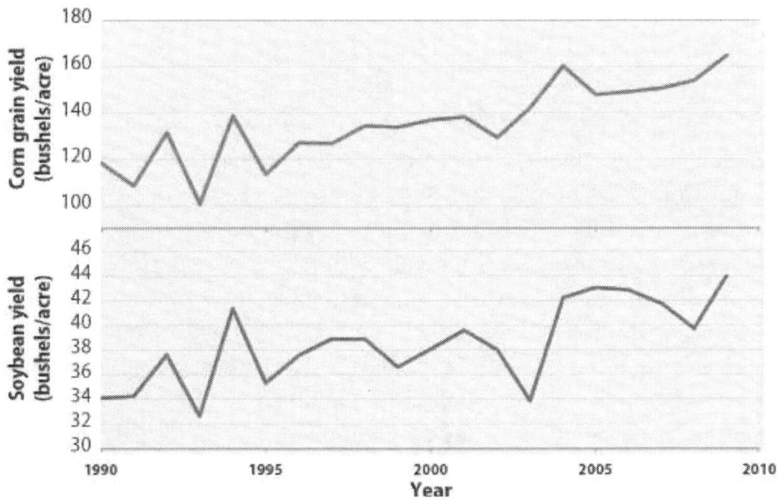

Data source: NASS, 2010a.

Figure 3-2. U.S. Corn and Soybean Yield.

3.2.2. Overview of Environmental Impacts

Corn and soybean production entails the use of pesticides, fertilizer, water, and fuel/energy, in addition to drainage systems in some areas. Each of these can affect the environment. Changes in land cover, vegetation, and habitat have additional impacts on the environment. Because corn stover is a byproduct of corn production, this report considers the incremental environmental impacts from corn stover separately from those of grain-only harvesting.

Cultivation of row crops such as corn and soybeans may lead to high levels of soil erosion, nutrient loss, and pesticide and water use if not managed adequately (Groom et al., 2008, Table 1). Agricultural conservation practices may be used to reduce or minimize the impact of row crop agriculture on the environment. These practices include: (1) controlled application of nutrients and pesticides through proper rate, timing, and method of application; (2) controlling erosion in the field (e.g., reduced tillage, terraces, grassed waterways); and (3) trapping losses of soil at the edge of fields or in fields through practices such as cover crops, grassland and riparian buffers, controlled drainage for tile drains, and constructed/restored wetlands (Dinnes et al., 2002; Blanco-Canqui et al., 2004; Blann et al., 2009; NRCS, 2010a).

The effectiveness of conservation practices, however, depends upon their adoption. The U.S. Department of Agriculture (USDA) Conservation Effects Assessment Project (CEAP) recently released a major study quantifying the effects of conservation practices commonly used on cultivated cropland in the upper Mississippi River basin. It found that, while erosion control practices are commonly used, there is considerably less adoption of proper nutrient management techniques to mitigate nitrogen loss to water bodies (NRCS, 2010a).

Even if conservation and BMPs are reliably implemented, they cannot be expected to always lead to rapid improvements in environmental quality. A case study in the Chesapeake Bay (CENR, 2010) found that the implementation of BMPs since 2000 has significantly lowered loadings of nitrogen (72 percent of sites showed downward trends), total phosphorus (81 percent of sites), and sediment (43 percent of sites). However, lower nutrient input has not yet improved dissolved oxygen levels overall in the Chesapeake Bay, with the exception of small-scale reductions in hypoxic zones.

3.2.3. Current and Projected Cultivation

In 2009, U.S. farmers planted 86 million acres of corn, harvesting 13.1 billion bushels (NASS, 2010a). Approximately 4.5 billion bushels (or 34.9 percent of corn grain consumed annually) were used for corn starch ethanol between September 2009 and August 2010 (ERS, 2010c, 2010d), up from 12.4 percent in 2004–2005 (see Figure 3-3; ERS, 2010b, 2010c).[9] Corn is grown throughout the United States, but the vast majority of the crop is grown in 11 states: Illinois, Indiana, Iowa, Kansas, Michigan, Minnesota, Missouri, Nebraska, Ohio, South Dakota, and Wisconsin. Figure 3-4 shows a map of planted acres by county in 2010.

EISA establishes 15 billion gallons as the maximum amount of corn starch ethanol that can contribute to meeting the 36 billion gallon per year renewable fuel target in 2022. Domestic production, which totaled 10.9 billion gallons in 2009 (U.S. EIA, n.d.[b]), is expected to meet this target in 2015 through a combination of increased corn yield, increased acreage dedicated to ethanol production, including more continuous corn, and, potentially, improved efficiency in converting corn starch to ethanol (Malcolm et al., 2009). Imported production totaled approximately 200 million gallons in 2009. The USDA estimates that planted corn acreage will remain at 89 to 90 million acres through 2019, despite increasing demand for biofuel in the United States (USDA, 2010a). In the RFS2 analysis, EPA estimates that in order to produce 15 billion gallons of corn starch ethanol per year by 2022, the percentage of corn bushels dedicated to ethanol could rise from the current 35 percent to 41 percent in 2022 (U.S. EPA, 2010a).

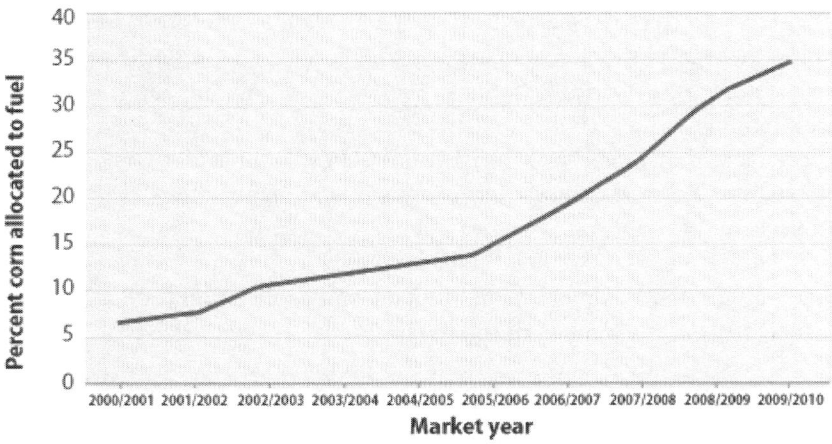

Note: Market year is September to August.
Sources: ERS, 2010b, 2010c.

Figure 3-3. Percent of Corn Grain Allocated to Ethanol for Fuel.

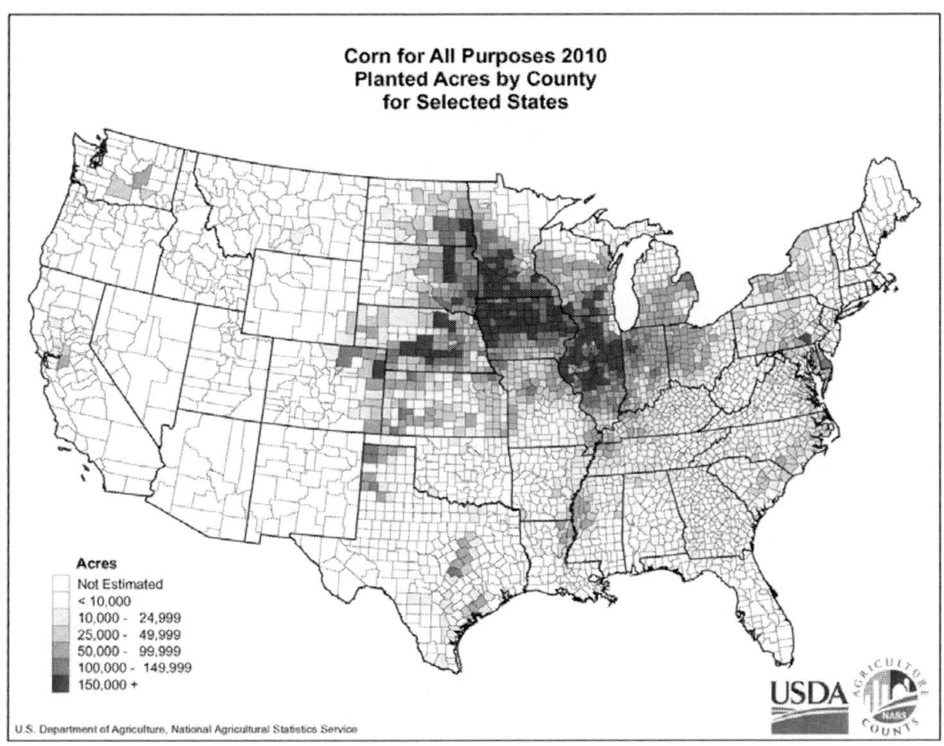

Source: NASS, 2010b.

Figure 3-4. Planted Corn Acres by County.

Corn stover—the stalks, leaves, husks, and cobs that are not removed from the fields when the corn grain is harvested—provides another potential feedstock for meeting EISA requirements. In the RFS2 Regulatory Impact Analysis (RIA), U.S. EPA (2010a) estimated that 7.8 billion gallons of ethanol could be produced from corn stover by 2022. Most corn stover harvesting for biofuel is expected to be from the major corn-producing states. As of July 2010, there was no commercial production of cellulosic ethanol from corn stover.

After corn, soybean is the second largest agricultural crop (in terms of acreage) in the United States. In 2010, American farmers planted 77.7 million acres of soybeans and harvested 3.4 billion bushels (NASS, 2010a). Soybean oil is the principal oil used for commercial production of biodiesel in the United States, responsible for about half of total biodiesel production. The rest comes from various other vegetable oils such as canola oil as well as waste fats, tallow, and greases (see Table 3-2 for a more detailed breakdown) (U.S. EIA, 2010). In harvest year 2008/2009, biodiesel accounted for approximately 5.5 percent of U.S. soybean consumption.[10] Almost 2 billion pounds of soybean oil (USDA, 2010b) yielded about half of the 505 million gallons of biodiesel produced in calendar year 2009 (U.S. EIA, n.d.[c]). This was a significant decline from the production total in 2008 of 683 million gallons from soybeans (U.S. EIA, n.d.[a]). Nonetheless, USDA expects biodiesel to account for approximately 7.7 percent of soybean consumption in 2012/2013 and hold relatively steady through 2019. USDA estimates that soybean acreage will level off at approximately 76 million acres through 2019 (USDA, 2010b).

In terms of cultivation, soybeans are typically grown in rotation with corn, in the same locations. Figure 3-5 shows that soybean production is centered in the upper Midwest and along the Mississippi River Valley, with Iowa, Illinois, Indiana, Minnesota, and Nebraska representing the top soybean-producing states.

There is some concern that the demand for corn and soybeans as biofuel feedstocks may lead to high prices of these commodities, inducing farmers with land currently enrolled in USDA's CRP to return to intensive agricultural production (e.g., Secchi et al., 2009). The CRP provides farmers with financial incentives to set aside a certain portion of their cropland in order to conserve or improve wildlife habitat, reduce erosion, protect water quality, and support other environmental goals. Biomass produced from the cultivation and harvesting of corn, corn stover, or soybeans on CRP lands is considered a renewable source of energy as defined in RFS2. The Food, Conservation, and Energy Act of 2008 (known as the Farm Bill) capped CRP acreage at 32 million acres, reducing enrollment by 7.2 million acres from the 2002 Farm Bill with the potential for making more acreage available for the production of row crops (Figure 3-1). Historically, land entering and exiting the CRP program has been more vulnerable to erosion than other cultivated land, but also less productive (ERS, 2008). So while the conversion of CRP land to intensive feedstock production is possible, the likelihood of such a land use conversion is uncertain given practical economic and agronomic considerations.

A recent USDA analysis estimates that in order to meet the volumetric requirements of RFS2, total cropland will increase 1.6 percent over 2008 baseline conditions by 2015, with corn acreage expanding 3.5 percent and accounting for most of the overall cropland increase (Malcolm et al., 2009). While corn acreage is expected to expand in most regions, USDA estimates that traditional corn-growing areas would likely see the largest increases—up 8.6 percent in the Northern Plains, 1.7 percent in the Corn Belt, and 2.8 percent in Great Lakes States (Malcolm et al., 2009). Other modeling studies have also projected an increase in the areal extent of cropland in the United States in response to the global demand for bioenergy (e.g., BRDI, 2008; Keeney and Hertel, 2009; Beach and McCarl, 2010; Hertel et al., 2010a, 2010b; Taheripour et al., 2010). These studies project conversion of pasture and other idle agricultural lands, as well as forest land, to intensive crop production in response to demand for biomass to produce ethanol and biodiesel. Not all land converted in these modeling studies is assumed to be used for biomass production, since the complex economics of agriculture, including future prices of cellulosic feedstocks, directly and indirectly affect land use changes.

Table 3-2. 2009 Summary of Inputs to U.S. Biodiesel Production[a]

Input		2009 Total (Million Pounds)	Percentage of Total
Feedstock inputs	Vegetable oils (canola, corn, cottonseed, palm, soybean, and other vegetable oils)	2,385	60.0%
	Animal fats (poultry fat, tallow, white grease, and other animal fats)	1,040	26.1%
	Recycled feedstock (yellow grease and other recycled feedstock)	169	4.2%
Other inputs	Alcohol	328	8.2%
	Catalysts	56	1.4%

[a] This table's contents must be considered as estimates due to withholding of confidential business information for some of the input categories.

Source: U.S. EIA, 2010.

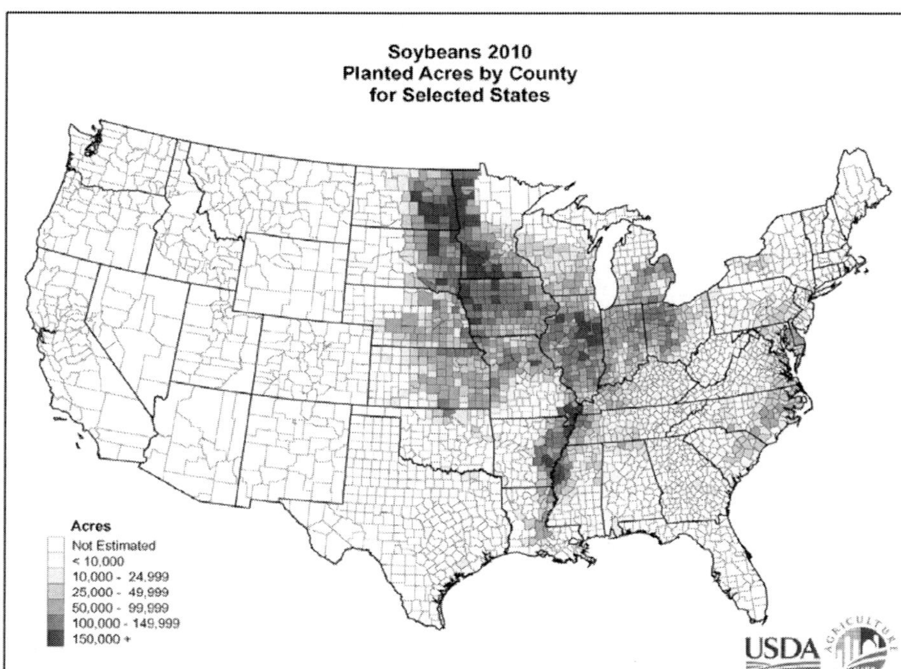

Source: NASS, 2010b.

Figure 3-5. Planted Soybean Acres by County.

3.2.4. Water Quality

Water quality impacts from increased corn and soybean production for biofuel are caused by pollution from nutrients, sediment, and pesticides, as well as biological contaminants such as pathogens that are released when animal manure is applied as fertilizer. Multiple studies examine corn production scenarios to meet EISA targets and find that increased nitrogen inputs to the Gulf of Mexico and other U.S. coastal waters are likely, and these inputs can worsen hypoxic conditions if crops are not grown under improved agricultural conservation practices and expanded nutrient BMPs (Donner and Kucharik, 2008; Malcolm et al., 2009; Rabalais et al., 2009).

3.2.4.1. Nutrient Loading

Corn has the highest fertilizer use per acre of any of the biofuel feedstocks, and it accounts for the largest portion of nitrogen fertilizer use among all feedstocks discussed in this report (U.S. EPA, 2010a). By one estimate, which surveyed 19 U.S. states, approximately 96 percent of corn acreage received nitrogen fertilizer in 2005, with an average of 138 pounds per acre (NASS, 2006). A study in Iowa found that each acre of corn requires about 55 pounds of phosphorus (as P_2O_5) for optimal production (Iowa State University, 2008). Assuming a yield of 154 bushels per acre (NASS, 2010c) and an ethanol conversion rate of 2.7 gallons per bushel (Baker and Zahniser, 2006), this results in 0.33 pounds of nitrogen and 0.13 pounds of phosphorus applied per gallon of ethanol produced. Nitrogen discharged from corn and soybean crops via runoff, sediment transport, tile/ditch drainage, and subsurface flow averages 24 to 36 percent of the nitrogen applied (and can range from 5 percent in drought years to 80 percent in flood years) (Dominguez-Faus et al., 2009).

Nutrients are applied to fewer soybean acres than corn and at much lower rates because soybean is a legume (U.S. EPA, 2010a). Legumes have associations in their roots with bacteria that can acquire atmospheric nitrogen and convert it into bioavailable forms, reducing the need for external addition of nitrogen fertilizer. However, losses of nitrogen and phosphorus from soybeans can occur at quantities that can degrade water quality (Dinnes et al., 2002; Randall et al., 1997). In 2006, USDA's NASS estimated that nitrogen and phosphorus fertilizers were applied to 18 percent and 23 percent of soybean acreage, respectively, with an average of 16 pounds of nitrogen and 46 pounds of phosphate applied per acre fertilized (NASS, 2007). The quantity of nitrogen fertilizer applied to soybean fields ranged from 0 to 20 pounds per acre, while the quantity of phosphate ranged from 0 to 80 pounds per acre. As with corn, the conversion of idled acreage to soybeans is estimated to result in losses of nitrogen and phosphorus from the soil through cultivation (Simpson et al., 2008).

Corn requires less fertilizer when grown in rotation with soybeans. Therefore, crop rotation provides an effective strategy for reducing the amount of fertilizer and pesticide applied to fields, and therefore could decrease runoff and leaching of the pollutants to water. Some studies of nitrate leaching from corn-soybean rotation cropping systems are inconclusive about whether these systems increase or decrease leaching rates compared to continuous corn systems (Klocke et al., 1999; Zhu and Fox, 2003). However, a more recent study estimated that only 2 to 40 percent of the total nitrogen leached from fields planted alternately with corn and soybeans came from the fields when they were planted with soybeans (Powers, 2005). This implies that most of the nitrogen leaching was due to corn production. In general, the total amount of nitrogen lost from corn fields tends to be higher than losses from soybean fields (Powers, 2005). However, it is important to consider several factors that cause variability in leaching rates for both corn and soybeans, including geography, soil type, hydrology, and tillage methods (Powers, 2005).

The removal of corn stover could lead to loss of soil surface cover if Natural Resources Conservation Service (NRCS) guidelines are not followed, thereby increasing runoff of nitrogen and phosphorus to surface waters including wetlands (Kim and Dale, 2005). Even partial removal of corn stover can result in nutrient losses to water due to increased runoff (Kim and Dale, 2005; Lal, 2004). In addition, corn stover removal can lead to the loss of soil nutrients needed for corn growth, and higher fertilizer rates are likely to be required to sustain crop productivity, increasing the likelihood of increased runoff and transport of nonpoint-source pollutants (Blanco-Canqui and Lal, 2009a, 2009b).

Nutrients—Surface Water Impacts

Increased production of row crops, especially corn due to biofuel demand, will likely increase nitrogen and phosphorus loading to surface waters (Malcolm et al., 2009). Excessive levels of nutrients in a body of water often cause accelerated algae growth, reducing oxygen levels and light penetration. Low dissolved oxygen (i.e., hypoxia) can kill many organisms, reducing population abundances and overall species diversity in the affected area (Pollock et al., 2007; Breitburg et al., 2009; Levin et al., 2009). This nutrient enrichment (eutrophication) can cause serious deterioration of both coastal and inland water resources. According to a 2008 report by the National Research Council, excess nutrients and sediment from the high corn- producing Midwest are the primary sources of water quality degradation in the Mississippi River basin and the Gulf of Mexico (NRC, 2008). Further, the National Summary of Impaired Waters (U.S. EPA, 2010b)[11] documented that in 2008, nationwide, approximately 50 percent of the 3.5 million miles of stream and rivers and 66 percent of the over 41 million acres of lakes and reservoirs in the United States were impaired due to nutrient enrichment. Increased corn and soybean production for biofuels could exacerbate this situation due to the nutrients from additional fertilizer used for increased acreage or switching to continuous corn, or from increased extent and density of subsurface tile drainage. A report by Donner and Kucharik (2008) predicts that the average annual flux of dissolved inorganic nitrogen (DIN) to the Gulf of Mexico could increase by 10 to 18 percent if EISA targets for non-advanced biofuels are met through either greater CRP to corn conversion or greater soybeans to corn conversion than the 2007 baseline; DIN flux would increase to 34 percent if all 36 billion gallons come from corn. This is in contrast with a USDA report that predicts an increase 1.8 percent nationally of nitrogen runoff into estuaries by 2015 due to agricultural acreage expansion and intensification (Malcolm et al., 2009). Despite the predicted increases in nutrient influx associated with row crops, there are ongoing efforts to address hypoxia in the Gulf of Mexico. These include the work of the Mississippi Basin/Gulf of Mexico Task Force and the USDA's Mississippi River Basin Healthy Watersheds Initiative (MRBI). The Task Force established a 30 percent reduction goal in nutrient loading to the Gulf to reduce the hypoxic zone to less than 5000 km^2 (Mississippi River/Gulf of Mexico Watershed Nutrient Task Force, 2008). The MRBI is a multiyear program that began in 2009. It is managed by NRCS and is intended to reduce the nutrient loading into the Gulf of Mexico. Twelve states are receiving support to adopt or expand conservation efforts in priority watersheds (NRCS, 2010b).

Mitigating the loss of nitrogen and other nutrients to water bodies is a research priority for USDA. Since drainage systems are a key conduit for nutrient loading, new research is focusing on alternative surface and subsurface drainage solutions. Subsurface tiles/pipes or artificial ditches are drainage systems that remove water from the soil subsurface to allow for crops to be planted. While these systems can move water from soils to surface water or wells, they can also quickly transport nutrients and pesticide runoff from fields without any of the attenuation that would occur if these contaminants were moving through wetlands or soils (U.S. EPA, 2010a). An interagency Agricultural Drainage Management Systems Task Force, formed in 2003 and recently expanded, is working to reduce the loss of nitrogen and phosphorus from agricultural lands through drainage water management (CENR, 2010).

One emerging conservation practice that addresses water quality degradation is the creation of wetlands on the perimeter of fields in order to receive surface runoff and filter out nutrients prior to the runoff's discharge into streams and rivers. Surface water runoff control, a set of conservation practices used to stop water erosion, reduces the overland losses of nutrients to the surrounding environment, but increases infiltration and loss of soluble nitrogen and phosphorus. A third strategy, lowering the water table during planting and harvesting, has been predicted to lower nitrogen losses in the Chesapeake watershed by 40 percent (CENR, 2010). The use of cover crops has also been shown to reduce nitrogen loss from fields (Dinnes et al., 2002). Other strategies, such as planting perennial grasses over subsurface tile drains or placing wood chips in drainage ditches, are also being explored. Implementing strategies such as these on agricultural lands that contribute a disproportionate share of nitrogen loads will maximize the environmental benefit of their application (CENR, 2010).

Significant opportunities exist for further increasing efficient use of nitrogen simply by increasing the number of growers who follow all nutrient BMPs. For instance, NRCS found that only about 38 percent of all cultivated cropland acres in the upper Mississippi River basin are already under the complete suite of nutrient management practices: proper source, rate, timing, and place of application (NRCS, 2010a). Given the extensive focus on the nutrient management issue at USDA and within the agricultural community, the proportion of acres benefitting from such nutrient management practices could increase.

Nutrients—Coastal Waters Impacts

Nutrient enrichment is a major concern for coastal waters across the United States, including the Gulf of Mexico, Chesapeake Bay, other estuaries, and the Great Lakes. For example, almost 15 percent of the coastal waters in the Gulf of Mexico and Northeast have poor water quality as measured by nutrient concentrations, extent of hypoxia, and water clarity (U.S. EPA, 2008b). The number of U.S. coastal and estuarine ecosystems documented as experiencing hypoxia increased from 12 in 1960 to over 300 in 2008 (out of 647 coastal ecosystems analyzed) (see Figure 3-6) (CENR, 2010). While these impacts are due to a number of types of nutrient inputs, such as lawn fertilizers, other agricultural uses, atmospheric deposition, and wastewater discharges, increased corn and soybean production for biofuel will likely increase nutrient loading in those watersheds where increased production occurs (SAB, 2007; Rabalais et al., 2007 as cited in CENR, 2010).

Hypoxia in the Gulf of Mexico is a long-standing environmental and economic issue that threatens commercial and recreational fisheries in the Gulf (U.S. EPA, 2010a). The primary cause of hypoxia in the Gulf of Mexico is excess nitrogen and phosphorus loadings from the

Upper Midwest flowing into the Mississippi River, suggesting that increased corn and soybean production may exacerbate the problem (U.S. EPA, 2010a). U.S. Geological Survey (USGS) SPARROW[12] modeling of the sources of nutrient loadings to the Gulf of Mexico estimated that agricultural sources contributed more than 70 percent of the delivered nitrogen and phosphorus to the Gulf of Mexico (Alexander et al., 2008). Corn and soybean production accounted for 52 percent of nitrogen delivery and 25 percent of phosphorus delivery. Modeling of the upper Mississippi River basin using SWAT[13] modeling indicated that, on average, it contributes 43 percent of the nitrogen load to the Gulf of Mexico, and 26 percent of the phosphorus load (SAB, 2007). One study estimated that corn production contributes between 60 and 99 percent of the total nitrogen load to the Mississippi River from eastern Iowa watersheds (Powers, 2007). Other studies have also determined that the majority of nitrate in the Mississippi River originates in the Corn Belt (Donner et al., 2004; Goolsby et al., 1999).

Nitrogen from fertilizers can also volatilize and then return to waters through atmospheric wet or dry deposition of either reduced or oxidized forms (e.g., NH_x, NO_y). Atmospheric nitrogen from all sources, including power plant emissions, is estimated to contribute up to approximately 27 percent of the nitrogen loading to both the Gulf of Mexico (Alexander et al., 2008), and to the Chesapeake Bay (Paerl et al., 2002).

A USDA study projects that reaching 15 billion gallons per year of ethanol from corn starch (i.e., not including stover) will result in an average 1.7 percent increase (over the 2008 baseline used in the USDA report) in nitrogen loads to surface water nationally by 2015, with the greatest contributions in nitrogen load occurring in the Corn Belt (1.3 percent) and Northern Plains (3.5 percent) (Malcolm et al., 2009). Another study used the Terrestrial Hydrology Model with Biochemistry to generate several corn-based scenarios to reach the 2022 non-advanced biofuels target. In these scenarios, dissolved inorganic nitrogen exported by the Mississippi and Atchafalaya Rivers increases 10 to 18 percent, which is 39 to 43 percent greater than the federal hypoxia reduction target established for the Gulf of Mexico (Donner and Kucharik, 2008).

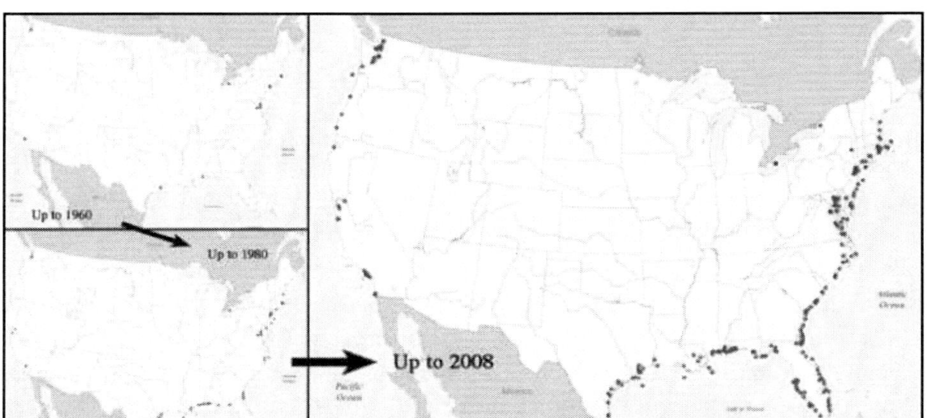

Note: Map does not display one hypoxic system in Alaska and one in Hawaii.
Source: CENR, 2010.

Figure 3-6. Change in Number of U.S. Coastal Areas Experiencing Hypoxia from 1960 to 2008.

Ecological features such as wetlands and riparian buffers play an important role in absorbing nutrients. Conserving wetlands where they exist, or creating artificial vegetated riparian buffers between waters and croplands, is a way to mitigate the impacts of nutrient loading. Riparian buffers and filter strips prevent potential pollutants in agricultural runoff (sediment, nutrients, pesticides, pathogens) from reaching surface waters. While the effectiveness of these buffers can vary depending on many factors, including slope, width, vegetation used, and how well they are maintained, studies have shown that they can remove up to 78 percent of phosphorus, 76 percent of nitrogen, and 89 percent of total suspended solids (Schwer and Clausen, 1989; Dosskey, 2001; Richardson et al., 2008).[14]

Nutrients—Ground Water Impacts

Ground water can be used for public and private drinking water supplies, and fertilizers can increase the concentration of nitrate in ground water wells, especially shallow wells (less than 200 feet deep). USGS sampled 495 wells in 24 well networks across the United States in predominantly agricultural areas from 1988 to 2004 and found significant changes in concentrations of nitrate in eight of the well networks. In seven of those eight networks, USGS found significant increases in nitrate concentrations; in three of those seven, nitrate concentrations exceeded the federal drinking water standards of 10 mg/L of nitrate-nitrogen (Rupert, 2008). Increased corn production for biofuels could worsen the problem of contaminated well water because of additional nitrogen inputs from fertilizer used to grow more corn. USDA projects that reaching 15 billion gallons per year of ethanol from corn will result in a 2.8 percent increase in nitrogen leaching to ground water, with the greatest increases occurring in the Great Lakes states and the Southeast; this increase occurs with a 1.6 percent increase in corn acreage (Malcolm et al., 2009). Similar estimates for soybean production were not identified.

Fertilizer application management strategies aim to reduce nitrogen leaching by maximizing the efficiency of applied fertilizer. Such strategies focus on collecting precise information on soil nutrient content in order to better inform application rates. USDA reports that phosphorus accumulation on farms has reached levels that often exceed crop needs (ARS, 2003). Better information on these conditions could help reduce nutrient runoff that leads to eutrophication.

3.2.4.2. Sediment

Nutrients and sediment are the two major water quality problems in the United States, and much attention has been focused on these issues in the Mississippi River basin and the Gulf of Mexico (NRC, 2008). As modeled in a 2010 NRCS study, cropped areas in the upper Mississippi River basin lose one ton of sediment per acre per year, with 15 percent of cropped acres experiencing more than 4 tons of sediment lost per acre in one or more years (NRCS, 2010a). The National Summary of Impaired Waters stated that in 2008 over 70,000 miles of streams and rivers and over 1.2 million acres of lakes and reservoirs in Mississippi River basin states are impaired because of sediments or turbidity (U.S. EPA, 2010b).[15] Nelson et al. (2006) reported that row crops, such as corn and soybean, result in higher erosion rates and sediment loads to surface waters, including wetlands, than non-row crops that might be used as biofuel feedstock, such as grasses. Sedimentation rates in agricultural wetlands can be higher than in natural grassland landscapes; increased sedimentation may, depending on sediment depths, cover viable seeds sufficiently to prevent germination (Gleason et al., 2003).

Studies evaluating the impacts of RFS2 over the next decade have provided estimates for the increases in sediment loads. A study conducted for EPA estimated that annual sediment loads to the Mississippi River from the upper Mississippi River basin would increase by only 0.52 percent between 2005 and 2015 and 0.30 percent by 2022 (assuming corn stover remained on the field following harvest) (Aqua Terra, 2010). A USDA study estimates that nationally, sediment loads in 2015 will be 1.6 percent greater with implementation of RFS2 than without (assuming ethanol production from corn starch only) (Malcolm et al., 2009).

Removal of corn stover from fields for use in biofuel production could increase sediment yield to surface waters and wetlands, but erosion rates can be highly variable depending on soil type, slope, management of fields, and the proportion of stover harvested (Cruse and Herndl, 2009; Kim and Dale, 2005). Results of SWAT modeling of the upper Mississippi River basin (Aqua Terra, 2010) indicated that leaving corn stover on fields helps reduce soil erosion and sediment transport, even when the amount of land in corn production increases. However, the amount of soil erosion that agricultural cropland experiences is a function of many factors, including not only residue left on the field, but also field operations (field preparation, tillage, etc.) in preparation for the next crop, timing of field operations, and other site-specific factors noted above (U.S. EPA, 2010a).

Conservation tillage practices, including no-till, strip-till, ridge-till, and mulch-till,[16] can reduce erosion by leaving at least 30 percent of the ground covered by crop residue and by limiting soil disturbance. According to USDA, 41 percent of planted acreage in the United States uses conservation tillage as a mitigation strategy (ARS, 2006). In 2002, the USDA Agricultural Research Service (ARS) studied the effect of ridge tillage on Northern Corn Belt plantations. The study showed that ridge tillage not only reduced erosion and sediment loading but also increased profitability, reduced fuel and labor use, and reduced economic risk relative to conventional tillage for a corn and soybean rotation (ARS, 2006). Additionally, these alternative tillage approaches can reduce trips across the field, lowering fuel use and improving the energy balance of the resulting biofuel. The use of conservation tillage, in combination with BMPs such as cover crops, may partially compensate for the increase in erosion potential caused by cover stover removal (Blanco-Canqui and Lal, 2009b). Depending on the soil type, these practices may allow a percentage of stover to be harvested sustainably (Blanco-Canqui and Lal, 2009b).

3.2.4.3. Pesticides

According to the National Summary of Impaired Waters (i.e., waters that do not meet the water quality standards) (U.S. EPA, 2009a, 2010c), pesticides were a cause of impairment for approximately 3.1 percent (372,009 acres) of threatened or impaired lakes, reservoirs and ponds and 3.5 percent (16,980 miles) of threatened or impaired streams and rivers in the U.S. (U.S. EPA, 2010b). Approximately 2 percent of the causes of impaired waters were attributed to pesticides and, of those, over three fourths were attributed to pesticides that are no longer registered for use by EPA (U.S. EPA, 2010b). Atrazine is a pesticide commonly used in corn production and lost from agricultural lands in the upper Mississippi River Basin (NRCS, 2010a). It was cited as a cause for approximately 1 percent of the total number of impaired or threatened stream and river miles in the U.S. (U.S. EPA, 2010b).

USDA estimates that insecticides were applied to 16 percent of the 2006 soybean-planted acreage (NASS, 2007). USDA also estimates that herbicides were applied to 98 percent of the planted soybean acreage in 2006. Soybean production releases less pesticide to surface and ground water per unit of energy gained than corn agriculture (Hill et al., 2006).

Growing continuous corn (rather than growing it in rotation with other crops) can increase population densities of pests such as the corn rootworm, resulting in increased pesticide applications to control these pest species (Whalen and Cissel, 2009) or the introduction of new varieties of genetically engineered crops (Bates et al., 2005; Glaser and Matten, 2003). A USDA study projects that cropland dedicated to continuous corn will increase by more than 4 percent (above a 2008 baseline) by 2015 to reach the target of 15 billion gallons of ethanol per year from corn (Malcolm et al., 2009).

While effective pest control may be critical to achieving the yield gains that underpin EISA biofuel projections and targets (Perkins, 2009), there are risks associated with the use of pesticides. The Federal Insecticide Fungicide and Rodenticide Act (FIFRA) registration process is intended to minimize these risks. Many factors contribute to the relative risks of pesticides to the environment, including fate and transport characteristics, method of application, depth to ground water, and proximity to receiving waters.

Integrated pest management (IPM) practices may help reduce pesticide use by tailoring treatment to pest infestation cycles, and by more precisely targeting the amount and timing of applications. IPM focuses on extensive monitoring of pest problems, comprehensive understanding of the life cycles of pests and their interaction with the environment, and precise timing of pesticide applications to minimize pesticide use. In addition to providing environmental benefits of lower pesticide use, IPM often lowers chemical pesticide expenses and pest damage to crops, as well as preventing the development of pesticide-resistant pests. The use of cover crops is an IPM practice that can dramatically reduce chemical application and soil erosion. USDA research in the Midwest in 2006 demonstrated that autumn-planted small grain cover crops reduced soil erosion, nitrate leaching, and suppressed weeds (Teasdale et al., 2007).

National adoption of IPM strategies varies. Corn and soybean growers reported scouting for weeds, insects, and diseases on 50 percent of acres or more in 2000, but reported adjusting planting or harvest dates to manage pests on less than 20 percent of acres (Weibe and Gollehon, 2006).

3.2.4.4. Pathogens and Biological Contaminants

The use of animal manure as a fertilizer has been tied to an increased risk of viruses and bacteria leaching into the water supply. Pathogens such as *Salmonella* sp., *Campylobacter* sp., and *Clostridium perfringens*—along with additives such as livestock antibiotics and hormones—may be released into surface or ground water when manure is applied to fields (Brooks et al., 2009; Lee et al., 2007b; Unc and Goss, 2004). The USDA Report to Congress on use of manure for fertilizer and energy reports that approximately 12 percent of corn and 1 percent of soybeans are fertilized with manure (MacDonald et al., 2009).

The flow paths by which pathogens can contaminate ground or surface water are the subject of current research. Transport through soil has been shown to remove harmful bacteria in some cases, though this may depend on soil characteristics, the hydrologic regime, and the pathogens in question (Malik et al., 2004; Unc and Goss, 2004). Contamination rates likely are greater where there is higher runoff relative to infiltration, a high water table, or a direct surface– ground water connection. Implementation of manure management practices, such as covering or storage at elevated temperatures before application, can reduce runoff and the presence of pathogens. In addition, applying manure during times of low runoff potential can reduce the risk of water contamination (Moore et al., 1995; Guan and Holley, 2003).

3.2.5. Water Quantity

3.2.5.1. Water Use

Agricultural production fundamentally depends on water. In many of the top corn- and soybean-producing states, agricultural water demand is met by natural rainfall; in other states, favorable yields are achieved using irrigation. In USDA farm census years between 2002 and 2008, approximately 14 to 15 percent of corn and 7 to 9 percent of soybean acres annually harvested were irrigated (NASS, 2009a, 2009b). In the Great Plains and the Midwest, where the majority of corn and soybean production takes place, farmers who irrigate rely largely on ground water (Kenny et al., 2009); for instance, in Nebraska in 2008, 95 percent of irrigated corn received ground water from wells (NASS, 2009b). When used to enhance yields, crop irrigation is by far the most significant use of water in the ethanol and biodiesel supply chains (see Figure 2-2), and it tends to be much higher than water use for most other non-renewable forms of energy on an energy content basis (King and Webber, 2008; Gerbens-Leenes et al., 2009; Wu et al., 2009).

Geography and the type of land/crop conversion will determine water use impacts from increased corn or soybean production to meet biofuel demands. Water use could increase if land in pasture or other low- or non-irrigated uses is converted to irrigated row crop production, especially corn in places like the Great Plains, where water demand for irrigation is even higher than for soybeans on a per area basis (NRC, 2008; NASS, 2009b). In predominantly rain-fed locations like much of the Midwest, water supply impacts are less likely to occur. Future assessment of biofuel feedstocks will also need to consider restrictions on water use due to competing demands for water resources (Berndes, 2002).

In some parts of the country, water demands for corn are met by natural rainfall, while in other places supplemental irrigation is required to achieve favorable yields. For instance, in Iowa in 2007, less than 1 percent of the more than 14 million acres planted in corn was irrigated. In contrast, approximately 60 percent of Nebraska's 9.5 million acres of corn was irrigated in the same year (NASS, 2009a). On fields that are irrigated, rates of application on a per area basis also vary from place to place. In 2008 in the United States overall, an average of 1 acre-foot (325,851 gallons) of water was used on an acre of irrigated corn (NASS, 2009b). In Iowa and Illinois, the rate of corn irrigation was half that, while in Nebraska the rate was 0.8 acre-feet (260,680 gallons) per irrigated acre. Above-average rates of irrigation are generally found in the western United States (NASS, 2009b).

Several studies have attempted to calculate the amount of irrigation used to produce corn ethanol on a per gallon basis. Wu et al. (2009) averaged irrigation water used over all irrigated and non-irrigated acres on a regional scale and found approximately 7 gallons of water were required per gallon of ethanol in states like Iowa and Illinois, and over 300 gallons of water/gallon ethanol were required in Nebraska and the Dakotas. Chiu et al. (2009), using similar methodology, were largely in agreement. Another study, which focused on irrigated acres only, found that up to 1,000 gallons of water were required per gallon of irrigated ethanol in the Great Plains (Dominguez-Faus et al., 2009). A handful of studies have also accounted for water use via crop evapotranspiration (ET) (see the text box) to produce a "water footprint" of ethanol production in the United States (e.g., Mubako and Lant, 2008;

Dominguez-Faus et al., 2009). Estimates of water use per gallon of ethanol tend to be higher in these calculations, because both irrigated and non-irrigated corn loses water to the atmosphere via ET. Taking into account the volume of corn starch ethanol produced per state and the area of corn currently needed as feedstock to achieve those volumes, as well as irrigation practices, total water use for corn ethanol can add up to very large amounts. Chiu et al. (2009) suggest approximately 5 billion gallons of irrigation water could be used in a single season in places like Iowa and Illinois versus 300 billion gallons in Nebraska.

WATER USE: IRRIGATION VS. EVAPOTRANSPIRATION

The agricultural water cycle involves inputs and outputs of water. Water enters into agricultural fields via precipitation and/or irrigation from ground or surface water sources. It exits via surface runoff, infiltration into soil layers below the rooting zone, and ET. ET is the combination of evaporation from the surface of the ground and transpiration. Transpiration is the process by which plants take up water from the soil and release water vapor through pores, called stomata, in their leaves.

Crop water use and its implications for water availability for other human and ecological demands can be understood by examining various parts of the agricultural water cycle, including irrigation and ET. Some crops in some regions require no irrigation to achieve favorable yields. However, when water is withdrawn for irrigation, this results in at least the temporary and sometimes functionally permanent reduction in the availability of water for other uses. As water is being withdrawn and applied, all of it is temporarily unavailable for other uses. A fraction of irrigation water may run off into surface water bodies or infiltrate into the soil or shallow groundwater aquifers, and if of sufficient quality, it becomes available once more for human surface withdrawal or ecological communities on a relatively short time scale. When irrigation water is pumped from deep regional aquifers, it may not return to (or recharge) those aquifers for centuries, which restricts that supply of water to meet other demands well into the future. Another important facet of the agricultural water cycle is the amount of water that exits fields via ET. The total amount of water that crops and other vegetation evapotranspire depends on many factors, including the species or variety of plants, cultivation and/or irrigation practices, weather, and soil properties. Consequently, calculating changes in ET attributable to a shifting biofuels landscape and estimating their magnitude and impact requires accounting for these factors. For example, Hickman et al. (2010) show that corn evapotranspires less water over its growing season than switchgrass and Giant Miscanthus, though whether this is an overall trend would have to be evaluated in combination with studies of other cultivars, seasons and locations. Water that is evapotranspired is lost to the atmosphere, where it is unavailable for human withdrawal or ecological use and where it can affect regional climate trends and feedbacks (e.g., VanLoocke et al., 2010). Therefore, measuring both the water withdrawn for irrigation of biofuel feedstocks and their water loss to the atmosphere are important to understand the influence of feedstock production on water availability, and some studies seek to reflect this (e.g., Dominguez-Faus et al., 2009).

The presence of a crop residue layer, such as corn stover, shields the soil surface, reducing evaporation while also maintaining soil organic matter, a critical component of the water-holding capacity of the soil. The harvesting of corn stover is likely to have little or negligible impact on water use above and beyond corn cultivation if undertaken in the most productive corn-growing regions of the United States, where corn stover is not functionally necessary for retention of soil moisture. However, under warmer conditions, corn growth can be enhanced by higher available water resulting from maintaining crop residues (Blanco-Canqui and Lal, 2009b). If corn stover is removed from dry corn cultivation areas with supplemental irrigation (e.g., in states like Nebraska), loss of soil moisture that would have otherwise been retained by corn stover cover could necessitate additional irrigation. The opposite, however, has been demonstrated in colder, wetter soils where heavy crop residue layers can delay corn emergence and lower crop yields (Liu et al., 2004). In such cases and locations, it is advantageous to remove at least some of the corn stover.

Water for soybean cultivation usually comes from natural precipitation and sometimes irrigation. In the leading soybean-producing state of Iowa in 2007, 8.6 million acres of soybeans were grown, of which less than 1 percent was irrigated (NASS, 2009a). Nebraska, on the other hand, grew 3.8 million acres of soybeans in 2007, of which over 40 percent was irrigated (NASS, 2009a). On soybean fields that were irrigated in 2008, rates of application on a per area basis averaged 0.7 acre-feet (228,095 gallons) in the United States overall (NASS, 2009b). In Iowa, Illinois, and Nebraska the rates of soybean irrigation were 0.4, 0.5, and 0.6 acre-feet, respectively, while in Arkansas the rate was above the national average at 0.9 acre-feet (293,265 gallons) of water per irrigated acre (NASS, 2009b).

Focusing solely on irrigated soybeans, Department of Energy (DOE) (2006) estimated an average nationwide rate of about 6,000 gallons of irrigation water to produce a gallon of biodiesel. A volume of biodiesel with the energy equivalent of a gallon of ethanol (which is less energy-dense than biodiesel), would require about 4,000 gallons of irrigation. A more recent study brings the irrigation volume in the range of 1,500 to 3,000 gallons of water per volume of biodiesel equivalent to the energy in a gallon of ethanol (Dominguez-Faus et al., 2009). It is important to note that these rates do not account for the more than 90 percent of soybeans that are grown without irrigation. An average value based on irrigated and non-irrigated soybean acres would likely be much smaller, unless it also included ET, which has been estimated at about 2,000 gallons or more per volume of biodiesel equivalent to the energy in a gallon of ethanol in the United States (Dominguez-Faus et al., 2009; Gerbens-Leenes et al., 2009).

3.2.5.2. Water Availability

Because agriculture accounts for such a large share of water use in the United States (37 percent of withdrawals nationwide in 2005, and a much larger percentage in some parts of the country, according to Kenny et al., 2009), changes in agricultural production could impact future water availability. For example, land conversion to irrigated corn from typically non-irrigated pasture, marginal, or CRP land could create more demand for water, adding to existing water constraints and potentially creating new ones in places like the Great Plains states.

To a large extent, the current capacity to produce biodiesel from soybeans resides in states with rain-fed soybean cultivation. Such strategic siting of biodiesel production facilities minimizes both demands for irrigation water for biodiesel feedstock and potential conflicts over water required for other purposes, such as power generation, public water use, ecosystems, and recreation. However, if biodiesel production develops in places requiring greater soybean irrigation such as the Great Plains, water availability could be reduced. This is especially true if irrigated soybean cultivation replaces other low or non-irrigated land uses. In the case of both corn and soybeans, because the vast majority of irrigation withdrawals in the Great Plains are from ground water wells (NASS, 2009b) that tap underground aquifers like the High Plains aquifer, ground water availability is likely to be affected more directly than surface water.

Both surface water and ground water withdrawals can negatively impact aquatic life. Surface water withdrawals can reduce flood flows (or peak flow regimes), as well as reduce total flow (or discharge) during summer months when irrigation requirements are high and surface water levels are low (Poff and Zimmerman, 2010). Ground water availability over the past several decades has been notably diminished in some places in the United States by withdrawals for irrigation (Reilly et al., 2008). The water level in the southern portion of the High Plains aquifer, in particular, has dropped 37 feet since extensive irrigation development in Texas in the 1930s and 1940s; recent annual water level declines in the states overlying the High Plains aquifer (e.g., Kansas, Nebraska, South Dakota, Oklahoma) are modest in comparison, but steady (McGuire, 2009). The consequences of ground water withdrawals that exceed recharge rates could include reduced water quality, prohibitive increases in the costs of pumping, reduced surface water levels through hydrological connections, and subsidence (Reilly et al., 2008). Withdrawals from hydrologically connected aquifers can lower base flow to rivers and streams that depend on ground water to maintain year-round stream flow. In some areas, stream flow has been reduced to zero because of ground water depletion, but in other areas, minimum stream flow during the summer has been sustained because of irrigation return flow to streams (Bartolino and Cunningham, 2003).

Options to mitigate the challenges outlined above do exist. Locating corn ethanol and soybean biodiesel production in regions that do not require irrigation is one option, although it may lead to displacement of row crop acreage for food or feed to more irrigation-intensive regions. Irrigation in locations that require it could be minimized by using crop varieties bred for high water use efficiency and/or drought tolerance or by installing more efficient irrigation delivery. Such strategies over the past decade have allowed nationwide irrigation delivery per acre to remain relatively flat for both corn and soybeans at the same time yields per acre have risen (Keystone Alliance, 2009). In the upper Mississippi River basin, specifically, the USDA estimates that the 2 percent of land that does receive irrigation has achieved 46 percent reduction in irrigation per acre, resulting from more efficient irrigation delivery and application techniques (NRCS, 2010a). It may also be possible to use recycled or reclaimed water to irrigate, thus reducing reliance on freshwater withdrawals (NRC, 2008).

3.2.6. Soil Quality

3.2.6.1. Soil Erosion

Soil erosion can have substantial negative effects on soil quality by preferentially removing the finest, uppermost soil particles, which are higher in organic matter, plant nutrients, and water-holding capacity than the remaining soil. The soil erosion impact of growing corn or soybeans for biofuel will vary, largely depending on the particular land use/land cover change and tillage practices. Conversion of uncultivated land, such as CRP acreage or pasture, to corn or soybeans for biofuels is the land use change scenario most likely to increase erosion and sedimentation. The USDA CEAP report on the upper Mississippi River basin found that for land in long-term conserving cover, like CRP, soil erosion and sediment loss were almost completely eliminated (NRCS, 2010a). Moreover, CRP acreage in riparian areas slows runoff, promoting the retention of sediment, nutrients, and other chemicals. The USDA Farm Service Agency (FSA) estimated that, in 2008, CRP land collectively prevented 445 million tons of soil from eroding (FSA, 2009). The soil-erosion effects of converting former or current pasture land to corn will vary depending on prior erosion rates. Pasture land in the United States' Southern Piedmont region, for example, can exhibit soil stability equal to forested or conservation-tilled land (Franzluebbers et al., 2000); converting this type of land to conventional corn production will increase soil erosion. In contrast, if much of the increase in corn or soybean production comes from a shift from other crops, the effect on soil erosion is likely to be much smaller. There have been substantial improvements over the last two decades in corn and soybean soil loss indicators (Keystone Alliance, 2009). By 2015, USDA predicts an increase in sheet erosion by 1.7 percent (in accord with the 1.6 percent increase in acreage above the 2008 baseline), and an increase in wind erosion by 0.7 percent. There is variation across regions in magnitude and whether the changes are driven by land use or management changes (Malcolm et al., 2009). Allocation of a higher percentage of corn or soybeans for biofuel production from land already in corn and soybean production should not alter soil erosion rates.

Tillage practices can mitigate soil erosion on current agricultural lands. Conventional tilling[17] breaks up soil aggregates, increasing erosion by wind and water (Lal, 2003). In contrast, conservation tillage—defined as practices that maintain at least 30 percent of the ground covered by crop residue (Lal, 1997)—can considerably reduce soil erosion (Cassel et al., 1995; Shipitalo and Edwards, 1998). No-till agriculture, a type of conservation tillage, disturbs the soil only marginally by cutting a narrow planting slit. According to an NRCS report, conservation tillage is practiced on 91 percent of all crop acreage in the upper Mississippi River basin, with 28 percent in no-till and only 5 percent in continuous conventional tillage (NRCS, 2010a). Conservation tillage practices may also partially mitigate the impact of converting CRP acreage to biofuel corn production (Follett et al., 2009). A majority of CRP acreage in areas of the Midwest is classified as highly erodible land, where tillage practices are influenced by the conservation compliance provisions of the 1985 Food Security Act (Secchi et al., 2009). For example, corn-soybean rotations with no-till cultivation can be required in order to maintain eligibility for certain USDA benefits and programs (Secchi et al., 2009).

Finally, removal of corn stover beyond a certain threshold may substantially increase soil erosion rates. Crop residues remaining after harvest reduce erosion both directly through the physical shielding of soil particles and indirectly through the addition of organic matter, promoting aggregation. Thus, stover removal is likely to be most problematic on erosion-prone soils. Due to this and cost concerns, a recent study suggested that only approximately 30 percent of corn stover[18] would be available for sustainable harvesting in the United States if erosion rates were to be kept lower than soil loss tolerances (T-values) as defined by the USDA NRCS (Graham et al., 2007). Because of wind erosion, the potential for corn stover removal in the western plains states may be particularly limited (Graham et al., 2007). Site cultivation practices may partially compensate for the effects of residue removal. If no-till agriculture were universally adopted, sustainably harvested corn stover supplies could increase from approximately 30 to an estimated 50 percent (Graham et al., 2007). Yet, even with no-till management, corn stover removal rates at or higher than 25 to 50 percent, depending on location, have been shown to increase erosion potential (Blanco-Canqui and Lal, 2009a).

3.2.6.2. Soil Organic Matter

Soil organic matter is critical to soil quality because it retains plant nutrients and water, facilitates carbon sequestration, promotes soil structure, and reduces erosion. The impact of corn and soybean production for biofuel on soil organic matter will depend on the land use history of the cultivated acreage. Corn production will negatively impact soil quality on acreage where organic matter has accumulated over time—for example, grasslands. Placing previously undisturbed soils into cultivation can result in carbon losses of 20 to 40 percent during the first five to 20 years of continuous conventional tillage (Davidson and Ackerman, 1993). Reduced losses would be expected in cases where conservation tillage is used (Follett et al., 2009) or from soils already depleted in organic matter. Increased corn or soybean production on currently cultivated land will have a smaller effect on soil organic matter, particularly where substantial amounts of crop residues are returned to the soil or a cover crop is used (Drinkwater et al., 1998; Lal, 2003; Adviento-Borbe et al., 2007). While soil quality degrades over time, yields and production can be maintained by the use of fertilizers, both commercial and organic.

Tillage practices may influence soil organic carbon levels as well. Meta-analyses have concluded that no-till or reduced tillage increases soil carbon levels (West and Post, 2002; Ogle et al., 2005). However, recent studies—more limited in scope—have suggested that no-till practices may increase carbon in the upper layers of the soil, but decrease amounts at lower depths compared to conventional tillage, with no difference in overall carbon storage (Baker et al., 2007; Blanco-Canqui and Lal, 2008). More studies of tillage effects on soil carbon deeper in the soil profile are needed. Determining the effect of tillage on soil carbon can be especially important for greenhouse gas modeling. For example, Kim et al. (2009) estimated that grassland conversion to corn for ethanol production would take 18 years to provide greenhouse gas benefits with conventional tillage, whereas with no-till it would take only four years.

The amount of corn stover needed to maintain soil organic matter levels is likely higher than that needed to control erosion (Wilhelm et al., 2007). Thus, the 30 percent harvesting estimate by Graham et al. (2007) may not leave enough stover on the field to maintain soil organic matter. Stover removal rates of 25 to 75 percent have been shown to decrease soil organic matter across several soil types even under no-till management (Blanco-Canqui and Lal, 2009a). There is concern that high stover removal rates may decrease soil carbon sequestration and concomitantly lower crop yields (Karlen et al., 2009). Corn stover removal at about 50 percent on a non-irrigated field in Nebraska led to significant declines in both grain and stover production (Varvel et al., 2008). Whatever the removal rate for a particular site, it has been recommended that soil erosion and organic matter content should be periodically monitored to allow stover removal rates to be adjusted accordingly (Andrews, 2006). The effects of crop residue removals on crop yields have been shown to be highly variable depending on soil type, climate, topography, and tillage management, among other characteristics (Blanco-Canqui and Lal, 2009b). Since the effects of management practices on soil organic matter often take many years to detect, predicting the impacts of different levels of residue removal on soils will take a combination of both modeling and monitoring of long-term, residue-removal field trials.

3.2.7. Air Quality

Air quality impacts during cultivation and harvesting of corn and soybeans are associated with emissions from combustion of fuels by farm equipment and from airborne particles (dust) generated during tillage and harvesting. Soil and related dust particles become airborne as a result of field tillage, especially in drier areas of the country. In addition, emissions result from the production and transport of fertilizers and pesticides used in corn and soybean production, and the application of fertilizers and pesticides to each crop. Air emissions associated with cultivation and harvesting of corn and soybeans for biofuel will mostly occur in sparsely populated areas (Hill et al., 2009). Subsequent stages in the biofuel supply chain (see Figure 2-2), including feedstock logistics and biofuel production, distribution, and use, also affect air quality and are discussed in Section 4.

Cultivating and harvesting corn and soybeans requires a range of mechanized equipment that use different fuels, including diesel, gasoline, natural gas, and electric power (U.S. EPA, 2010a). Generally, equipment used to produce corn and soybeans consumes more diesel than for most other crops, while the rate of gasoline consumption is somewhat less than that of other crops. Primary emissions from fuel use include nitrogen oxides (NO_x), volatile organic compounds (VOCs), carbon monoxide (CO), sulfur dioxide (SO_2), and coarse and fine particulate matter (PM_{10} and $PM_{2.5}$). Gasoline use may also result in benzene, formaldehyde, and acetaldehyde emissions. With respect to corn stover, additional fuel use depends on the method of stover harvest. For example, methods that can simultaneously collect grain and stover will use less fuel than those requiring multiple passes with a harvester. For this reason, one-pass harvesters are currently being developed and tested (Shinners et al., 2009).

Cultivation also affects the release of CO_2, NO_x, and methane (CH_4) from the soil. Conservation tillage practices, including no-till, are generally assumed to sequester greater amounts of soil organic carbon than conventional tillage, reducing CO_2 emissions—though recent studies have questioned this finding (see Section 3.2.6.2). Additionally, there are uncertainties regarding tillage and the release of NO_x and CH_4 from the soil. Time from the initiation of tillage practices appears to be a determining factor (Six et al., 2004). In a meta analysis study, the implementation of no-till resulted in an initial increase in both N_2O and CH_4 relative to conventional tillage; within 20 years, however, cumulative emissions were substantially lower under no-till, although N_2O estimates were highly variable (Six et al., 2004).

Emissions are also associated with generation of electricity used for irrigation water pumping. Irrigation power needs are estimated to range from 3 to 11 kilowatt-hours (kWh) per irrigated acre, depending on the region, with a national average of 8 kWh per irrigated acre. For soybean cultivation, electricity use is estimated to be 4.6 kWh per acre (Sheehan et al., 1998a; Pradhan et al., 2008; Hill et al., 2006). Emissions associated with this use depend on the source of the electricity consumed. Coal is the predominant fuel source for electricity in the Midwest, accounting for 71.3 percent of generation in the 12 primary corn-producing states. Coal-fired power plants are significant sources of SO_2, NO_x, CO_2, and mercury emissions.

Corn, with a moisture content over 18 to 20 percent, may require some drying to reach a water content appropriate for storage (South Dakota State University, 2009). Grain driers use liquefied petroleum gas (LPG) and electricity. LPG and electricity use depend on grain moisture content at harvest. For example, typical Midwest grain harvest conditions and yields require 20 gallons of LPG per acre harvested. The exact amount depends on grain moisture conditions at harvest.

Pesticides are commonly used on both corn and soybeans, with corn having more intensive application rates (NRC, 2008, as cited in U.S. EPA, 2010a, Table 3-3) than soybeans. Corn has the highest nitrogen fertilizer use per acre of any biofuel feedstock. Because soybeans are legumes, they require much lower amounts of fertilizer, particularly nitrogen (NASS, 2006, 2007). Soybeans can acquire nitrogen from the atmosphere and therefore require less external nitrogen fertilization than corn.

Air emissions associated with fertilizer manufacturing and transport include NO_x, SO_x, VOC, CO, and particulate matter (PM_{10} and $PM_{2.5}$), while pesticide production and blending may result in emissions of 1,3-butadiene, benzene, and formaldehyde (U.S. EPA, 2010a).

Application of fertilizers and pesticides may result in releases to the air. The primary pollutants associated with the releases to air are benzene and acrolein. The results described in U.S. EPA 2010a are consistent with another study, which found increases in benzene, formaldehyde, acetaldehyde, and butadiene emissions, although that study included feedstock transport and thus is not directly comparable (Winebrake et al., 2001). The application of inorganic and organic fertilizers can increase NO_x, ammonia (NH_3), and CH_4 emissions from the soil (Mosier et al., 1996; Jarecki et al., 2008; Janzen et al., 2003; Das et al., 2008). Emissions of CO, NO_x, and SO_2 increased with the use of corn stover as a feedstock in a hypothetical system (i.e., a simulation based on corn stover life cycle data), with higher NO_x emissions mainly due to denitrification of increased amounts of nitrogen fertilizers added to farm soils (Sheehan et al., 2004).

3.2.8. Ecosystem Impacts

3.2.8.1. Terrestrial Biodiversity

Overall, row crops provide habitat for a less diverse set of species than pasture or CRP lands (Fletcher et al., 2010). Using a meta-analysis approach, Fletcher et al. (2010) found that bird abundances are significantly lower in row crops, particularly for species of conservation concern. However, no-till fields have a greater diversity and abundance of species, including birds, invertebrates, and small mammals (Warburton and Klimstra, 1984). A variety of conservation management practices on cropland can improve wildlife habitat, including providing nesting and winter cover (Brady, 2007). Yet additional conversion of pasture or CRP lands will contribute to additional habitat loss and landscape homogenization (Landis et al., 2008; Fletcher et al., 2010).

While land use/land cover conversion in general contributes to a loss of landscape diversity and increases in fragmentation, there is little evidence to date of widespread land use changes due to biofuel production. Species' responses to habitat fragmentation are complex (Ewers and Didham, 2006), which makes it difficult to discern effects due to biofuel production. Habitat isolation can alter dispersal success and population structure within fragments; this has longer-term consequences for genetic, morphological, and behavioral traits of species (Ewers and Didham, 2006). Breeding bird surveys in Iowa found that the abundance of nesting species increases in diverse landscapes with mosaics of crop and non-crop habitats, as compared with crop monocultures (Brady, 2007). If landscape diversity decreases (especially if CRP land is converted to corn), migratory birds may lose habitat and will likely decline in numbers. On CRP lands, several grassland bird species have increased in abundance, and it is estimated that without the 7.4 million acres of CRP in the Prairie Pothole region of the United States, over 25 million ducks would have been lost from the annual fall migratory flights between 1992 and 2004 (Dale et al., 2010).

The removal of corn stover residues from agricultural corn fields for ethanol production also has potential consequences for biodiversity by reducing habitat and food sources (Brady, 2007). Increased crop residue amounts on fields generally result in a greater diversity of small mammals (Brady, 2007), while removing crop residues has been shown to negatively affect both terrestrial and soil organisms (Lal, 2009; Johnson et al., 2006). Similar to no-till practices, maintaining crop residues like corn stover on fields can increase the diversity of beneficial soil- dwelling invertebrates that can improve soil quality (Brady, 2007).

Increased corn production can also impact other aspects of biological diversity and associated ecosystem services. In Iowa, Michigan, Minnesota, and Wisconsin, biological control of soybean aphids declined due to lower habitat diversity as the proportion of corn in the local landscape increased, resulting in increased expenditures for and application of pesticides, and reduced yields (Landis et al., 2008). This results from reduced landscape diversity that decreases habitat availability for many insects and animals in the local region (Landis et al., 2008). Similarly, intensification of soybean production and pesticide use may also threaten biodiversity and nearby plants and animals (Artuzi and Contiero, 2006; Koh and Ghazoul, 2008; Pimentel, 2006). Also, agricultural herbicides affect the composition of local plant communities, which then affects the abundance of natural enemy arthropods and the food supply of local game birds (Taylor et al., 2006).

3.2.8.2. Aquatic Biodiversity

The impact of increased corn and soybean cultivation on ecosystems and biodiversity depends, in large part, on where crop production occurs and what management techniques are used. Much of the Midwestern landscape uses some type of surface or subsurface drainage to convey water away from fields. These drainage waters carry sediments, nutrients, and pesticides into surface and ground waters (Blann et al., 2009). Approximately half of the nitrogen lost from croplands travels through subsurface drainage and nearly two-thirds of it is subsequently released to surface waters, where it is combined with the 21 percent of nitrogen directly lost to runoff (NRCS, 2010a). In surface waters these inputs result in eutrophication and increased turbidity. Eutrophication can occur as fertilizer application increases nutrient loadings (nitrogen and phosphorus) in surface waters such as streams, rivers, lakes, wetlands, and estuaries (Carpenter et al., 1998; Pollock et al., 2007; Breitburg et al., 2009; U.S. EPA, 2010a). Increased phosphorus concentration has been correlated with declines in invertebrate community structure (Carpenter et al., 1998), and high concentrations of ammonia nitrogen are known to be toxic to aquatic animals (Kosmala et al., 1999; Faria et al., 2006). Severe oxygen depletion and pH increases, both of which are correlated with eutrophication, can inhibit growth and lead to mortality in fish and invertebrates (Carpenter et al., 1998; Pollock et al., 2007; U.S. EPA, 2010a). In addition, as aquatic systems become more enriched by nutrients, algae growth can cause a shift in species composition. Nutrient enrichment in estuaries leads to hypoxia, which limits biodiversity and threatens commercial and recreational fisheries (Wang et al., 2007a; U.S. EPA, 2010a). Cumulative effects of hydrologic and water quality changes due to agricultural drainage have led to declines in intolerant, sensitive species; shifts in aquatic community composition; and homogenization of aquatic faunal assemblages to more tolerant, generalist species (Blann et al., 2009).

Crop production not only releases nutrients to water bodies, but also sediments. Cultivation practices or corn stover harvest rates leading to soil erosion can increase wetland sedimentation, which may, depending on sediment depths, cover viable seeds sufficiently to prevent germination (Gleason et al., 2003). Row crops also release more sediment into wetlands than perennial grasses (Nelson et al., 2006). An increased input of sediments into aquatic ecosystems can increase turbidity and water temperatures and bury stream substrates, limiting habitat for coldwater fish (U.S. EPA, 2006a).

In addition to nutrients and sediments, agricultural drainage and runoff can contain a variety of pesticides and other pollutants that are transported into water bodies (Blann et al., 2009). For example, Malcolm et al., (2009) project a four percent increase in continuous corn production in response to biofuel demand, which is likely to lead to more herbicide application. One such herbicide is atrazine, commonly used in the United States and predominantly on corn. As part of the EPA's 2003 Memorandum of Agreement, atrazine registrants are required to conduct watershed monitoring to ensure protection of aquatic ecosystems. If any of the watersheds show levels of atrazine above the Agency's level of concern for two years, the registrants must initiate watershed-based management activities in concert with state or local watershed programs to reduce atrazine exposure (U.S. EPA, 2010g). To date the results show that for most locations monitored, atrazine levels are below EPA's current level of concern. Although EPA concluded in 2007 that atrazine does not adversely affect amphibian gonadal development based on a review of laboratory and field studies (FIFRA Scientific Advisory Board, 2007), the Agency has begun a comprehensive reevaluation of atrazine's ecological effects, including potential effects on amphibians, based on data generated since 2007.

Fungicide pollution from runoff events also has been shown to impact algae and aquatic invertebrates in areas where soybeans are intensively grown (Ochoa-Acuna et al., 2009). Manure application can also lead to runoff that contains pathogens such as *Salmonella* sp., *Campylobacter* sp., and *Clostridium perfringens,* along with additives such as livestock antibiotics and hormones (Unc and Goss, 2004; Lee et al., 2007b; Brooks et al., 2009).

Conservation practices that are implemented and encouraged for erosion control and to reduce nutrient losses—such as grass or riparian buffers, constructed/restored wetlands, or enrollment in CRP—can ameliorate some of the ecosystem impacts described above. Compared with the 56.6 percent of nitrogen lost from conventionally farmed croplands, conservation practices reduce overall nitrogen loss by 18 percent on average, although this depends on location and specific practices (NRCS, 2010a). Lands used for these management strategies can serve as habitat for a variety of species and have been shown to improve species diversity and abundances in agricultural landscapes, especially if these lands are left uncultivated for long periods (van Buskirk and Willi, 2004; Brady, 2007). Agriculturally dominated landscapes with a diversity of natural or non-crop habitats can also enhance the abundance and diversity of predators of insect pests (Bianchi et al., 2006). However, the value of such habitats varies by species, habitat composition (e.g., habitat type, plant species richness), and landscape structure (e.g., habitat heterogeneity, surrounding land uses, connectivity) (Jeanneret et al., 2003), and specialist species are less likely to benefit from these types of habitats and heterogeneous landscapes (Filippi-Coadccioni et al., 2010).

3.2.8.3. Invasive Plants

Though neither corn nor soybeans are native to the United States, modern varieties of corn and soybeans under production today in the United States pose little risk of dispersing seeds or regenerative plant parts or creating hybrids with related plants that will become weeds or invasive plants in the future. Corn and soybeans rarely overwinter successfully in major production areas, but on occasion, seed from the previous year's crop can emerge in the following year and the plants persist through a single growing season as a weed. Such populations of plants do not become a chronic problem, however, because they do not sustain themselves (Owen, 2005). To date, no cases of invasive corn or soybeans have ever been reported in natural areas in the United States.

The extensive cultivation of row crops that are genetically engineered to resist glyphosate, a commonly used herbicide, may result in indirect effects on other weed species and invasive plants. One study correlated the increased use of this herbicide with the appearance of glyphosate resistance in at least 10 agricultural weeds in the United States; loss of effectiveness of glyphosate could encourage the use of more toxic herbicides (NRC, 2010a).

3.2.9. Key Findings

Much of the environmental impact of corn starch ethanol and soybean biodiesel production depends on the types of land put into cultivation. To date, most additional acreage has originated from lands currently in crop production. Expanding corn crop production to CRP or pasture will likely have the greatest environmental impacts. Between September 2009 and August 2010, approximately 35 percent of corn consumed domestically was converted into ethanol biofuel (ERS, 2010c, 2010d). Corn acreage has increased over 2005 levels in part due to ethanol demand, and planted acreage is expected to increase from 2008/2009 levels of 86 million acres to 90 million acres in 2019 to meet the EISA target of 15 billion gallons per year (USDA, 2010c). The most plausible scenario at this time is that these additional acres will be conventionally managed, tilled corn in predominantly rain-fed areas, replacing conventionally grown soybeans or other row crops.

Currently, biodiesel accounts for approximately 5.5 percent of the soybean consumption; USDA expects this percentage to increase to 7.7 percent by 2012 and hold steady through 2019. Greater diversion of soybeans to biodiesel production will not result in additional impacts due to land use change. USDA also expects that soybean acreages will hold steady at 76 million acres (USDA, 2010b), though this number may increase to meet the EISA target. Moreover, it may be necessary to increase acreage, yield, or the proportion of the soybean harvest that is devoted to biodiesel in order to meet EISA targets (FAPRI, 2010a). The most plausible scenario for soybeans at this time is that an increased proportion of conservation-tilled soybeans will be diverted for biodiesel production.

The use of corn stover for ethanol production may not increase acreage dedicated to corn. The most plausible scenario at this time is a 40 percent removal rate from conventionally managed, tilled corn grown in predominantly rainfed areas as a separate harvest.

Increasing production of corn for ethanol and soybeans for biodiesel will likely have implications for water quality. Increased corn and soybean production could increase nutrient, sediment, and pesticide loadings to water bodies, including the Gulf of Mexico, Great Lakes, and Chesapeake Bay, although fewer negative impacts are expected with soy production. Private drinking water wells could see increases in nitrate and public drinking water systems could see increases in their costs to lower nitrate levels. However, some of the potential increased nutrient loadings from corn grown for ethanol might be reduced if farmers expand their use of conservation practices. Increased risk of pathogens entering surface waters from application of animal manure fertilizers is also possible. Removal of corn stover could lead to loss of soil surface cover, thereby increasing runoff of nutrients, phosphorus in particular, and sediments to surface waters; harvesting corn stover may reduce soil nutrient availability, leading to increased fertilizer applications.

The magnitude of water availability impacts from increased corn or soybean production for biofuel will vary geographically. If corn replaces other crops in the Midwest (the most plausible scenario), water availability will be minimally impacted. However, if corn cultivation replaces perennial grasses such as those on CRP land, it may reduce ET, leading to increases in water availability. Increased corn and soybean production in areas requiring irrigation, such as the Great Plains, will increase water usage, potentially decreasing water availability. Removal of corn stover for ethanol will not affect water availability in most parts of the United States.

Negative soil quality impacts from biofuel feedstock production can arise from converting acreage with perennial vegetation cover to conventional corn and/or soybeans likely increasing soil erosion, sedimentation, and nutrient losses. In contrast, allocation of a higher percentage of corn or soybeans for biofuels from land already in production is likely to have much smaller impacts. High stover removal rates are of particular concern with regard to loss of soil and organic matter, which in turn can decrease soil carbon sequestration and adversely impact crop yields. Impacts can be reduced through conservation practices, particularly no-till, yet even with this management practice risks to soil organic matter from high stover removal remain.

An increase in the production of corn and soybean for biofuel will likely lead to increased pollution from fossil fuels associated with cultivation and harvesting and from airborne particles (dust) generated during tillage and harvesting. Air emissions also result from the production and transport of fertilizers and pesticides used in corn and soybean cultivation, and the application of fertilizers and pesticides for each crop. Increasing their use will likely increase the volume of emissions.

Ecosystem health/biodiversity impacts include degradation of aquatic life due to eutrophication and herbicide runoff, impaired aquatic habitat due to sedimentation from soil erosion, and decreases in landscape diversity. Conversion of CRP lands, which are predominantly grasslands, may lead to declines in grassland birds, ducks, and other wildlife that use these lands as habitat. Corn and soybean typically are not invasive in the U.S. corn- and soybean-growing regions.

For a more comprehensive, qualitative comparison of the environmental impacts of corn, soybean, and corn stover, including a discussion of the most plausible impacts, see Section 6.

3.2.9.1. Key Uncertainties and Unknowns

Uncertainties and a scarcity of data exist in many key areas concerning environmental impacts of biofuel feedstock production. In particular:

- The impacts of additional corn and soybean production are determined by two highly uncertain factors: where the production occurs and the types of management practices employed. In particular:
  - Increased corn and soybean yields may partially offset the need for increased acres in production to achieve EISA goals in 2022. However, the extent to which yield increases will be sustained is currently unknown, and thus the extent to which increased production of corn and soybeans will occur on marginal lands, CRP, and/or via continuous corn production on existing lands now in rotation with other crops is also uncertain.
  - The extent to which conservation practices are currently implemented on cropland nationally is relatively unknown, and the potential for future improvements, including improvements in yield; management of nutrients, pesticides, drainage, and energy use; and erosion control systems, is also uncertain.
- The ability to track impacts will depend on the quality and consistency of monitoring fertilizer and pesticide usage, such as data provided by USDA's NASS. USDA tracking of fertilizer application rates ended in 2005.
- The ability to evaluate current and future water shortages associated with ethanol and biodiesel production is limited by the available data. Annual measurements of the extent of irrigation and amounts of surface and ground water used are not systematically collected nationwide, forcing researchers to use incomplete information to calculate crude water use estimates. The availability of fresh water for a particular use is determined by many factors, including rainfall, soil water retention and ground water recharge, water demand for competing uses, water appropriation policy, and water contamination; attribution of water shortages to a specific use may be difficult to measure without improvements in data collection (Alley et al., 2002; Reilly et al., 2008).
- The uncertainties regarding the effect of corn and soybean production on soil quality arise predominantly from uncertainties regarding the amount and type of land converted to corn or soybeans as a result of biofuel demand. Secondarily, uncertainties regarding the effect on soil quality are caused by lack of detailed land management data. For example, more frequent and detailed data—including geographical location—on tillage practices employed would substantially reduce uncertainties surrounding the soil quality response of producing biofuels.
- The key uncertainties with respect to air quality impacts of increased corn and soybean production are similar to those for water quality with respect to fertilizer and pesticide use and application. In addition, NO_x and NH_3 emission rates from fertilized soil are highly uncertain and variable—they rely on microbial conversion of fertilizer to nitrate, which in turn is influenced by environmental conditions. Similarly, estimates of NO_x emissions from the soil due to tillage practices are highly variable, and a source of additional uncertainty. It is also uncertain how extensively cover crops and tillage practices (both of which can reduce fugitive dust emissions) are

employed. For corn stover, there are a range of assumptions regarding cropping practices, harvest techniques, and farm inputs that require more study.

- Ecosystem health and biodiversity, including fish and wildlife, are heavily impacted by uncertain environmental factors such as nutrient and sediment runoff. Nutrient loadings from row crop production into surface waters depend on many different factors, including changes due to weather, and are therefore widely variable (Powers, 2007). Regardless, the ability to reduce chemical exposure of biota can be beneficial to the ecosystem and local biodiversity. In addition to resolving uncertainties about those factors, more studies are needed on landscape- level associations between corn and soybean production and terrestrial and aquatic biodiversity, as well as biodiversity-related services such as pollination and natural pest control. While conservation practices can improve habitat and water quality, there is considerable uncertainty with respect to where these practices occur, exactly how effective they are, and which species benefit.

3.3. Perennial Grasses

3.3.1. Introduction

Perennial grasses are herbaceous plants that grow in successive years from the same root system. Seed production for most perennial grasses is lower than for grain crops such as corn, resulting in starch and sugar contents that are inadequate for commercial production of biofuel ethanol. However, an active field examining physical, chemical, and biological processes for conversion of cellulose into fermentable sugars has emerged over the past three decades, making perennial grasses an attractive feedstock. Cultivation of perennial grasses as biofuel feedstocks has many potential environmental advantages over traditional row crops such as corn and soybeans. However, major technological challenges exist for the development of these more advanced biofuel conversion technologies, and the realization of these benefits depends largely on where and how these crops are eventually grown. Currently, no commercial-scale facilities for converting perennial grasses to cellulosic ethanol are operating in the United States. However, several switchgrass cellulosic ethanol production facilities are under development (RFA, 2010).

The predominant perennial grasses for biofuels are likely to be monocultures of switchgrass (*Panicum virgatum*) or Giant Miscanthus (*Miscanthus x giganteus*). Other grasses have also been explored and will not be thoroughly reviewed here, including *Arundo donax, Phalaris arundinacua, Sorghum bicolor,* diverse mixtures of native species (see the text box in Section 3.3.9), as well as various "cane" hybrids.

The selection of switchgrass as a model biofuel feedstock in the United States resulted from decades of research by the USDA and DOE, including the Oak Ridge National Lab (ORNL), following the oil crisis of the 1970s. The initial screening program examined 37 potential feedstocks, across a range of soil and management regimes, eventually resulting in the selection of switchgrass in 1991 as the most promising overall feedstock for future study. A recent review of the program is provided by Wright and Turhollow (2010), and all ORNL and subcontractor reports have been made publically available on the Biofuels Feedstock Information Network website.[19] Switchgrass was selected as the most promising model feedstock, partially due to funding constraints and partially because it met certain environmental, management, and economic criteria. In short, switchgrass was selected because of its perennial life history (as opposed to annual) and low demands for inputs after the first year, because of its reduced environmental damages and economic costs, and because its yields were generally high and more consistent from year to year, even though other feedstocks had higher yields in some locations for some years (Wright and Turhollow, 2010). Following this screening, switchgrass was intensively studied for 10 years, as summarized by McLaughlin and Kszos (2005). Several more recent updates exist in the literature (e.g., Wullschleger et al., 2010).

Giant Miscanthus has been studied across Europe extensively since the 1980s (also in response to the oil crisis of the 1970s) under the auspices of several national and multinational programs. A review of these programs is provided in Lewandowski (2000). Giant Miscanthus had not been considered as a feedstock in the United States until recently (Heaton et al., 2004a, 2008). It has similar economic and environmental advantages to switchgrass. However, it is likely more productive: it may be one of the most productive land plants in temperate regions, though it is more difficult to establish (Heaton et al., 2008; Lewandowski et al., 2000).

Switchgrass is a native grass of North America that was widespread across much of the Great Plains before European arrival (Parrish and Fike, 2005). Switchgrass is well adapted to disturbances such as fire and grazing (Knapp et al., 1986) and has historically been grown in the United States as forage for grazing livestock (Parrish and Fike, 2005). Several researchers note that many studies have been published on switchgrass as a forage crop, but practices for forage versus biofuel production can be quite different (e.g., harvest frequency; Heaton et al., 2004b). Two major subtypes of switchgrass have been identified in the wild, an upland and a lowland type, that differ in some key characteristics such as water use (Parrish and Fike, 2005). Research from DOE and USDA across 15 states indicates a yield for switchgrass production averaging from 4 to 10 tons per acre (McLaughlin and Kszos, 2005), in agreement with more recent compilations (Wullschleger et al., 2010). There is considerable variation across sites, ecotypes, management, and other factors. Farm-scale studies have demonstrated that ethanol yield from switchgrass ranges from approximately 240 to 370 gallons per acre, compared to an average of 330 gallons per acre for corn grain (Schmer et al., 2008; assumes 0.0456 gallons per pound for conversion of cellulosic biomass and 0.048 gallons per pound for conversion of corn grain).

Giant Miscanthus is a grass native to Asia, a rare but naturally occurring hybrid from two parental species (*Miscanthus sinensis* and *Miscanthus sacchariflorus*) that have long been used as forage in Asia (Stewart et al., 2009). Giant Miscanthus is sterile and does not produce viable seed (Hodkinson et al., 2002). Thus, all individuals currently studied are genetically identical to the original specimen brought from Japan in the 1930s. This horticultural specimen was soon propagated and transported to Europe, and then later to the United States. Research from Europe indicates variable but high productivities for Giant Miscanthus across Europe (2 to 25 tons per acre), with higher levels in warmer wetter regions such as 13 to 14 tons per acre in Italy (Lewandowski et al., 2000). The first replicated field trials for bioenergy production from Giant Miscanthus in the United States occurred in Illinois from 2002 to 2004, and included side-by side comparisons with switchgrass. Researchers reported yields of 13 tons per acre for Giant Miscanthus and 4.5 tons per acre for switchgrass (Heaton et al., 2008). Since then, other studies have also demonstrated high yields in the United States (Propheter et al., 2010).

Considerable genetic variation for both of these species (for the parental lines of Giant Miscanthus) has yet to be explored as a way to optimize feedstock production and biofuel refining (Keshwani and Cheng, 2009; Vogel and Masters, 1998; Sarath et al., 2008; Carroll et al., 2009; Demura and Ye, 2010). But promising traits, including low lignin and ash content and late or absent flowering periods, indicate ample potential for high crop yields and efficient conversion to ethanol (Jakob et al., 2009). Though standard irrigation, fertilizer, and pesticide use practices for large-scale production have yet to be developed, the potential for biofuel production from these feedstocks is promising.

3.3.2. Overview of Environmental Impacts

Because current production of perennial grasses for biofuels is negligible, current impacts from their production under the RFS2 are also considered negligible. This is likely to change as research and markets develop. As production of biofuel from perennial grass becomes technologically and economically viable, demand for perennial grass will increase. This will result in conversion of qualifying land to perennial grasses, the location and extent of which will depend on region-specific agricultural and economic conditions. Perennial grass production will likely require traditional agricultural activities, including pesticide, fertilizer, water, and fuel/energy usage. The intensity of these activities relative to the land management practices they are replacing will determine the extent to which perennial grass production impacts water quality, water availability, air quality, soil quality, and biodiversity. Interestingly, even though conversion from traditional row crops to perennial grasses is not expected to be a widespread land use change in response to the RFS2, much of the literature on the environmental effects from cellulosic biofuel production focuses on this transition rather than more likely transitions such as from lands under the CRP. EPA is relying on peer-reviewed literature for this first triennial report; thus, the remaining sections include caveats where appropriate to highlight this mismatch.

3.3.3. Current and Projected Cultivation

Perennial grass species, including switchgrass, have historically thrived in the Midwest and are generally well suited to grow as a biofuel feedstock over much of the continental United States (see Figure 3-7). Current production of switchgrass and Giant Miscanthus as biofuel feedstocks in the United States is limited to research field trials in several geographic locations (reviewed in Heaton et al., 2008; Wright and Turhollow, 2010; and Wullschleger et al., 2010). However, the vast majority of the nearly 31 million acres of lands in the CRP, which satisfy the eligibility requirement for EISA, are dominated by perennial grasses, often switchgrass (Adler et al., 2009; FSA, 2010).

Large areas of the eastern United States could support high-yield production of switchgrass (Thomson et al., 2009; Wullschleger et al., 2010). Economic models have projected future cultivation of switchgrass for biomass on CRP as well as existing cropland (de la Torre Ugarte et al., 2003; Walsh et al., 2003). However, the policy and economic assumptions used in those studies are no longer current. More recent economic analysis of commercial-scale switchgrass production suggests that displacement of crops such as corn and soybean is, at present, unlikely (Nelson et al., 2006; Vadas et al., 2008; Jiang and Swinton, 2009; James et al., 2010). The vast majority of land enrolled in the CRP uses switchgrass or other native or introduced grasses (FSA, 2010) and commercial-scale switchgrass trials frequently take place on CRP or other marginal agricultural land (e.g.,

Perrin et al., 2008; Wright and Turhollow, 2010). There are no similar studies for Giant Miscanthus in the United States.

Projected cultivation of perennial grasses, in terms of location and management practices, is highly uncertain. In the near term, it seems likely that perennial grasses for cellulosic ethanol will be produced on lands not already under active cultivation of high-value row crops.

3.3.4. Water Quality

Perennial grasses, often grown as a conservation practice along the margins of agricultural fields to reduce sediment and nutrient runoff into surface water and wetlands, are expected to have fewer water quality impacts than conventional agricultural crops (Keshwani and Cheng, 2009; Blanco-Canqui et al., 2004; Blanco-Canqui, 2010). This will depend, however, on the agricultural intensity of the perennial grass cropping system (e.g., the extent of fertilizer and pesticide use) and the land use that is replaced. In general, both switchgrass and Giant Miscanthus have demonstrated positive responses to fertilization and water supplements, more so for switchgrass with nitrogen and for Giant Miscanthus with water (Heaton et al., 2004b). These responses vary substantially (Heaton et al., 2004a; Lewandowski et al., 2000; Wullschleger et al., 2010; Wang et al., 2010). However, inputs for managed perennial grass cropping systems will likely be higher than for unmanaged lands. Table 3-3 shows inputs needed to grow perennial grasses compared to agricultural intensity metrics associated with growing conventional crops.

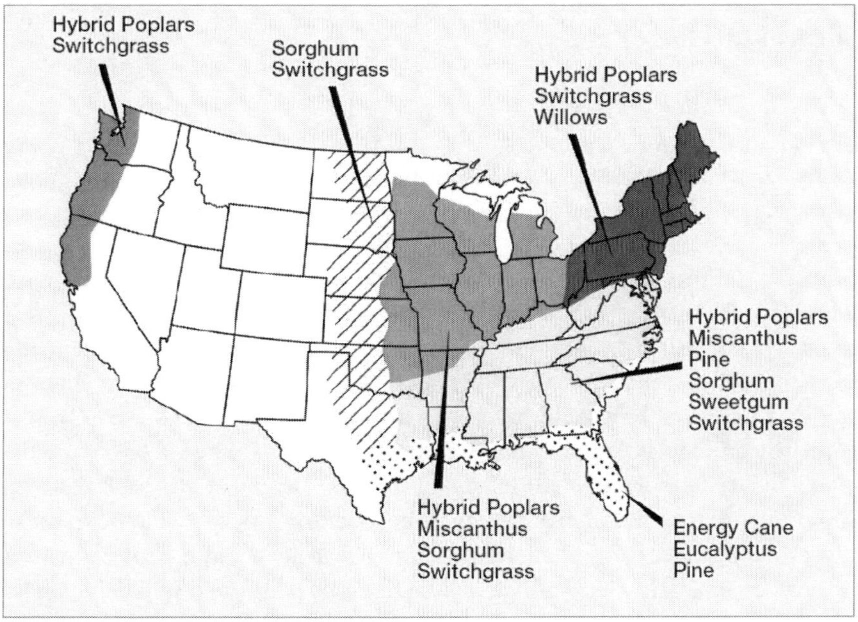

Source: Dale et al., 2010, updated from Wright, 1994.

Figure 3-7. Generalized Map of Potential Rain-Fed Feedstock Crops in the Conterminous United States Based on Field Plots and Soil, Prevailing Temperature, and Rainfall Patterns.

Table 3-3. Example Comparison of Agricultural Intensity Metrics for Perennial Grass, Short-Rotation Woody Crops and Conventional Crops

Metric[a]	Perennial Grass[d]	Short-Rotation Woody Crops[d]	Corn[c]	Soy[c]	Wheat[c]
Erosion[b] (T ac^{-1} yr^{-1})	0.1	0.9	9.7	18.2	6.3
Fertilizer (N | P | K) (lb ac^{-1} yr^{-1})	45|54|54	54|13|13	120|54|71	18|40|62	54|31|40
Herbicide (lb ac^{-1} yr^{-1})	0.22	0.35	2.73	1.63	0.16
Insecticide (lb ac^{-1} yr^{-1})	0.02	0.01	0.34	0.14	0.02
Fungicide (lb ac^{-1} yr^{-1})	0.0009	0.0001	0.0007	0.0009	0.0089

[a] All metric units converted to English units. T/ac^{-1}/yr^{-1} converted from Mg/ha^{-1}/yr^{-1}; lb/ac^{-1}/yr^{-1} converted from kg/ha^{-1}/yr^{-1}.

[b] Conditions for average erosion rates: for corn and soy, 4 percent slope clay loam soil; for wheat, 4 percent other soil; for perennial grass, after establishment; for short-rotation woody crops, 5 percent slope.

[c] Fertilizer levels for corn, soy, and wheat are the approximate national average (USDA, 1991). Herbicide, insecticide, and fungicide levels for corn, wheat, and soy are mean annual projections (USDA, 1991).

[d] Herbicide, insecticide, and fungicide levels for perennial grass and short-rotation woody crops are from ORNL (1991). Unpublished estimates from field experiments of the Biofuels Feedstock Development Program, 1978–1991.

Source: Ranney and Mann, 1994.

3.3.4.1. Nutrient Loading

Nutrients—Surface Water Impacts

Several factors affect losses of nutrients to surface waters, including fertilizer application rates, irrigation, nutrient uptake rates by the crop, and soil and landscape properties. Results from over a decade of research from DOE recommend application rates for switchgrass grown for biofuels ranging from 37 to 107 pounds per acre per year, varying by region (McLaughlin and Kszos, 2005). These findings are generally upheld by a more recent comparison of 18 publications across 17 states (Wullschleger et al., 2010).

A meta-analysis of Giant Miscanthus in Europe and its response to management suggests that nitrogen addition at approximately 90 pounds per acre per year stimulates growth by approximately 15 percent after the third year of cultivation (Miguez et al., 2008). It should be noted that this magnitude is small compared to responses of other crops. Similar detailed studies in the United States are generally lacking (but see Heaton et al., 2008, 2009). A recent study reported that Giant Miscanthus can fix atmospheric nitrogen, which could explain the relatively weak responses to nitrogen fertilizer and lead to a large benefit to its use as a feedstock (Davis et al., 2010). These findings remain to be confirmed in other studies.

Relative to annual row crops such as corn, production of switchgrass and Giant Miscanthus requires less fertilizer and reduces surface and subsurface nutrient losses (McIsaac et al., 2010). Both species are inherently efficient in their nitrogen use, because they store carbohydrates and nutrients in their roots at the end of the growing season (Beaty et al., 1978; Beale and Long, 1997; Parrish and Fike, 2005; Heaton et al., 2009). Therefore, the practice of harvesting the above-ground biomass after translocation of nutrients to below-ground storage structures reduces the need for fertilization in subsequent growing seasons. In the only field study to date comparing corn, Giant Miscanthus, and switchgrass under comparable conditions at the field scale, McIsaac et al. (2010) found that soil nitrate levels under mature stands of switchgrass and Giant Miscanthus were lower than under fertilized corn (184 pounds per acre) by 97 and 93 percent respectively. Less is known when these crops are grown under identical management conditions or for other fertilizer inputs. However, switchgrass is considered efficient in its use of potassium and phosphorus (Parrish and Fike, 2005), while Giant Miscanthus may require additional potassium inputs in some circumstances (Clifton-Brown et al., 2007).

In total, because of lower nutrient inputs and high efficiencies for switchgrass and Giant Miscanthus, and lower water requirements especially for switchgrass, conversion of row crops to perennial grass production will likely reduce surface water impacts from nutrient loading. If perennial grasses are grown for biomass on CRP acreage, however, nutrient loading to waterways may increase.

Nutrients—Coastal Waters Impacts

As mentioned above, switchgrass and Giant Miscanthus cropping systems are expected to require fewer fertilizer additions than traditional row crops, and have been shown to reduce chemical oxygen demand in runoff when used as filter strips (Keshwani and Cheng, 2009). This will reduce their impact on the hypoxic zones of U.S. coastal waters if they replace row crops.

3.3.4.2. Sediment

Switchgrass and other perennial grasses are frequently used as an erosion control management practice to reduce sediment loads from row crops (Hill, 2007; McLaughlin and Walsh, 1998; U.S. EPA, 2009a). Perennial grasses such as switchgrass have been shown to reduce erosion by 99.2 percent when compared to an average of corn, wheat, and soybeans (see Table 3-3). Similar results are expected for Giant Miscanthus, which has been shown to produce more root biomass in field comparisons with switchgrass (Heaton et al., 2008). Therefore, because of their perennial root structure and assuming conservation-oriented agricultural practices, switchgrass and Giant Miscanthus production is not expected to increase sediment loads to surface waters except possibly during the planting stages.

3.3.4.3. Pesticides

Perennial grasses are generally less susceptible to pests than traditional row crops (Oyediran et al., 2004; Keshwani and Cheng, 2009). Switchgrass plantings use approximately 90 percent less pesticide than row crops (Keshwani and Cheng, 2009). However, herbicides are used initially to establish and maintain switchgrass and Giant Miscanthus plantings for harvest (Lewandowski et al., 2000; Keshwani and Cheng, 2009). Information relevant to potential pesticide use for Giant Miscanthus in the United States is generally lacking; however, researchers in Europe have reported that pesticide requirements are lower than for row crops (Lewandowski et al., 2000).

Switchgrass and Giant Miscanthus have been found to be susceptible to insects such as the corn leaf aphid, sugarcane aphid (Bradshaw et al., 2010), and fall armyworm (Prasifka et al., 2009), as well as to nematodes (Tesfamariam et al., 2009) and pathogens (Parrish and Fike, 2005; Garrett et al., 2004; Christian et al., 2001; Lewandowski et al., 2000). However, disease levels are considered generally low for both species compared to row crops and especially corn studies. The lack of commercial perennial grass production as biofuel feedstock therefore makes it difficult to predict how much pesticide would be needed for this application and what the environmental impacts would be. However, it is likely that chemical inputs (e.g., herbicides) would be needed during the establishment phase (Parrish and Fike, 2005; Lewandowski et al., 2000). Research has found that herbicides that are safe for corn application can be safely applied to Giant Miscanthus (Bullard et al., 2001). In non-commercial production, pesticide releases from perennial grass plantings are much less than from corn or soybeans (Hill et al., 2006). Most species are likely to be more susceptible to pests when grown in monocultures than in polycultures (Hooper et al., 2005).

As an example, cellulosic feedstock production may impact the spread of the western corn rootworm (WCR), whose soil-borne larval stage is estimated to be responsible for more than $1 billion in annual losses in the U.S. Corn Belt (Rice, 2003). Recent research reported that WCR is able to use Giant Miscanthus and several North American grasses as a host, though not as effectively as corn (Oyediran et al., 2004; Spencer and Raghu, 2009). Similar information on WCR use of switchgrass as a host is not available, though perennial grasses generally are more resistant to pests than corn (Lewandowski et al., 2003; Oyediran et al., 2004).

3.3.4.4. Pathogens and Biological Contaminants

The reviewed literature does not directly discuss the effect of perennial grass plantings on pathogens in runoff or the potential for pathogen loads associated with perennial grass management (i.e., from manure used as fertilizer). Since perennial grasses require fewer inputs and take up more impurities from surface water, fewer contaminants are expected from its growth compared to row crops.

3.3.5. Water Quantity

3.3.5.1. Water Use

Switchgrass and Giant Miscanthus are both C4 grasses (like corn) that use water efficiently and are adapted to warmer environments. Neither appears to require water inputs to attain high yields, except in arid regions, when summers are dry, and in very dry years (Beale et al., 1999; Lewandowski et al., 2000; Heaton et al., 2004a; McLaughlin and Kszos, 2005; Wang et al., 2010). However, both species have been found to increase yields with higher water inputs, more so for Giant Miscanthus than switchgrass (Heaton et al., 2004a). Thus, it is unclear whether switchgrass or Giant Miscanthus, when grown as bioenergy crops, will be irrigated or not and to what degree, though it is assumed that both will require fewer water inputs than row crops. Generally, studies that calculate water use for ethanol produced from switchgrass often assume that the feedstock is rain-fed, requiring no irrigation, and is capable of tolerating moisture deficits (e.g., Dominguez-Faus et al., 2009; Wu et al., 2009). In the first comparative study to date of corn (*Zea mays*), switchgrass, and Giant Miscanthus, Hickman et al. (2010) found that the cumulative ET and water use over the growing season was higher for Giant Miscanthus than corn, with switchgrass as intermediate. This was mostly because the perennial grasses had a higher leaf area and a longer growing season (Dohleman et al., 2009; Hickman et al., 2010). High biomass production for Giant Miscanthus did not offset this water usage in terms of efficiency. Thus, the common presumption that perennial grasses will use less water depends on the details of how and where they are grown, and what land use they replace.

The upland and lowland types of switchgrass differ in their water use. The upland type tends to tolerate dry conditions, while the lowland type requires more water (Parrish and Fike, 2005). Switchgrass farmers may be able to minimize potential irrigation withdrawals (and ET) by cultivating the upland type of switchgrass. Given the clonal origins of most Giant Miscanthus studied to date, it is likely that the range of sensitivities to water stress will be low, though the same may not be said of its parent species (*M. sinensis* and *M. sacchariflorus*), which are known to have high genetic variability (Hodkinson et al., 2002).

3.3.5.2. Water Availability

Depending on where perennial grasses are grown, whether irrigation is required, and what crops they replace (if any), perennial grass production could improve or worsen water availability. If perennial grasses replace more water-dependent crops, ground water availability could be improved in places like Nebraska, where aquifers provide 85 percent of the water to agriculture (Kenny et al., 2009; NASS, 2009a). On the other hand, if ground water–irrigated perennial grasses replace unmanaged CRP land, water availability would be expected to be reduced.

Changes in ET as a result of the cultivation of perennial grasses could either increase or decrease field-level or local water supplies. Higher cumulative ET for Giant Miscanthus and switchgrass compared to corn (Hickman et al., 2010) suggests that growing perennial grasses may decrease surface runoff and subsurface infiltration, and increase ET (VanLoocke et al., 2010). This has been a growing concern in Europe (Richter et al., 2008) and the United States (VanLoocke et al., 2010). These impacts, along with albedo effects from a longer growing season, may lead to increases in local and regional humidity and to a local cooling (Georgescu et al., 2009). A modeling study for an Iowa watershed reported that converting corn-soybean to perennial grasses (i.e., switchgrass) would increase ET and reduce water yields measured through annual stream flow by about 25 percent (Schilling et al., 2008). A recent regional analysis of the impacts of widespread cultivation of Giant Miscanthus in the Midwest found increases in ET and decreases in surface runoff by 1.6 inches per year or more when Giant Miscanthus is planted over 25 percent or more of the region (VanLoocke et al., 2010). Dramatic changes were predicted when cover exceeded 50 percent. However, no significant changes were found when Giant Miscanthus covered 10 percent of the land area (though coverages between 10 percent and 25 percent were not examined). These simulations may not reflect likely scenarios. In addition, given the high productivity for these grasses, such high levels of cultivation may be unlikely to be needed to meet RFS2 standards (VanLoocke et al., 2010). In either case, these changes could have large impacts on ecosystem services tied to the hydrologic cycle (Brown et al., 2005). Much more work is needed to determine recommended practices for growing perennial grasses as a bioenergy feedstock to determine subsequent impacts on water quantity.

3.3.6. Soil Quality

3.3.6.1. Soil Erosion

Both switchgrass and Giant Miscanthus have extensive root systems that prevent the erosion of soil. In addition, unlike corn and soybeans, these perennial grasses are not planted on an annual basis, reducing the frequency of soil disturbance. Currently, switchgrass can be planted in conventional tillage and no-till systems, whereas Giant Miscanthus is planted in tilled fields (Heaton et al., 2008; Parrish and Fike, 2005). This one-time tillage can increase erosion risk, particularly in Giant Miscanthus, where plant growth is slow the first year following planting and does not provide substantial ground cover (Lewandowski et al., 2000). In subsequent years, however, Giant Miscanthus stands generally have high yields and dense root mats (Heaton et al., 2008; Lewandowski et al., 2000), and likely provide substantial erosion control benefits relative to annually planted crops. Erosion control by switchgrass has received more study than that of Giant Miscanthus. Switchgrass has been extensively planted on CRP acreage for erosion reduction, and planting switchgrass in riparian zone grass barriers and vegetation strips has been shown to substantially reduce runoff, sedimentation, and nutrient loss (Eghball et al., 2000; Blanco-Canqui et al., 2004; FSA, 2009). Switchgrass intensively managed for biofuel feedstock production, however, may increase nutrient losses relative to switchgrass plantings intended as erosion control.

3.3.6.2. Soil Organic Matter

In general, soil organic matter increases more under perennial than annual species because of the continuous accumulation of plant material (Sartori et al., 2006). Soil carbon is a primary constituent of soil organic matter. If perennial grasses replace annual crops, perennials will likely increase soil organic matter (Bransby et al., 1998; Schneckenberger and Kuzyakov, 2007; Blanco-Canqui, 2010). McLauchlan (2006) reviewed changes in soil organic matter following agricultural abandonment, and concluded that soil organic carbon accumulated with the cessation of agriculture and the establishment of perennial vegetation—although the perennial vegetation was not subjected to periodic harvesting. Where perennials are planted on degraded soils with low organic matter content, soil erosion can be reduced and carbon stocks restored (Clifton-Brown et al., 2007; McLaughlin and Kszos, 2005). For example, using stable isotope data, switchgrass was predicted to increase soil carbon by approximately 12 percent on a degraded soil over a decade of production and harvesting (Garten and Wullschleger, 2000).

Switchgrass planted on CRP acreage or former cropland eligible for CRP enrollment has been shown to increase soil organic matter. When grown on former cropland eligible for CRP enrollment, switchgrass, with annual harvesting, significantly increased soil organic carbon (in the top 11.8 inches of soil) by an average annual value of 981 pounds per acre over a five-year period (Liebig et al., 2008). In this particular study, the 10 switchgrass sites ranged along a north–south transect from southern Nebraska to northern North Dakota, and the fields received an average of 172 pounds of nitrogen per hectare per year. In another study conducted on CRP acreage, switchgrass production increased soil organic matter, but only with the application of fertilizers (Lee et al., 2007a).

Besides the influence of nitrogen fertilizer, the magnitude of soil organic matter accumulation under these perennials can depend, in part, on harvest frequency, soil type, and site preparation (Blanco-Canqui, 2010). Harvesting biomass reduces the amount of plant matter available for soil organic matter; however, switchgrass production for bioenergy generally results in the accumulation of soil organic matter (Anderson-Teixeira et al., 2009). Relative to reported values for corn, soil carbon increased under Giant Miscanthus cultivation when its above-ground vegetation was harvested annually; however, this result varied according to soil type, with carbon increasing in a loamy soil but not in a sandier textured soil

(Schneckenberger and Kuzyakov, 2007). Finally, the effect on soil organic matter of preparing land for these biofuel feedstocks has received little attention to date (Anderson-Teixeira et al., 2009). Soil preparation for perennial grass cultivation will be much less frequent than for row crops (e.g., every decade vs. annually) (McLaughlin and Kszos, 2005). Consequently, the effects on soil organic matter will be smaller. Nevertheless, the amount of soil carbon lost from this conversion and the time needed for perennials to regain that carbon requires further study.

3.3.7. Air Quality

As mentioned earlier, little is known overall about the extent to which fertilizer, herbicides, and pesticides will be used to increase perennial grass production. The production of fertilizer requires fossil fuel inputs, resulting in air pollutant emissions, including SO_x and NO_x (U.S. EPA, 2010a). Studies indicate that NO_x emissions should decrease when perennial grasses replace row crops as a biofuel feedstock (Wu and Wang, 2006). However, if they are instead grown on marginal or CRP land that does not receive nitrogen inputs, total NO_x emissions will increase. Nitrogen fertilizer rates are based on field trials, which are not extensive (Wu and Wang, 2006) and may differ from on-farm conditions (Hill et al., 2009). Much less is known about phosphorus (P_2O_5) fertilizer requirements for either species, though the prevailing assumption is that switchgrass and Giant Miscanthus will likely use as much phosphorus as row crops, or less (Ranney and Mann, 1994; Lewandowski et al., 2000; Parrish and Fike, 2005). These reductions would translate to lower SO_2 emissions if perennial grasses replace row crops, and overall increased SO_2 emissions if perennial grasses replace lands not receiving chemical inputs (Wu and Wang, 2006). As described earlier in Section 3.3.4.3, perennial grasses are expected to require less pesticide and herbicide than row crops, except when initially establishing perennial grass plantings when inputs can be comparable (Lewandowski et al., 2000; Parrish and Fike, 2005). The lack of experience with commercial perennial grass production as a biofuel feedstock precludes firm conclusions about potential air quality impacts.

As with corn and soybeans, harvesting of perennial grasses will involve use of farm equipment, and thus is expected to generate NO_x and PM emissions. However, VOC, CO, NO_x, PM_{10}, $PM_{2.5}$, and SO_2 emissions associated with switchgrass production have been reported as being lower than emissions from corn or soybean production (Wu and Wang, 2006; Hess et al., 2009b). Thus, similar to above, overall effects depend on whether perennial grasses replace row crops or unmanaged (or relatively less managed) lands.

3.3.8. Ecosystem Impacts

3.3.8.1. Biodiversity

Research generally indicates that perennial grasses support a greater diversity of species, including birds, small mammals, and invertebrates, than row crops (Herkert, 2007; Semere and Slater, 2007a; Fargione et al., 2009; Dale et al., 2010). However, active management and harvesting of perennial grasses for feedstock production is likely to negatively impact at least some of these species (e.g., Murray and Best, 2003; Murray et al., 2003; Kaufman and Kaufman, 2008). Much of the scientific literature on biodiversity and biofuels compares row crops with unmanaged lands or lands enrolled in the CRP, though highly managed perennial grass feedstock cultivation systems are unlikely to resemble these unmanaged areas. Specific biodiversity responses will depend on location, perennial grass species, and agricultural and conservation management practices.

Many studies indicate that a greater diversity of birds and mammals is supported by perennial grasses and CRP lands than by row crops such as corn or soybean (Herkert, 2007; Fargione et al., 2009; Dale et al., 2010; Fletcher et al., 2010). Planting switchgrass in tallgrass ecoregions can increase local grassland bird diversity (Roth et al., 2005). Switchgrass can also serve as suitable breeding habitat for a variety of bird species and increase their abundance, depending on the timing of harvest and other management practices (George et al., 1979; Murray et al., 2003). CRP lands planted using native grass mixtures that include switchgrass generally have greater bird abundances, diversity, activity, and breeding success in the winter than non-native grass monocultures and several migratory non-game species of management concern (Thompson et al., 2009). However, if land currently enrolled in the CRP subsequently is harvested for switchgrass, some bird species may decline (Murray and Best, 2003; Murray et al., 2003).

Research from Nebraska and Iowa shows that populations of white-tailed deer are not likely to decline following conversion of land from corn to native grassland (i.e., dominated by switchgrass), but may experience a contraction of home ranges to areas near row crops, increasing crop losses and the potential for disease transmission among wildlife (Walter et al., 2009). Research has also shown a greater abundance and diversity of beneficial insects in switchgrass than corn, although additional research is needed on whether these species provide increased pollination or pest suppression services and whether switchgrass also provides habitat for insect pests (Gardiner et al., 2010; Landis and Werling, 2010). Overall, little research exists on the habitat value of managed switchgrass fields for other terrestrial species, such as small mammals, reptiles, and many invertebrates.

In terms of some specific harvest practices, increases in avian diversity are insensitive to whether switchgrass is strip harvested or completely harvested (Murray and Best, 2003). However, field studies have shown that different species prefer habitats under different management regimes, suggesting that switchgrass cultivation under a mosaic of field ages and management regimes will maximize total avian diversity over a large landscape (Murray and Best, 2003; Roth et al., 2005). Other factors also affect habitat quality, such as stand density and the use of rotating unharvested areas as refuges of wildlife habitat. Research from Iowa found that dense switchgrass fields managed for biomass often supported generalist species, and did not support species of management concern (Murray and Best, 2003).

Impacts on aquatic ecosystems have received even less attention thus far. Stream flow in Iowa rivers has been augmented by agricultural drainage since the mid-20th century, when row crops came to dominate this area (Schilling et al., 2008), and conversion back to perennial vegetation could bring surface water availability to more historical levels (see Section 3.3.5.2). Characteristics of surface water flow are important determinants of aquatic biological community health (Poff and Zimmerman, 2010). More research is needed to determine how large-scale perennial grass production and management practices may affect aquatic ecosystems (Powlson et al., 2005).

Research on how Giant Miscanthus cultivation might affect biodiversity in the United States is virtually nonexistent, and it is uncertain whether research from Europe would directly translate to U.S. communities and ecosystems. Nonetheless, studies in the United Kingdom have shown that Giant Miscanthus can provide improved habitat compared to row crops for many forms of native wildlife—including ground flora, invertebrates, small mammals, and bird species—due to the low intensity of the agricultural management system (Semere and Slater, 2007a, 2007b). Research from the United Kingdom shows that non-crop plants from a wide range of families (*Poaceae, Asteraceae,* and *Polygonaceae*) coexist within young Giant Miscanthus cropping systems due to a lack of herbicide applications. These plots support a greater diversity of bird populations than annual row crops, but less than with short-rotation woody crops (SRWCs) such as willow (Bellamy et al., 2009; Sage et al., 2010). These effects are likely to be transient as fields mature and crop height and coverage become more homogeneous and dense (Bellamy et al., 2009; Fargione et al., 2009). Similar patterns may be likely for the United States.

3.3.8.2. Invasive Plants

The risk that switchgrass or Giant Miscanthus (or any other perennial grass feedstock) will become an agricultural weed or invasive plant depends on their specific biology and their interaction with the environments in which they are grown. Invasive plant traits include rapid growth rate, ability to grow in dense stands, efficient resource use, tolerance to a wide range of environmental conditions, tolerance to disturbance, resistance to pests and diseases, and rapid and widespread abilities to disperse and establish (Barney and DiTomaso, 2008). Unfortunately, the traits that make a species potentially invasive often overlap with those of favorable biofuel feedstocks (Raghu et al., 2006; Barney and DiTomaso, 2008). Properties of the environment that facilitate invasion include a moderate climate, fertile areas, low levels of pests and/or disease, and low biodiversity (Stohlgren et al., 1999; Fridley et al., 2007), properties exhibited by many agricultural locations in the United States. Thus, the potential for invasion exists.

Weed risk assessments are formalized procedures for determining invasion risk. They are designed to predict invasive and non-invasive species/varieties and distinguish between them based on a set of questions about their history of invasiveness in other places, biological traits, and suitability for the environment into which they will be introduced. The most widely accepted approach, developed for Australia (Pheloung et al., 1999), has been recently adapted and tested in Florida (Gordon et al., 2008), Hawaii (Daehler and Carino, 2000; Buddenhagen et al., 2009), and several areas in Europe (Crosti et al., 2010; Krivanek, 2006), with good predictive results.

This weed risk assessment approach, modified for California, indicated that switchgrass could become invasive if introduced to that state (Barney and DiTomaso, 2008). Similar potential for switchgrass invasion was found for Hawaii (Buddenhagen et al., 2009). Conversely, Parrish and Fike (2005) anecdotally reported no records of switchgrass escaping cultivation in Australia, Europe, and the Pacific Northwest of the United States. While it may be possible for improved switchgrass varieties to become weedy anywhere in the United States, unimproved varieties of switchgrass are considered non-invasive in their native range. Switchgrass varieties for biofuel production are being bred for rapid growth, tolerance to low fertility soils, and the ability to grow in dense stands (Parrish and Fike, 2005; Rose et al., 2008; Sarath et al., 2008; Das and Taliaferro, 2009; Yang et al., 2009), all of which could increase invasive potential. On the other hand, breeding for traits like sterility can be used to reduce the risk of invasion. Reapplying the aforementioned weed risk assessment in California, assuming switchgrass bred for seed sterility, yielded a non-invasive result (Barney and DiTomaso, 2008).

Little experimental information exists about the ability of Giant Miscanthus to disperse from cultivation and persist as a weed or invade natural areas, but testing in Europe since the mid-1980s has not resulted in any known escapes (Lewandowski et al., 2000). A modified version of the Australian weed risk assessment recommended no restrictions on planting current varieties of Giant Miscanthus in the United States, because the plant produces no living seeds and is therefore unlikely to spread (Barney and DiTomaso, 2008). Using the weed risk assessment developed for Florida, Gordon et al. (2008) recommended no restrictions on Giant Miscanthus.[20] An earlier study noted that Giant Miscanthus can spread vegetatively and could undergo genetic changes to produce seeds once more, potentially enhancing invasive potential (Raghu et al., 2006). However, the likelihood of such an event remains unknown. Relatives of Giant Miscanthus, especially *Miscanthus sinensis,* have been grown in the United States for landscaping and horticultural purposes for decades (Stewart et al., 2009). Several researchers have highlighted the potential invasiveness of *M. sinensis* (Raghu et al., 2006; Barney and DiTomaso, 2008; Davis et al., 2010; Kaufman and Kaufman, 2008).

Another way a feedstock could escape cultivation is by crossing with free-living populations of compatible plants. Switchgrass, being native to the United States east of the Rockies and highly outcrossing, could transfer traits from conventionally bred or genetically modified cultivated varieties to wild populations. The effect on wild populations would depend on the frequency of crossing, the traits transferred, and the environment, but could include outcomes ranging from extinction of wild switchgrass populations to negligible or benign effects to enhanced invasiveness of crop-wild hybrids (Ellstrand et al., 1999). Because there are no wild populations of Giant Miscanthus in the United States and it is sterile, there is no possibility of escape through outcrossing.

Some other grass species that have been considered for use as biofuel feedstocks currently invade wetlands, including giant reed (*Arundo donax*) (Bell, 1997) and reed canarygrass (*Phalaris arundinacea*) (Lavergne and Molofsky, 2004). *Arundo donax* has been evaluated by several weed risk assessment protocols and has a high likelihood of becoming invasive in Florida, where production is being proposed (Barney and DiTomaso, 2008).

While feedstock cultivation poses the greatest risk for invasive impacts, reproductive parts from feedstocks could also be dispersed during transport from the field to storage or ethanol-processing facilities. Roads, railroads, and waterways can act as man-made corridors for non-native and invasive plants. Harvested switchgrass possesses living seed and Giant Miscanthus can reproduce vegetatively from plant cuttings, both of which may be dispersed during feedstock transport.

Several mitigation options for reducing the potentially negative environmental impacts from perennial grass production have been suggested. A prominent approach is to apply a weed risk assessment and reject planting species or varieties that are predicted to be invasive (Pheloung et al., 1999). Such an approach, though recently adopted generally by the U.S. Invasive Species Advisory Committee (National Invasive Species Council, 2009), is not formally a part of current RFS2 regulations surrounding cultivation of biofuel feedstocks. USDA's Biomass Crop Assistance Program (BCAP) does define as ineligible for some forms of crop assistance payments "any crop that is invasive or noxious or has the potential to become invasive or noxious" as determined by local, state, and federal entities (USDA, 2010e). Another option is to avoid cultivation of feedstocks with a history of invasiveness, especially in places that are climatically similar to where invasion has already occurred.

Breeding feedstocks to limit their dispersal into other fields or natural areas could reduce the probability of invasion. Another strategy for managing potential invasiveness is cleaning harvesting machinery and vehicles used to transport harvested feedstock, which would help to decrease unintended dispersal. Though prevention is most desirable, early detection and rapid response mechanisms could also be put into place to eradicate persistent populations of feedstock species as they arise, but before they have the chance to spread widely (DiTomaso et al., 2010). Such early detection and rapid response mechanisms might involve local monitoring networks and suggested mechanical and chemical control strategies (timing and application rate of herbicides, for example) devised by local agricultural extension scientists for specific feedstocks.

NATIVE GRASSLANDS AS A BIOFUEL FEEDSTOCK

Recent research has suggested using mixtures of native perennials as a biofuel feedstock on marginal or infertile lands (Tilman et al., 2006, 2009; Campbell et al., 2008; Weigelt et al., 2009). This practice is limited by several technological and management hurdles, but it is associated with many environmental benefits not found to the same degree in other feedstocks discussed in this report. Termed "low-input high-diversity" (LIHD) mixtures, they are essentially composed of several plant species that perform different functions within the community (e.g., high root mass to prevent soil erosion, nitrogen fixation to reduce fertilizer inputs) potentially at different times (e.g., spring versus fall) or the same function in a different manner (e.g., root growth and soil carbon sequestration at shallow versus deeper soil depths). LIHD mixtures, by definition, have more plant biodiversity than other monoculture-based feedstocks. This higher plant biodiversity is often associated with a variety of benefits, including higher stability of production, higher quality of habitat for wildlife, lower potential for invasion of the community, reduced need for chemical inputs (fertilizers, pesticides) (Fargione et al., 2009; Hooper et al., 2005; Loreau et al., 2002; Reiss et al., 2009). Also, monoculture crops are expected to require more active management (e.g., to prevent losses from pests) than polycultures such as LIHD (Hill et al., 2006; Tilman et al., 2009; Weigelt et al., 2009). When systems are viewed as a composite of many co-occurring processes (e.g., primary production, soil stabilization, and decomposition), polycultures sustain higher levels of multiple processes, sometimes termed "ecosystem multifunctionality" (Hector and Bagchi, 2007; Fornara and Tilman, 2009; Zavaleta et al., 2010). Diverse mixtures also often produce more biomass than their average constituent species grown in monoculture; however, the productivity of the most productive constituent species is in many cases similar to or higher than that of the mixture (Cardinale et al., 2006, 2007; Loreau et al., 2002). Although it seems likely that highly productive feedstocks (e.g., switchgrass and Giant Miscanthus) managed for maximum yield will produce more biomass for biofuel production than LIHD mixtures, there are no direct field-scale comparisons between LIHD and other feedstocks. The only comparison to date found that switchgrass grown on fertile lands recently in production across the Midwestern corn belt (Nebraska, South Dakota, North Dakota) out-produced LIHD grown on unproductive land in Minnesota that had been abandoned from agriculture (Schmer et al., 2008).

Production of a feedstock composed of a mixture of species will likely face greater technological and management hurdles than production of single-species feedstocks. For example, a mixture of species, having variable tissue densities and arrangements in the cropping system, may be more difficult to harvest, transport, and process into biofuel than a relatively uniform feedstock grown from a single species. Much more research is needed in this area to determine the potential role of LIHD as a biofuel feedstock on marginal or infertile lands.

3.3.9. Key Findings

Current environmental impacts from the production of perennial grasses as a biofuel feedstock are considered negligible because no large-scale commercial operations are yet in existence. The potential benefits of using perennial grasses as a biofuel feedstock instead of traditional row crops such as corn, however, are substantial. These can include reduced soil erosion, enhanced soil structure and carbon sequestration, reduced nitrogen loading and sedimentation to waterways, reduced hypoxia in coastal areas, and greater support for populations of non-crop plants, as well as animals and soil biota (Fargione et al., 2009; Hill, 2007; Williams et al., 2009). These benefits occur because perennial grasses are likely to require fewer chemical inputs than traditional row crops, and they have perennial roots that enable longer planting intervals and less soil disturbance (Parrish and Fike, 2005; Keshwani and Cheng, 2009). However, the magnitude and even the presence of these advantages depends on whether perennial grasses replace traditional row crops versus lands that are managed less intensively (e.g., CRP acreage or pasture), as well as how grasses are managed. If perennial grasses replace lands that received little or no inputs, and are grown with chemical amendments to increase production in large-scale operations that resemble current row crops, overall environmental costs may be significant. Studies highlighted above that incorporate agro-economic considerations suggest that the most plausible land use change is conversion of unmanaged lands (e.g., CRP) to perennial grass production, rather than conversion of row crops such as corn and soybean, due in part to the high market value of these row crops. The invasion potential for unimproved varieties of switchgrass and Giant Miscanthus over most of the United States is considered low, though little can be predicted about the invasion potential of improved varieties that may be developed. In total, realizing the potential benefits of perennial grasses as a biofuel feedstock will necessitate careful consideration of land use changes and management practices, as well as widespread implementation of BMPs where possible.

For a more comprehensive, qualitative comparison of the environmental impacts of perennial grasses, including a discussion of the most plausible impacts, see Section 6.

3.3.9.1. Key Uncertainties and Unknowns

- Because no commercial-scale facilities exist for converting perennial grasses to cellulosic ethanol, many uncertainties remain about how perennial grasses will affect environmental conditions when grown as feedstock at commercial scales. This holds for all impact categories documented in this report (soil carbon, leaching, biodiversity etc.), as well as for yields that may be lower in non- experimental plots that are managed less intensively. This highlights the need for large-scale studies comparing perennial grass cultivated under a variety of management regimes with row crops and other feedstocks.

- Environmental impacts from cultivation of perennial grasses will probably be largely driven by land use changes, which remain poorly understood, and management practices.
- Most existing literature on switchgrass focuses on its ecology and uses as forage, or focuses on dynamics in the context of CRP, which may or may not resemble switchgrass grown for high yields as a biofuel feedstock. In short, this literature might not be completely applicable.
- It is unclear how the abundant genetic potential for both switchgrass and Giant Miscanthus (parental lines) can be used to increase their feasibility as feedstocks. If researchers can develop novel cultivars of these plants with significantly improved yields, there may be even greater benefits for perennial grasses as biofuel feedstocks.
- Little is known about region-specific recommendations for fertilizer and pesticide use for increasing perennial grass production. Precision management strategies (e.g., minimal fertilization, irrigation, and pest management at specific times) may increase productivity without deleterious ecological impacts.
- The water requirements of different grass species in different areas of the country are not documented; however, widespread cultivation of perennial grasses may negatively impact regional hydrologic cycles and related ecosystem services (e.g., aquifer recharge), while improving others (e.g., nutrient runoff).
- The role of nitrogen fixation in explaining the productivity of Giant Miscanthus requires further study and may have large ramifications on the potential use of Giant Miscanthus as a feedstock.
- The potential invasiveness of current varieties of switchgrass and Giant Miscanthus over much of the United States is low, but the invasion risk of improved varieties remains unknown. Studies to evaluate feedstocks for biological characteristics associated with invasiveness should be conducted, and methods to synthesize this information (e.g., weed risk assessments) should be applied to local situations for anticipating and preventing potential invasions.
- It remains uncertain whether the continual removal of above-ground biomass will deplete soil nutrients over the long term, particularly on marginal soils. On these soils, it may be particularly critical to harvest after translocation of nutrients back into the root systems.
- More landscape-level research is needed to understand how the distribution of multiple land use systems across a large landscape (e.g., row crops interspersed with perennial biofuel grasses and native habitat) will affect local and regional biodiversity.

3.4. Woody Biomass

3.4.1. Introduction

Woody biomass is an attractive energy source because extensive amounts may be available domestically and, if managed correctly, the production of this feedstock can provide environmental benefits. Woody biomass includes trees (e.g., removed or thinned from forests); forest residues (e.g., limbs, tree tops, and other materials generally left on site after logging); short rotation woody crops (SRWCs, i.e., fast-growing tree species, such as willow [*Salix* sp.] and hybrid poplar [*Populus* sp.], cultivated in plantation-like settings); and milling residues. Thinning is a common forestry practice that removes trees within a forest stand, stimulating stand growth by reducing plant competition (e.g., Reukema, 1975). Rotation refers to the length of time between tree establishment and harvesting. Currently, woody biomass is burned for electricity generation in select locations in the United States, and pulp and saw mills use residues to produce heat, steam, and electricity. In 2008, about 10 percent of the renewable electricity generated in the United States came from woody biomass (White, 2010). Commercial-scale, cellulosic-ethanol plants using this feedstock are not yet in operation, but demonstration and development facilities exist.

Estimates of woody biomass available domestically differ widely and vary by price paid per ton of feedstock. At $40 to $46 per dry ton, it has been estimated that approximately 4 billion gallons of second-generation biofuel could be made from woody biomass (BRDI, 2008). EPA's RFS2 RIA notes that at $70 per ton, 40 to 118 million dry tons are potentially available for biofuel production in 2022 (U.S. EPA, 2010a). At a currently demonstrated conversion rate of 80 gallons of ethanol per dry ton, up to 9.4 billion gallons of ethanol could be produced from 118 million dry tons (Foust et al., 2009). Additionally, the conversion rate of biomass to ethanol or other biofuels will likely improve in the future.

Not all woody biomass is eligible under the RFS2 requirements. The RFS2 limits the origin of woody biomass to "planted trees and tree residue from actively managed tree plantations on non-federal land cleared at any time prior to December 19, 2007" (U.S. EPA, 2010d).

Both forest harvesting residues and thinning operations are expected to be the predominant sources of woody biomass for future biofuel use, but SRWCs might be important as well at higher feedstock prices (Perlack et al., 2005; U.S. EPA, 2010a; White, 2010). In the following sections, the potential environmental impacts of the use of harvest residues, thinning, and SRWCs are discussed in more detail. For comparison purposes, the environmental impacts of SRWCs are considered in relation to annual row crops. Economic analyses, however, conducted before the establishment of RFS2 guidelines, suggest that the most likely sources of land for SRWC plantations are CRP or fallow agricultural lands, rather than prime agricultural acres or grasslands (Volk et al., 2006; Walsh et al., 2003). Additionally, SRWCs replacing row crops, grasslands, or unmanaged forests are ineligible as a biofuel feedstock according to the RFS2.

3.4.2. Overview of Environmental Impacts

Current environmental impacts of production and use of woody biomass as a biofuel feedstock are negligible, since no large-scale, commercial operations are yet in existence to create demand for this feedstock. In the case of forest thinning and residue removal, there are direct environmental effects of biomass removal, as well as an effect from operation of forestry machinery. In the case of SRWCs, traditional forestry and agricultural activities undertaken during feedstock cultivation and harvesting, such as pesticide or fertilizer application, irrigation, and fuel/energy use, have the potential to impact the environment. If planted on degraded soil, SRWCs can improve both soil and water quality. The choice of tree species can influence the risk of establishment and impact of invasive species. All these activities can alter water quality and availability, soil and air quality, and biodiversity, with resulting effects on ecosystems. The extent of the impacts depends on each activity's intensity and the management practices in use.

3.4.3. Current and Projected Production Areas

Woody biomass is likely to be produced in major forest harvesting areas, predominantly in places such as the upper Lake States, the Southeast, and the Pacific Northwest (see Figure 3-8). Since forest residues and biomass thinning will be collected as a byproduct of harvesting operations, their usage is unlikely to produce land use/land cover changes (Williams et al., 2009). To reduce the threat of catastrophic wildfires, thinning or other biomass removal may occur in certain locations of high wildfire risk (USDA, 2005; Gorte, 2009). SRWCs will follow a similar geographic pattern to that illustrated in Figure 3-7, since these are restricted to existing tree plantations.

3.4.4. Water Quality

3.4.4.1. Nutrients

Cultivation of woody biomass can affect water quality, primarily through nutrient runoff and sedimentation. The effects of removing harvest residues on nutrient loads vary depending on topography, soil nutrient content, and the chemistry of the residues themselves (Titus et al., 1997). In a review analysis on logging impacts in boreal forests, Kreutzweiser et al. (2008) found that harvesting with residue removal relative to harvesting-only generally increased soil nutrient losses, but they failed to observe a clear trend in the export of nutrients to aquatic systems. Residue removal has been suggested as a management technique to reduce nitrogen in forests that receive high atmospheric deposition, such as in the northeastern United States (Fenn et al., 1998; Lundborg, 1997). Under these circumstances, residue removal might decrease nitrogen loads to waterways (Lundborg, 1997). Compared to forest residue removal, moderate forest thinning typically does not affect loss of soil nutrients to ground or surface waters (Baeumler and Zech, 1998; Knight et al., 1991). Bäumler and Zech (1999) found that thinning increased stream ion concentrations, particularly ammonium (NH_4^+), but the effects returned to pre-thinning conditions within a year.

As stated in the RFS2, SRWCs are only considered renewable biomass if cultivated on non-federal, previously managed forest lands or existing forest plantations. For comparative purposes, however, nutrient losses from SRWCs are, in general, considerably less than in annually cropped systems (Table 3-3). In willow plantations, the recommended fertilizer application rate is 89 pounds of nitrogen per acre (100 kilograms per hectare) every three years, which equates on an annual basis to approximately 22 percent of the average rate for corn production (Keoleian and Volk, 2005; NASS, 2006). Initially after planting, SRWC plantations can exhibit losses of nitrogen at rates comparable to conventional corn production, yet following this establishment phase, nitrogen losses decline to low levels (Aronsson et al., 2000; Goodlass et al., 2007; Randall et al., 1997). A comparison of nutrient exports from a short-rotation poplar stand and an undisturbed forest found no difference (Perry et al., 1998), and measurements of nitrogen in ground water and leaching from established willow plantations generally show little eutrophication potential for aquatic ecosystems (Keoleian and Volk, 2005). In coppiced systems, where trees are harvested at the ground level and re-grow from the stump, the harvesting of the above-ground portion of the tree appears to have little impact on nitrogen leaching (Goodlass et al., 2007). Losses can be substantially higher when the stand is replanted (Goodlass et al., 2007). Longer rotation lengths would likely improve nutrient retention on site and reduce losses to waterways.

Pesticides might be used with SRWCs; for purposes of comparison, it is noted that the amount used would be significantly less than that for corn or soybeans (Table 3-3; Ranney and Mann, 1994).

3.4.4.2. Sediment

Forest soils generally exhibit low erosion rates and thus small sediment losses to surface waterways (Neary et al., 2009). Forested riparian buffers reduce sediment and nutrient runoff into adjacent streams compared to row crops or pasture land (Zaimes et al., 2004). Additionally, erosion rates at harvested sites are relatively short-lived and decline once vegetation is re established (Aust et al., 1991; Miller et al., 1988). However, harvesting residues left on-site physically shield soil particles from wind and water erosion, and promote soil stability through the addition of organic matter. Thus, relative to harvesting-only operations, combined harvesting and removal of residues could increase erosion and associated sediment loading to surface waters, especially on steeper slopes (Edeso et al., 1999). Relative to undisturbed forests, thinning can temporarily increase erosion on steeper slopes or in semi-arid areas with high risk of wind erosion (Cram et al., 2007; Whicker et al., 2008).

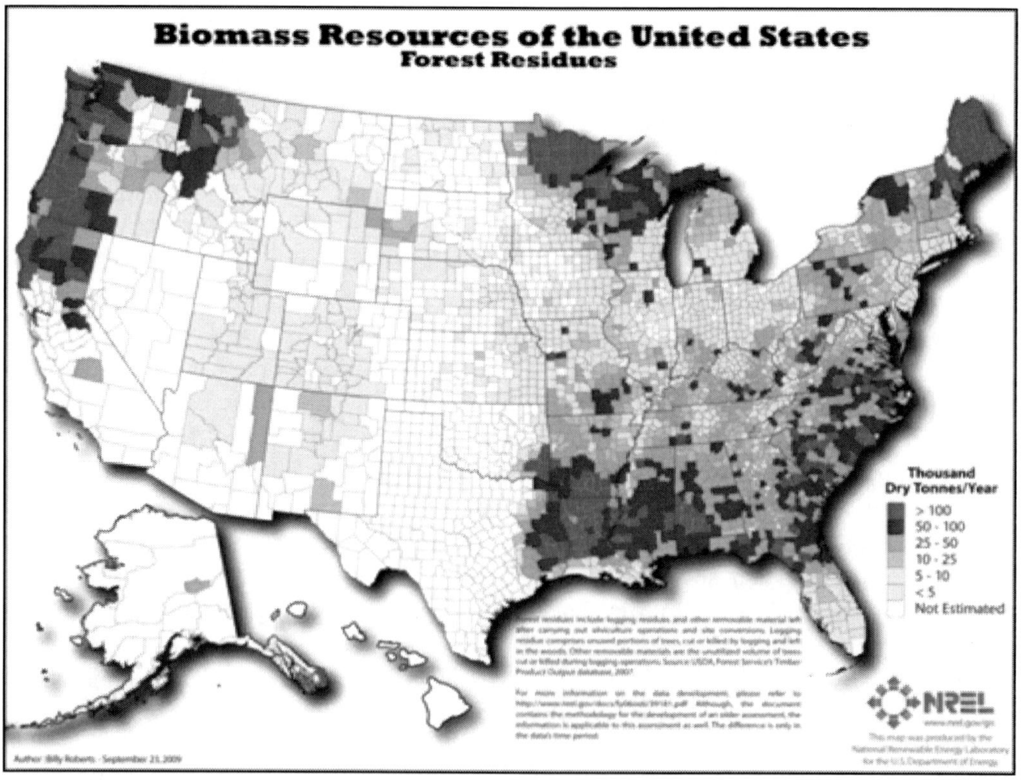

Source: Milbrandt, 2005.

Figure 3-8. Estimated Forest Residues by County.

3.4.4.3. Forestry Best Management Practices

According to the 2006 EPA National Assessment Database, forestry practices are a relatively small source of water quality impairment to streams and other surface waters (though the majority of states did not report this information) (U.S. EPA, 2006d). Louisiana, for example, reported that about 3.5 percent of impaired stream miles were likely due to forestry practices, compared to those attributed to such sources as agriculture (15.3 percent) and municipal discharges (11.5 percent) (U.S. EPA, 2006d).

BMPs have been demonstrated to be effective in reducing water quality impairment from forestry activity, and may help ameliorate impacts from increased demand for woody biomass as a result of biofuels (Shepard, 2006). Overall, there are generally high implementation rates for forestry BMPs, although actual practices and rates of implementation vary by state (NCASI, 2009). BMPs include establishing stream-side management/buffer zones, minimizing the construction of roads and stream crossings, using portable stream crossing structures, and choosing low-impact equipment that is of the appropriate size and scope for the site (Phillips et al., 2000; Aust and Blinn, 2004; Shepard, 2006). These BMPs have been shown to decrease erosion and sedimentation to waterways. Improved logging road construction and maintenance practices, for example, have been linked to decreases in sedimentation to streams in Oregon watersheds (Reiter et al., 2009). In a Texas watershed study, McBroom et al. (2008) found that modern harvesting techniques with proper use of BMPs significantly reduced erosion compared to past logging activities.

3.4.5. Water Quantity

3.4.5.1. Water Use

The use of harvest residues from mature stands of trees and thinning does not require additional water use at the feedstock production stage.

For the most part, growth of SRWCs will likely occur in areas with high water availability, such as the Northeast, Southeast, and Northwest. Because they are usually not irrigated, trees require less total water than row crops (Evans and Cohen, 2009). However, they can still have a large impact on regional water availability due to their much higher ET rate. For example, conversion of natural pine savanna and low-intensity pasture to plantations of slash pine (*Pinus elliotii*) and loblolly pine (*Pinus taeda*) in the Southeast could result in nearly 1,000 gallons of additional water consumed per gallon of ethanol (Evans and Cohen, 2009). Further, in certain locations and in some years, additional irrigation water may be required to maintain high biomass accumulation (Hansen, 1988), though precision application systems can reduce the amount of water applied. However, since SRWCs are only considered renewable biomass if cultivated on previously managed forest lands or existing forest plantations, the risk of increased ET rates is very low.

3.4.5.2. Water Availability

The use of forest harvest residues should have little or no effect on water availability at the feedstock production stage. Forest thinning can increase streamflow, but data suggest that at least 20 percent of the basal area of stand may need to be removed before a change in flow is detectable (Troendle et al., 2010). Removal of woody biomass that has overgrown traditional savannah grassland and dry forest ecosystems, largely due to fire exclusion, could provide a benefit of increased streamflow for a period of time.

Plantations of SRWCs may reduce runoff into streams and rivers compared to traditional row crops like corn and soybeans, potentially benefiting water quality (Updegraff et al., 2004). However, some experts warn that reduced runoff coupled with high water requirements for SWRCs could reduce or eliminate stream flow (Jackson et al., 2005). In places with seasonal flooding, modulation of surface water flow closer to pre-agricultural development levels could possibly mitigate flooding risk (Perry et al., 2001). Like most feedstock production impacts, the positive benefits or negative effects will depend on the location and practices used.

3.4.6. Soil Quality

3.4.6.1. Soil Erosion

The soil erosion impacts of SRWCs will depend on harvesting and planting frequencies; impacts are lower when time between planting intervals is longer. Some species of SRWCs may require intensive soil preparation for successful establishment, and it is during this brief establishment phase that erosion rates can be high (Keoleian and Volk, 2005). For example, higher sediment losses were observed within the first three years of seedling establishment in sweetgum (*Liquidamber styraciflua*) plantations compared to no-till corn or switchgrass (Nyakatawa et al., 2006). The slow-developing canopy failed to provide adequate ground cover to protect against erosion as a result of rainfall (Nyakatawa et al., 2006). In established SRWC plantations, soil erosion rates are much lower than those of annually harvested row crops (Keoleian and Volk, 2005; Blanco-Canqui, 2010). The use of a cover crop can also significantly reduce erosion caused by SRWC establishment (Nyakatawa et al., 2006), and the soil erosion effects of SRWCs are likely to be lower under a coppicing system, which reduces the frequency of soil disturbance by keeping the root systems intact. Willows are generally managed by the coppicing system and harvested at three- to four-year intervals for a total of seven to 10 harvests (Keoleian and Volk, 2005). This allows 21 to 40 years between soil disturbances.

3.4.6.2. Soil Organic Matter

Harvesting of forest residues removes plant material that could otherwise become soil organic matter. A review study suggested that, on average, a complete, one-time removal of forest residues slightly decreased soil organic matter in coniferous forests, but did not affect levels in hardwood or mixed stands (Johnson and Curtis, 2001). A recent meta-analysis found no significant impact on soil carbon with a one-time harvest and residue removal (Nave et al., 2010). The importance of residues to soil organic matter can vary with forest type and soil. Leaving logging residues can be particularly important for soils with low organic matter content, and repeated harvesting of residues in the same location could lead to overall declines in soil organic matter (Thiffault et al., 2006). The addition of commercial or organic fertilizers can increase soil organic matter; therefore, this could be a management strategy to offset potential losses due to residue removal (Johnson and Curtis, 2001). Thinning of forests has been shown to reduce carbon in forest floor layers, but less evidence is available regarding its impact on soil (Grady and Hart, 2006; Jandl et al., 2007). The effect of thinning over the long term will depend on both the frequency and intensity of the specific thinning operations.

Production of SRWCs can add organic matter to the soil, sequestering carbon, but the net benefits of these crops depend upon time between harvests and prior land use. If frequently harvested short-rotation forests, particularly with residue removal, replace longer-rotation forests, then the overall effect on soil organic matter over time is likely to be negative (Johnson et al., 2010). Though a one-time harvest with residue removal might not by itself decrease soil organic matter (see above; Johnson and Curtis, 2001), repeated removals of biomass at a greater frequency than longer-rotation forests are likely to decrease soil organic carbon over the long term (Johnson et al., 2010). If SRWCs are planted on degraded, abandoned agricultural lands, there is much greater potential to enhance soil organic matter (Schiffman and Johnson, 1989; Huntington, 1995; Richter et al., 1999). Some initial soil carbon can be lost during forest establishment in a former agricultural field (Paul et al., 2002). The amount of time it takes for soil carbon to accumulate varies. In hybrid poplar plantations in Minnesota, it was estimated to take 15 years to meet the carbon levels of the agricultural field replaced (Grigal and Berguson, 1998). A review study suggested that on average it can take 30 years to exceed soil carbon levels of abandoned agricultural fields; though when the forest floor was also considered, carbon accumulation rates were higher, reducing the time needed to regain carbon from the initial forest establishment (Paul et al., 2002).

3.4.6.3. Soil Nutrients

Use of harvesting residues removes a potential source of soil nutrients that can be used by the regenerating forest. Residue removal might reduce nitrogen loads in forests that receive high atmospheric deposition (Fenn et al., 1998; Lundborg 1997), yet this risks depletion of calcium and other nutrients critical for plant growth (Federer et al., 1989). Harvesting with residue removal can lead to declines in soil nutrients and forest productivity, but in some cases, it can be sustainable for at least one rotation (McLaughlin and Phillips, 2006; Thiffault et al., 2006). Powers et al. (2005), in a survey of forest stands across North America, found that a one-time harvest with residual removal did not affect stand productivity. In a subset of those sites, located in North Carolina and Louisiana, Sanchez et al. (2006) observed a similar result, but did find declines of soil phosphorus availability in the Louisiana stands. The cumulative effects of repeated removals from the same site require further study. Application of commercial or organic fertilizers may be necessary to compensate for nutrients lost. Overall, residue removal may be less problematic on high-fertility soils than on coarser-textured, low-fertility soils (Page- Dumroese et al., 2010). Since thinning operations remove less biomass than harvesting with residue removal, the risk posed to soil nutrients by thinning is likely to be smaller (Luiro et al., 2010).

There is concern that continual harvesting of SRWCs will deplete soil nutrients over the long term (Adegbidi et al., 2001). Commercial fertilizers or organic waste products, such as municipal effluent, can be used to offset these losses (Stanton et al., 2002). Nutrient removal from such effluents by SRWCs could provide an additional environmental benefit, though it remains unclear how much nitrogen, other nutrients, or contaminants might leach from these systems if this technique is used.

3.4.7. Air Quality

Air quality impacts during harvesting of forest residues and thinning are associated with emissions from combustion of fossil fuels by logging equipment. Few data are available for evaluating air emissions from SRWCs such as hybrid poplar and willow. As with switchgrass,

SRWCs require less tillage (reducing fugitive dust emissions) and fewer applications of fertilizer relative to row crops (reducing emissions associated with fertilizer production and application).

However, some species such as poplar and willow that are potential feedstocks for cellulosic ethanol are known to emit biogenic VOCs such as isoprene (U.S. EPA, 2010a), although emissions of VOCs from these species are moderate compared to emissions from some others (Isebrands et al., 1999). Compared to non-woody crops that emit relatively little isoprene, these trees could affect ozone concentrations if planted extensively. This effect will be highly sensitive to environmental conditions, preexisting vegetative cover, and the presence of other atmospheric chemicals, especially NO_x (Hess et al., 2009a; U.S. EPA, 2006b).

3.4.8. Ecosystem Impacts

3.4.8.1. Biodiversity

Both positive and negative consequences of residue removal on forest biodiversity have been reported in the scientific literature. Forest residues or debris are habitat for many mammal, bird, amphibian, reptile, and invertebrate species; function as plant germination sites; and are positively related to the structure and composition of the understory plant community (Franklin et al., 2002; Scheller and Mladenoff, 2002; Waddell, 2002; Janowiak and Webster, 2010). Species diversity in forest ecosystems is strongly linked to structural diversity, of which forest residues and woody debris are a component (Janowiak and Webster, 2010). The extent and intensity of residue or debris harvesting and type of management employed will determine the level of impact on biodiversity, both in terms of species diversity and abundances (Janowiak and Webster, 2010). While understory cover and diversity often increase with increased levels of forest thinning (Thomas et al., 1999), plant species diversity is generally highest in old-growth forests and decreases under various forest management scenarios from selective cutting to clear cutting (Scheller and Mladenoff, 2002; Khanina et al., 2007). Similar results have been documented for amphibians (e.g., Karraker and Welsh, 2006) where abundance is greatest in older forests, and not significantly different in thinned forests for some species. Some small mammal species also increase in abundance under a variety of forest management practices (e.g., Homyack et al., 2005); and bird diversity and abundance are higher in thinned forests than more intensively managed areas (e.g., Kalies et al., 2010). The results of these studies suggest that a landscape with forest patches of different ages and careful management can support relatively high species diversity and abundances, although questions remain about species-specific responses, long-term conditions, and demographic consequences (Karraker and Welsh, 2006; Niemelä et al., 2007; Kalies et al., 2010).

Tree harvesting activities can impact aquatic biodiversity in a number of ways. For example, removal of woody biomass by harvesting of forest residues or thinning in riparian areas may reduce woody debris in headwater streams, an important component for aquatic habitat (Angermeier and Karr, 1984; Chen and Wei, 2008; Stout et al., 1993; Thornton et al., 2000). In addition, tree canopies over streams help maintain cooler water temperatures conducive to cold- water smallmouth bass, trout, or salmon populations (Binkley and Brown, 1993; U.S. EPA, 2006c). Thinning practices in riparian areas that are consistent with widely applied forestry BMPs are less likely to negatively impact aquatic communities, particularly when these practices do not significantly reduce riparian canopy cover and increase stream temperatures (Chizinski et al., 2010).

The biodiversity effects of SRWCs versus other managed forests will likely depend upon how much habitat complexity can develop during intervals between harvests. Several studies have documented that bird species diversity on woody biomass plantations is comparable to that of natural shrubland and forest habitats (Dhondt et al., 2007; Perttu, 1995; Volk et al., 2006), although this is not always the case (Christian et al., 1998). Bird and small mammal species found on SRWC plantations tend to be habitat generalists that can also use open habitats like agricultural lands, while birds and small mammal species in mature forests are more specialized and require forest cover (Christian et al., 1998). Changes in the type and amount of edge habitat also can alter species interactions, since edges can serve as dispersal barriers or filters, influence mortality, contribute to overall habitat use that maintains populations, and generate new interactions (e.g., predation, competition); again, these processes are largely species-specific (Fagan et al., 1999). Therefore, habitat edges can benefit some species and be detrimental for others. If there is enough time between harvests to allow understory plants to establish in SRWC plantations, bird species diversity can increase due to increases in habitat complexity (Christian et al., 1998). Post-logging studies of birds also show species-specific responses, with some species increasing in abundance immediately after logging and others increasing a decade after harvest (Schlossberg and King, 2009). For comparison purposes, there is some evidence that planting SRWCs can improve species habitat relative to agricultural crops (Christian et al., 1998).

3.4.8.2. Invasive Plants

Like perennial grasses, woody plants cultivated for biofuel feedstock could become invasive. This is based on documentation that trees used in forestry have become invasive, though one estimate suggests most invasive trees were introduced for landscaping, not production forestry (Reichard and Hamilton, 1997). Woody plant invasions can negatively affect biodiversity and water availability (Richardson, 1998).

Predictive frameworks based on past invasions help identify what conditions and woody plant traits make invasions more likely in the future. Different frameworks often consider different factors. A study of determinants of woody plant invasion in Central Europe concluded that long residence time (>180 years since introduction) and high planting intensity correlated with escape and naturalization, while long residence time and ability to withstand low temperatures correlated well with invasions (Pysek et al., 2009). If these factors hold true for the United States, large-scale, widespread planting of woody biofuel species may pave the way for invasion, though possibly not until the next century or beyond. In a study focused on North America, woody plants that were native outside the continent or not sterile hybrids were more likely to be invasive (Reichard and Hamilton, 1997). These results suggest that using species that are native to the United States, or that are sterile hybrids between two species, could reduce the risk of invasion. Finally, an assessment of risk factors in woody plant invasions of New England identified plants that were invasive elsewhere and had a high growth rate as more likely to be invasive (Herron et al., 2007), two factors that could be considered when promoting or discouraging particular species as feedstocks.

In at least one case, a predictive assessment has been applied specifically to a woody species discussed as a potential biofuel feedstock: *Eucalyptus grandis* in Florida (Rockwood et al., 2008). The predictive assessment was based on the Australian Weed Risk Assessment

(Pheloung et al., 1999) and modified for Florida growing conditions (Gordon et al., 2008). It has been validated with 158 species introduced to Florida and has proven to possess good predictive power. It correctly identified 92 percent of species independently determined to be invasive and 73 percent of species known to be non-invasive. *E. grandis* has been cultivated in Florida for more than two decades and has not invaded. However, because this species could be planted widely as a biofuel feedstock, when the predictive assessment was applied in 2009, *E. grandis* received a conclusion of "Predict to be invasive; recommend only under specific management practices that have been approved by the University of Florida's Institute of Food and Agricultural Sciences Invasive Plant Working Group." Approved management practices for four different cultivars include maintenance of a buffer around production areas and harvesting prior to seed maturation.[21]

As with biofuels crops and grasses, it is possible that varieties of woody species, both native and non-native to the United States, could be developed as biofuel feedstocks that have significantly different traits than either varieties in production now or those woody species used to develop predictive assessments. Some, like *E. urograndis* (*E. grandis x E. urophylla*) genetically modified for freeze tolerance, may be grown in locations where they never have before. Additionally, woody plants that have been selectively bred or genetically modified and are also reproductively compatible with related species in the natural environment could transfer those novel traits into wild populations. In all situations described, it would be important to assess the likelihood of invasion or the transfer of novel traits to wild populations specifically and carefully.

3.4.9. Key Findings

Current environmental impacts of production and use of woody biomass as a biofuel feedstock are negligible, since no large-scale, commercial operations are yet in existence to create demand for this feedstock. However, estimates suggest that the potential for biofuel production from woody biomass is substantial, with predominant sources coming from forest harvest residues, thinning, and SRWCs. The removal of forest harvest residues is the largest source of woody biomass assumed in the RFS2 RIA (U.S. EPA 2010a). The most plausible impacts from residue removal appear to be slightly negative for air and soil quality, especially with multiple removals on nutrient-poor soils in the case of the latter. In some cases, the application of fertilizers may be necessary to offset losses in soil nutrients. Other impacts appear to be relatively negligible, particularly regarding water quantity and invasiveness. The environmental impacts of moderate thinning regimes without residue removal appear to be relatively modest.

Although woody biomass plantings eligible under RFS2 can be grown only on non-federal, managed forested land, there are considerable benefits of planting SRWCs to replace row crops or on degraded, abandoned agricultural land. For example, woody biomass species require fewer inputs of fertilizer and pesticides than row crops, resulting in reduced runoff of these substances into surface and ground water. In contrast, SRWCs grown to replace traditional managed forest land can exhibit a range of environmental impacts, depending on management. SRWCs will have the most negative consequences if they significantly reduce forest replanting intervals and require high water and chemical inputs. Additionally, the introduction of non-native species or genetic modification that enhances invasiveness could increase the risk to native populations. Conversely, if native species are planted and managed with limited soil disturbance and low chemical and water inputs, SRWCs can provide environmental benefits relative to other short-rotation managed forests.

For a more comprehensive, qualitative comparison of the environmental impacts of woody biomass, including a discussion of the most plausible impacts, see Section 6.

3.4.9.1. Key Uncertainties and Unknowns

- Though there are commercial-scale power plants that generate electricity from woody biomass, this biomass source is not yet converted to liquid biofuel on any large scale. A mature cellulosic ethanol industry might demand more feedstock than these current power plants, so its environmental impacts could be greater and possibly differ. This uncertainty regarding future demand creates unknowns for any projection of this feedstock's environmental effects, positive or negative.
- Specific environmental impacts will vary depending on forest and soil type, topography, climate, and other factors.
- Lack of information about the amount and relative proportion of woody biomass that would come from harvest residues, thinning, and SRWCs to support large- scale operations creates substantial uncertainty. The potential effects of harvest residues and thinning are easier to assess because a body of literature from other forestry applications, such as whole-tree harvesting, does exist. Even so, uncertainties arise from variations in the percent of residues removed during harvesting and in the degree of thinning, which can range from small to large proportions of the existing stand.
- Quantifying impacts of SRWCs to ecosystems and biodiversity will depend on knowing where and under what agronomic conditions SRWCs are grown and how they are managed. Uncertainty about these factors limits understanding of the potential impacts of this feedstock.

3.5. Algae

3.5.1. Introduction

Algae are of interest as a biofuel feedstock because of their high oil content, ability to recycle waste streams from other processes, and minimal land requirements (U.S. EPA, 2010a). Algae production demands less land area per gallon of fuel produced than other feedstocks (Dismukes et al., 2008; Smith et al., 2009).

In the case of biofuels, the word "algae" typically refers to microalgae.[22] There are many different types of algae, methods to cultivate them, and processes to recover oil from them. Algae grown photosynthetically are limited to growth during daylight hours and require carbon dioxide. Heterotrophic algae, which do not use photosynthesis, can be grown continuously in the dark, but require a fixed carbon source such as sugars because they cannot use carbon dioxide directly (Day et al., 1991).

Basic research on algae as a biofuel feedstock was conducted by DOE under a program known as the Aquatic Species Program (ASP) (Sheehan et al., 1998). The ASP focused on the production of biodiesel from high-lipid-content algae grown in ponds, using waste CO_2 from coal-fired power plants. Over the almost two decades of this program, advances were made in the science of manipulating the metabolism of algae and the engineering of algae production systems. However, very little research focused on the environmental consequences of large-scale algae production. Research and pilot studies have shown that the lipids and carbohydrates in microalgae can be refined and distilled into a variety of biodiesel- and alcohol-based fuels, including diesel, ethanol, methanol, butanol, and gasoline (U.S. EPA, 2010a). Algae also have the potential to serve as feedstock for other types of fuels, including bio-oil, bio-syngas, and bio hydrogen. This section focuses on the use of algae for biodiesel, because biodiesel is the most likely near-term pathway for algae use as biofuel.

Cultivation of algae feedstocks can take place in photobioreactor facilities with closed-cycle recirculation systems or in open-system-style impoundments. *Open systems* use pumps and paddle wheels to circulate water, algae, and nutrients through shallow, uncovered containments of various configurations. *Closed systems* employ flat plate and tubular photobioreactors and can be located outdoors or indoors. Variations include hybrid (combined open and closed) cultivation and heterotrophic cultivation (which uses organic carbon instead of light as an energy source). Different algae cultivation strategies are being studied to determine which is most suitable for supporting large-scale biofuel production (Chisti, 2007; U.S. EPA, 2010a).

Harvesting requires that the algae be removed, dewatered, and dried. Dewatering is usually done mechanically using a screw press, while drying can use solar, drum, freeze, spray, or rotary techniques (U.S. EPA, 2010a). After harvesting, the biofuel production process begins: oil is extracted from the algae through chemical, mechanical, or electrical processes (U.S. EPA, 2010a). Algal oil can then be refined with the same transesterification process used for other biofuel feedstocks such as soybean oil.

While the different methods of algae cultivation and recovery will clearly have very different environmental impacts, such as energy consumption and chemical use and disposal, it is premature to draw definitive conclusions about these impacts, given the nascent state of cultivating algae for biofuel. Likewise, relatively few scientific studies have examined the environmental impacts of algal biofuel production, although the literature was developing rapidly as this report went to publication. The second triennial report to Congress will likely contain much more information—from both industry and academia—on the environmental impacts of algal biofuel production.

3.5.2. Overview of Environmental Impacts

Algae-based biofuel production systems are still being investigated at the pilot stage using smaller-scale prototype research facilities. The potential environmental and resource impacts of full-scale production are highly uncertain because much of the current relevant data is proprietary or otherwise unavailable, and many key parameters are unknown, including where and how algae will be produced and what species and strains of algae will be used as feedstocks.

Algae cultivation can require the use of pesticides, fertilizers, water, and fuel. Each of these activities, in turn, can impact air quality, water quality, and water availability (soil quality is likely not a concern). In addition to these impacts, there is potential for invasive algae strains to escape from cultivation (Flynn et al., 2010). Industrial oil extraction and biodiesel production, biodiesel and byproduct transport and storage, and biodiesel and byproduct end use also entail environmental impacts, which are discussed further in Section 4.

3.5.3. Current and Projected Cultivation

Land use is one of the primary drivers behind interest in algae as a biofuel feedstock. Relative to other feedstock resources, algal biomass has significantly higher productivity per cultivated acre (Chisti, 2007). Moreover, algae cultivation requires relatively little land (e.g., about 2 percent compared to soybeans) (Smith et al., 2009). Algae's lack of dependence on fertile soil and rainfall essentially eliminates competition among food, feed, and energy production facilities for land resources (Muhs et al., 2009). Because algae-based biofuel production facilities do not require specific land types, they may be sited closer to demand centers, reducing the need to transport significant quantities of either biofuel or feedstock from one region of the country (e.g., the Midwest) to another (e.g., coastal population centers). Despite these potential advantages, significant long-term research and development will be required to make microalgal biofuels processes economically competitive (Huesemann and Brenemann, 2009).

Proximity to input sources (such as carbon dioxide sources) and output markets, as well as the availability of affordable land, will likely drive algae production facility siting decisions. The U.S. Southwest is viewed as a promising location for economical algae-to-biofuel cultivation due to the availability of saline ground water, high exposure to solar radiation, and low current land use development. Based on pilot studies and literature on algae cultivation, likely areas for siting algae-based biofuels facilities also include coasts, marginal lands, and even co-location with wastewater plants (Sheehan et al., 2004; U.S. EPA, 2010a). Algae grown in conjunction with animal and human wastewater treatment facilities can reduce both freshwater demands and fertilizer inputs, and may even generate revenue by reducing wastewater treatment costs. U.S. companies are already using wastewater nutrients to feed algae in intensively managed open systems for treatment of hazardous contaminants (Munoz and Guieysse, 2006).

3.5.4. Water Quality

Scaled production of algae oil for biofuels has not yet been demonstrated; therefore, water quality impacts associated with large-scale use of algae-based biofuels are currently speculative. Wastewater is a key factor influencing water quality impacts of algae production facilities, including whether wastewater is used as a water source for algae cultivation, and whether wastewater is discharged from the algae cultivation site. Depending on the treatment requirements, release of wastewater could introduce chemicals, nutrients, additives (e.g., from flocculation), and algae, including non-native species, into receiving waters. Releases of nutrient-rich growth media could affect water quality by inducing higher productivity of native algae, which can contribute to eutrophication.

Co-locating algae production facilities with wastewater treatment plants, fossil fuel power plants, or other industrial pollution sources can improve water quality and utilize waste heat that contributes to thermal pollution, while reducing freshwater demands and fertilizer inputs (Baliga and Powers, 2010; Clarens et al., 2010). When these facilities are co-located, partially treated wastewater acts as the influent to the algae cultivation system. Algae remove nutrients as they grow, which improves the quality of the wastewater and reduces nutrient inputs to receiving waters. If fresh surface water or ground water is used as the influent, nutrients must be added artificially in the form of fertilizer.

Significant environmental benefits could be associated with the ability of algae to thrive in polluted wastewater. Algae can improve wastewater quality by removing not only nutrients, but also metals and other contaminants, and by emitting oxygen. Thus, algae can effectively provide some degree of "treatment" for the wastewater (Darnall et al., 1986; Hoffmann, 1998).

3.5.5. Water Quantity

3.5.5.1. Water Use

Water is a critical consideration in algae cultivation. Factors influencing water use include the algae species cultivated, the geographic location of production facilities, the production process employed, and the source water chemistry and characteristics. EPA has estimated that an open-system biofuel facility generating 10 million gallons of biofuel each year would use between 2,710 and 9,740 million gallons of water each year; a similar-scale photobioreactor-type facility would use between 250 and 720 million gallons of water annually (U.S. EPA, 2010a).

The harvesting and extraction processes also require water, but data on specific water needs for these steps are limited (U.S. EPA, 2010a). Compared to the water required for algae growth, however, demands are expected to be much lower.

3.5.5.2. Water Availability

The use of fresh water versus brackish, saline, or wastewater will largely determine the effects of algae cultivation on water availability. If fresh water is employed, algae production could exacerbate or create water availability problems, especially in promising locations like the Southwest, which are already experiencing water shortages. However, the water used to grow algae does not have to be high-quality fresh water. Algae can thrive in brackish water, with salt concentrations up to twice that of seawater, and can obtain nutrients from wastewater such as industrial, agricultural, coal plant, and ethanol plant effluent (U.S. EPA, 2010a). Thus, competition for freshwater resources may be mitigated by siting facilities in areas that can provide suitable brackish or wastewater sources. Additionally, co-locating algae production facilities with wastewater treatment plants can reduce, but not eliminate, water demands (Clarens et al., 2010). Evaporation losses from open ponds could still reduce water availability if wastewater could otherwise be treated and re-used. Relative to open ponds, the water availability impacts of algae production can also be mitigated in large part by using photobioreactors, which require less water and land area than open systems (U.S. EPA, 2010a).

3.5.6. Soil Quality

Very little peer-reviewed literature exists on the soil impacts of algae production. These impacts are likely to be negligible and have therefore not been the subject of much study.

3.5.7. Air Quality

The effects of algae-based biofuels on air quality have received little attention to date in peer-reviewed literature. As a result, additional research is needed to determine whether anything unique to algae production processes would raise concern about air emissions.

Open or hybrid open systems appear to have greater potential to impact air quality compared to enclosed photobioreactors, given the highly controlled nature of the latter systems. No studies are yet available, however, to characterize or quantify emissions associated with open systems used to produce algae for biofuel. Studies have measured air emissions of open-system algae ponds that are part of wastewater treatment systems (Van der Steen et al., 2003), but these studies may have very limited applicability to open systems for commercial-scale production of algae oil for biodiesel. Additional research will be required to estimate and characterize emissions from pumping, circulation, dewatering, and other equipment used to produce algae for biofuel.

3.5.8. Ecosystem Impacts

3.5.8.1. Biodiversity

Algae production is likely to have fewer biodiversity impacts than production of other feedstocks because algae typically require less land, fertilizer, and pesticide than do other feedstocks, and because algae production plants may be co-located with wastewater treatment plants. As mentioned above, the location of algae production facilities will be a key factor affecting the potential for impacts. Using wastewater to capture nutrients for algae growth could help reduce nutrient inputs to surface waters (Rittmann, 2008). Algae also require lower inputs of fertilizers and pesticides than other feedstocks, which may translate into fewer ecological impacts to aquatic ecosystems (Groom et al., 2008). Production facilities for algae that need sunlight to grow could be located in arid regions with ample sunlight (Rittmann, 2008); however, growing algae in areas with limited water resources could impact the amount of water available for the ecosystem because of draws on ground water. It is unknown what impacts an accidental algae release might have on native aquatic ecosystems, particularly if the algae released have been artificially selected or genetically engineered to be highly productive and possibly adaptable to a range of conditions.

3.5.8.2. Invasive Algae

The potential for biofuel algae to be released into and survive and proliferate in the environment is, at present, highly uncertain. It will depend on what species and strains of naturally occurring, selectively bred, or genetically engineered algae are used and how they are cultivated.

The risk of algae dispersal into the environment is much lower in closed bioreactor systems than open system production, though unintentional spills from bioreactors in enclosed production facilities are possible. High winds blowing across open systems may carry algae long distances, depositing them in water bodies, including wetlands. Wildlife that enter the ponds may also disperse algae to other water bodies. Closed systems, in addition to limiting algae dispersal, have the benefit of protecting algal media from being contaminated with other microbes, which could compete with the cultivation strains for nutrient resources.

Effluent from algal biomass dewatering processes may contain residual algae, which could thrive in receiving waters. Treatment strategies will need to be developed to prevent algae in effluent from contaminating the surrounding ecosystem.

The ability of cultivated algae to survive and reproduce in the natural environment is unknown: one theoretical study suggests that native algae would out-compete some, but not all, strains with the most desirable commercial characteristics (Flynn et al., 2010). Further empirical work is critical to determine competitive and hybridizing abilities of biofuel algae in the natural environment and to measure possible effects on algal community dynamics and ecosystem services.

3.5.9. Key Findings

Current environmental impacts of production and use of algae as a biofuel feedstock are negligible, since no large-scale, commercial operations are yet in existence. Due to the lack of data on commercial-scale, the future environmental impacts of algae production are highly uncertain. Nevertheless, some key findings can be ascertained from the established literature. Unlike other feedstocks presented in this report, algae production does not require large amounts of land. This means it could have a much smaller environmental footprint than other feedstocks; its influence on water quality and quantity will largely determine its environmental impacts. This influence, in turn, will depend on the type of water used and the production system. Algae production using wastewater effluent, for example, can treat high levels of nutrients. Thus, combining commercial-scale algae production with wastewater treatment plants may create synergies that increase algae yields, while decreasing the environmental impacts of both facilities. In contrast, using freshwater for production of algal biofuels will potentially decrease water availability in areas such as the Southwest, where water is already scarce. Algae production in open water ponds requires substantial amounts of water relative to the use of photobioreactor systems.

Though highly uncertain, the most plausible impacts of growing algae in open ponds using wastewater are likely positive for water quality, slightly negative for water quantity, and relatively negligible for other environmental end points. Open system cultivation systems may have a greater potential than photobioreactor systems to adversely affect air quality, but no studies have yet evaluated air pollutant emissions associated with these systems. Little is known about how increases in algal biofuel production might affect biodiversity. Algae require lower inputs of fertilizers and pesticides than other feedstocks, which may also translate into fewer ecological impacts to aquatic ecosystems. The ability of cultivated algae to escape into and survive in the natural environment is uncertain. Experts speculate that photobioreactor cultivation systems would be superior to open systems in preventing the escape of cultivated algae.

For a more comprehensive, qualitative comparison of the environmental impacts of algae, including a discussion of the most plausible impacts, see Section 6.

3.5.9.1. Key Uncertainties and Unknowns

- Very little is known about the environmental impacts of commercial-scale algae production.
- Most of the uncertainties related to the production of algae for biodiesel stem from a lack of knowledge about which technologies may be used in future commercial applications, where they will be located, and what species and strains of algae will be used.
- Water availability impacts from feedstock growth will depend on where the algae are grown, if open or closed systems are used, and whether water is recycled.

3.6. Waste-Based Feedstocks

3.6.1. Introduction

Diverse wastes, including construction debris, municipal solid waste (MSW), yard waste, food waste, and animal waste, have the potential to serve as biofuel feedstocks. Depending on the waste, conversion system, and product, potential exists for municipalities, industries, and farmers to transform a material with high management costs to a resource that generates energy and profits. Tapping into waste energy sources has many challenges, including dispersed locations and potentially high transport costs, lack of long-term performance data, the cost of converting waste to energy, and the possibility that the resulting biofuel might not meet quality or regulatory specifications for use (Bracmort and Gorte, 2010).

Use of wastes as biofuel feedstocks will vary based on their availability, the ability of conversion technologies to handle the material, and the comparative economics of their use for fuel versus power and other products. Types and quantities of wastes used will vary by region. A large number and variety of waste-based materials are being investigated and implemented as feedstocks for ethanol and biodiesel, mostly on local scales. For example, several states—e.g., Massachusetts (Advanced Biofuels Task Force, 2008; Timmons et al., 2008), California (Chester et al., 2007), and Ohio[23]—have explored waste availability and its potential to meet regional energy needs, either for power or for transportation fuel. Feedstocks may be converted to biofuel or used as an energy source to power a biorefinery.

3.6.2. Municipal Solid Waste

The biogenic portion of MSW (paper, wood, yard trimmings, textiles, and other materials that are not plastic- or rubber-based), could be a contributing feedstock for ethanol and other biofuels. Using 2005 data, the U.S. EIA calculated that 94 million tons (MT) (about 56 percent) of the 167.8 MT of MSW waste generated that year had biogenic BTU content (U.S. EIA, 2007). This estimate included food waste—the third largest component by weight, and a potentially viable biofuel feedstock in addition to biogenic material listed above. It is estimated by Shi et al (2009) that at least 21billion gallons of waste paper-derived cellulosic ethanol can be produced globally from MSW (Shi et al., 2009). While this is not a likely scenario (since, for example, some of the biogenic fraction—paper, wood, etc.—would be recycled or reused), it demonstrates that MSW could be a significant source for biofuel. In addition, there are significant environmental co-benefits associated with using MSW for biofuel, including diverting solid waste from landfills and incinerators, extending their useful life, and reserving that capacity for materials that cannot be recycled or reused.

3.6.3. Other Wastes

Several types of waste materials that currently present environmental and economic challenges have the potential to be harnessed as feedstock for biofuel. These materials include waste oil and grease, food processing wastes, and livestock waste (Antizar-Ladislao and Turrion- Gomez, 2008).

The DOE estimates that the restaurant industry generates 9 pounds of waste oil per person annually, and that the nation's wastewater contains roughly 13 pounds of grease per person per year (Wiltsee, 1998). Several municipalities and industries have implemented collection programs, and are converting these wastes to biodiesel.

Annually, the United States generates an estimated 48 million tons of food processing wastes (i.e., food residues produced during agricultural and industrial operations), not including food waste disposed and processed through wastewater treatment plants (Kantor et al., 1997). These wastes have potential as biofuel feedstocks.

The United States generates over 1 billion tons of manure, biosolids, and industrial byproducts each year (ARS, n.d.). The amount of manure generated at confined and other types of animal feeding operations in the United States is estimated to exceed 335 million tons of dry matter per year (ARS, n.d.). While much of this manure is applied to cropland and pasture as fertilizer, excess is often available and could be tapped as a biofuel feedstock. It has been estimated that around 10 percent of current manure production could be used for bioenergy purposes under current land use patterns once sustainability concerns are met (i.e., this manure is available after primary use of manure on soils to maintain fertility) (Perlack et al., 2005). Methane emissions from livestock manure management systems, which account for a significant percentage (10 percent or 17.0 million metric tonnes of carbon equivalent [3.0 teragrams] in 1997) of the total U.S. methane emissions, are another potential energy source (U.S. EPA, 1999).

Using any of these excess waste materials as biofuel feedstocks could create a higher-value use with significant environmental and economic benefits.

3.6.4. Environmental Impacts of Waste-Based Biofuel

Among other environmental benefits, using waste-based biofuels diverts waste from landfills (avoiding the generation of landfill methane) and diverts waste and trap greases (helping to avoid costly plant disruptions that contribute to combined sewer overflow). Biorefineries that use wastes, particularly MSW, tend to be located near the sources of those wastes; accordingly, they are also near dense populations of end-users of transportation fuels, which helps reduce the GHG life cycle footprint of waste-derived fuels (Antizar-Ladislao and Turrion-Gomez, 2008; Williams et al., 2009).

More information is needed to understand and evaluate the environmental effects of waste-based biofuels. Different wastes have different characteristics, including size, volume, heterogeneity, moisture content, and energy value. These characteristics will, to a large degree, determine feasible and appropriate collection, processing, and conversion methods, which in turn will determine net energy gain, as well as environmental impacts such as air and GHG emissions. Research is needed to compare the benefits and impacts of various technological options for converting MSW to biofuel, and to compare, on a regional basis, the environmental benefits and impacts of MSW to other biofuel feedstocks. Currently, data are lacking for such comparisons. Comparative life cycle assessments that consider both the direct impacts or benefits and indirect impacts or benefits (e.g., impacts of reduced landfilling of MSW) are needed to understand the true value of waste as an alternative feedstock.

3.6.4.1. Key Findings and Uncertainty

There have been comparatively few attempts at assessing the environmental impacts associated with the production and use of waste-based biofuels (Williams et al., 2009). In general, waste as a feedstock is expected to have a smaller environmental impact than conventional feedstocks. However, the choice of waste management options and the particular technology for energy recovery will influence the environmental medium impacted and the level of impact (Chester and Martin, 2009; Kalogo et al., 2007). As the number of waste conversion facilities increases, environmental monitoring and research will be needed to address the information gaps that currently limit environmental assessment.

3.7. Summary of Feedstock-Dependent Impacts on Specialized Habitats

EISA Section 204 requires an assessment of the impacts of biofuels on a variety of environmental and resource conservation issues, including impacts on forests, grasslands, and wetlands. This section provides an overview of impacts on these specific habitats.

3.7.1. Forests

Section 211(o) of the Clean Air Act limits (CAA) planting of SRWC and harvesting of tree residue to actively managed tree plantations on non-federal land that was cleared before December 19, 2007, or to non-federal forestlands; it limits removal of slash and pre-commercial thinning to non-federal forestlands. However, as described in Table 3-4, a variety of activities associated with producing woody biomass feedstock may have direct impacts on forests.

Table 3-4. Overview of Impacts on Forests from Different Types of Biofuel Feedstocks

Feedstock	Forest Impact	Report Section
Row crops	No direct impacts are likely since the conversion of forests to row crops is ineligible under RFS2; may have indirect impacts.	3.7.1
	Use of forested buffers as conservation practices to control erosion and nutrient runoff can increase forest habitat and connectivity for some species.	3.2.8
Perennial grasses	Most grass species are unlikely to have impacts.	3.3.4.2
Woody biomass	Use of forested buffers as a best management practice to control erosion and nutrient runoff can increase forest habitat and connectivity for some species.	3.4.4
	SRWC plantations may deplete soil nutrients with repeated, frequent harvesting, particularly on marginal soils, but may sustain levels with coppicing, longer rotations, and strategic use of cover crops.	3.4.4.1
	SRWC plantations can sustain high species diversity, although bird and mammal species in these plantations tend to be habitat generalists.	3.4.8.1
	Some tree species under consideration, like *Eucalyptus,* may invade forests in certain locations.	3.4.8.2
	Harvesting with residue removal can be sustainable for at least one rotation, yet multiple harvests and removals at the same site may pose a risk to forest growth, particularly on nutrient-poor soils.	3.4.6.2; 3.2.6.1
	Harvesting forest residues may decrease woody debris available for species habitat.	3.4.8.1
Algae	Unlikely to have any significant impacts.	3.5.1

While row crop cultivation is not expected to directly affect forests, there may be indirect effects. Recent economic modeling has predicted a net decrease in the acres of forested lands in the United States in response to an expansion of cropland needed to satisfy future demand for ethanol (Keeney and Hertel, 2009; Taheripour et al., 2010; Hertel et al., 2010a, 2010b). Woody biomass is the feedstock most likely to affect forests in place; algae and most perennial grasses are unlikely to have an impact on this habitat.

3.7.2. Grasslands

In addition to the restrictions on forested sources of renewable biomass mentioned above, Section 211(o) of the CAA more broadly limits the lands on which any biofuel feedstock can be produced to those that were cleared or cultivated at any time before December 19, 2007, either in active management or fallow and non-forested. Therefore, grassland that remained uncultivated as of December 19, 2007, is not included in the RFS2 Final Rule and would not be considered eligible for renewable biomass production. However, in the Midwest approximately 96, 59, and 50 percent, respectively, of the historical tallgrass, mixed grass, and short grass prairie have already been converted for human purposes (Samson and Knopf, 1994).

Table 3-5. Overview of Impacts on Grasslands from Different Types of Biofuel Feedstocks

Feedstock	Grasslands Impact	Report Section
Row crops	Conversion of grasslands (e.g., CRP) to row crops particularly impacts grassland- obligate species, potentially leading to declines, including declines in duck species.	3.2.8.1
	Higher proportions of corn within grassland ecosystems lead to fewer grassland bird species.	3.2.8.1
	Use of grassland buffer strips as conservation measures to mitigate erosion and nutrient runoff will increase grassland area and can provide habitat for some grassland species depending on management regimes.	3.2.4.1
Perennial grasses	Conversion of row crops to switchgrass and use of grassland buffer strips as conservation measures to mitigate erosion and nutrient runoff may improve grassland habitat for some species depending on management regimes.	3.3.4.1
	Commercial grassland production as a feedstock may require chemical inputs and can negatively impact water quality.	3.3.4.3
	Overall biodiversity will be impacted by harvesting and management regimes.	3.3.8.1
	Cultivation of perennial grasses outside their native ranges and/or introduction of more vigorous varieties of native species could lead to invasions of pastures and native grasslands.	3.3.8.2
Woody biomass	Unlikely to have significant impacts, since eligible woody biomass is restricted to managed forest lands under RFS2.	3.4.1
Algae	Unlikely to have any significant impacts.	3.5.9

Most of the lands that would be eligible for renewable biomass production under the CAA, because they were cultivated at some point prior to December 19, 2007, are now part of the CRP (see Section 3.2.3). Multiple studies project that some conversion of CRP to cropland will occur (Malcolm et al., 2009; Beach and McCarl, 2010). Because the vast majority of land enrolled in CRP uses native or introduced grasses (FSA, 2010), it is likely that the conversion of CRP lands to biomass production will impact grassland ecosystems (Table 3-5). However, cultivation of perennial grasses in conjunction with conservation practices could have some positive effects on previously degraded grasslands.

3.7.3. Impacts on Wetlands

Provisions in both the Food Security Act of 1985 (commonly known as the Swampbuster Program) and the Clean Water Act (Section 404 Regulatory Program) offer disincentives that limit the conversion and use of wetlands for agricultural production. These programs have reduced the rate of wetland losses due to agricultural development, although some wetlands are still lost due to partial drainage and indirect effects of altered volume and timing of runoff (Dahl, 2000; Blann et al., 2009). Continued losses of wetlands have consequences for the landscape mosaic of habitat, reducing connectivity for a variety of organisms (Blann et al., 2009). Current legislation also does not protect all types and sizes of wetlands, making some (e.g., small, isolated, or seasonal wetlands) more vulnerable to direct losses or indirect impacts of drainage (Blann et al., 2009). Therefore, the feedstocks assessed in this report are still expected to impact wetlands (Table 3-6).

Table 3-6. Overview of Impacts on Wetlands from Different Types of Biofuel Feedstocks

Feedstock	Wetlands Impact	Report Section
Row crops	Increased sediment, nutrients, chemicals, and pathogens from runoff flow into downstream wetlands and change wetlands community structure.	3.2.4
	Possible loss of or impacts on small, isolated, or seasonal wetlands due to conversion or partial drainage with consequences for other species.	3.2.8; 3.7.3
	Increase in wetland habitat if wetlands are constructed to control erosion or nutrient runoff.	3.2.4.1
Perennial grasses	Reduced sediment and nutrient loadings, leading to improved water quality (but dependent on specific management practice).	3.3.4.1; 3.3.4.2
	Some grass species under consideration may invade wetlands, including giant reed (*Arundo donax*) and reed canarygrass (*Phalaris arundinacea*).	3.3.8.2
	Possible loss of or impacts on small, isolated or seasonal wetlands due to conversion or partial drainage with consequences for other species.	3.3.8;
Woody biomass	SRWC plantations can initially increase runoff of nutrients and sediment to water bodies, yet these losses rapidly decline to low levels after the establishment phase.	3.4.4.1
Algae	Algal strains created may escape from cultivation, potentially affecting wetlands.	3.5.8.2

3.8. Genetically Engineered Feedstocks

Genetic engineering of crops has a history of research, development, and commercialization that extends back for more than 15 years. Along with the growth of this biotechnology industry, the United States established a coordinated framework for regulatory oversight in 1986 (OSTP, 1986). Since then, the relevant agencies (EPA, USDA, and the Food and Drug Administration) have implemented risk assessment programs that allow informed environmental decision-making prior to commercialization. These programs have been independently assessed over the years (NRC, 2000, 2001, 2002) and improvements made to ensure the safety of the products. At the same time, the methodology for biotechnology risk assessment has been scrutinized and general frameworks created to facilitate robust approaches and harmonize the processes internationally (Conner et al., 2003; Pollard et al., 2004; Andow and Zwalen, 2006; Raybould, 2007; Auer, 2008; Craig et al., 2008; Nickson, 2008; Romeis et al., 2008;). Nonetheless, there are environmental concerns associated with genetically modified organisms (GMOs) that are currently used as biofuel feedstocks, as well as anticipated concerns for GMOs that will be developed for the next generation of biofuel feedstocks.

Brookes and Barfoot (2006, 2008, 2009, 2010) conducted a series of extensive post-commercialization assessments of genetically engineered maize, soybeans, cotton, sugar beets, and canola varieties at 10-year intervals. In these analyses, the authors found consistent reductions in the amounts of pesticides used and a reduction in GHG emissions for agricultural systems where these GMO crops are grown. These results are supported by others (Brinmer et al., 2005; Knox et al., 2006), although regional differences in the reductions have been noted (Kleter et al., 2008). The results for corn and soybean are consistent with the general trends (Brookes and Barfoot, 2010). Assuming that current genetically engineered varieties of corn and soybeans receive continued regulatory oversight, no additional environmental concerns are anticipated with these organisms in their current genetic configuration, even with an increase in their production. However, as feedstocks for biofuel change to accommodate cellulosic technologies and algae production, the range of environmental considerations, including impacts from GMO varieties, will change as well (Wilkinson and Tepfer, 2009; Lee et al., 2009).

To harness the full potential of biomass, the genetic engineering of feedstocks has been recognized as a key technology (Gressel, 2008; Antizar-Ladislao and Turrion-Gomez, 2008; Sexton et al., 2009). The approaches being considered include increasing plant biomass by delaying flowering, altering plant growth regulators, and manipulating photosynthetic processes; modifying traits (e.g., herbicide tolerance, insect resistance) in non–row crop plants that reduce cultivation inputs; and modifying cellulose/lignin composition and other traits that result in cost reductions in bioprocessing (i.e., facilitating the biorefinery process) (Sticklen, 2007, 2009; Ragauskas et al., 2006; Gressel, 2008). These new varieties may have implications for the environment beyond what has been considered in first-generation biotechnology crops, and the scientific community has begun to examine whether and how well existing risk assessment procedures will work for bioenergy crops (Chapotin and Wolt, 2007; Firbank, 2007; Lee et al., 2009; Wilkinson and Tepfer, 2009; Wolt, 2009).

4. BIOFUEL PRODUCTION, TRANSPORT, STORAGE, AND END USE

4.1. Introduction

This section addresses potential environmental impacts of post-harvest activities of the biofuel supply chain (see Figure 4-1). These activities comprise feedstock logistics (Section 4.2) and biofuel production (including handling of wastes and byproducts) (Section 4.3), distribution (Section 4.4), and end use (Section 4.5).

Production of biofuel from feedstock takes place at biofuel production facilities through a variety of conversion processes. The biofuel is transported to blending terminals and retail outlets by a variety of means, including rail, barge, tankers, and trucks. Biofuel distribution almost always includes periods of storage. Once dispensed at the final outlet, biofuel is combusted in vehicles and other types of engines, usually as a blend with gasoline or diesel, or in some cases in neat form.

Biofuel production, distribution, and end use primarily affect air and water, with some consequences for aquatic ecosystems. Air emissions may be released by a variety of sources. Many factors affect the quantity and characteristics of these emissions, including the type and age of equipment used, and operating practices and conditions.

Air emissions associated with end use of ethanol combustion are relatively independent of feedstock or conversion process, whereas biodiesel emissions are highly dependent on feedstock type. As discussed later in the section, biofuel combustion may result in higher emissions of some pollutants compared to gasoline combustion, and lower emissions of others.

Biofuel production requires the use of water, which may contribute to ground water depletion or lower surface water flow, depending on the amount of water withdrawn and water availability. Potential water quality impacts include wastewater discharge during the conversion process and the potential for leaks and spills to surface and ground water during biofuel handling, transport, and storage. Additionally, phosphorus runoff from the manure of animals that have been fed an ethanol byproduct—for example, dried distillers grains with solubles, which have a high phosphorus content (Regassa et al., 2008)—may have the potential to impact water quality and aquatic ecosystem condition.

Possible air and water impacts associated with ethanol and biodiesel, as well as opportunities for mitigation, are discussed in Sections 4.2 to 4.5. Discussion focuses primarily on the impacts of corn ethanol and diesel from soybean oil, because these constitute the vast majority of biofuel produced and used in the United States as of July 2010.

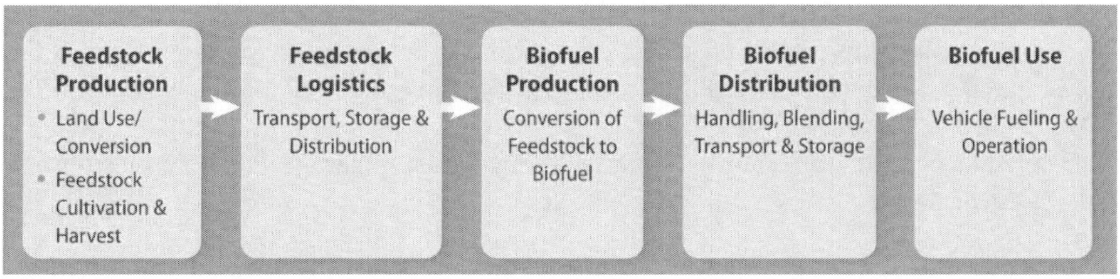

Figure 4-1. Biofuel Supply Chain and Use of Biofuel.

4.2. Feedstock Logistics

4.2.1. Handling, Storage, and Transport

Feedstock logistics comprise activities associated with handling, storing, and transporting feedstocks after harvest to the point where the feedstocks are converted to biofuel. Though alternative feedstock logistic systems have recently been proposed—e.g., the Uniform-Format: Solid Feedstock Supply System (Hess et al., 2009a)—this report considers the conventional system of transporting biomass directly to refinery without prior processing. The most significant environmental impacts of these activities are the emissions associated with energy use. Both greenhouse gases (GHGs) and criteria pollutant emissions result from the combustion of fuels used during transportation. In general, feedstock logistics may be optimized, and emissions reduced, by integrating feedstocks, processing facilities, and consumer demands at a regional scale to minimize transport distances.

4.2.1.1. Ethanol

Harvested corn is transported to a biorefinery, where it is converted to ethanol and a number of co-products. Air quality will be impacted by emissions from the combustion of fuels used for transportation vehicles and equipment.

4.2.1.2. Biodiesel

After harvest, soybeans or other vegetable oil seeds used as biodiesel feedstocks are transported from fields to the drying site, storehouse, or collection center, followed by transport to the biodiesel refinery. In the case of soybeans, mechanical crushing is typically used to separate soybean hulls from soybean oil. Air quality may be affected by emissions from the combustion of fuels used for transportation vehicles and equipment.

4.3. Biofuel Production

4.3.1. Biofuel Conversion Processes

4.3.1.1. Ethanol

As of November 2009, there were 180 corn starch ethanol facilities in the United States with a combined capacity of 12 billion gallons per year (bgy) (U.S. EPA, 2010a).[24] At that time, 27 of these (representing 1,400 million gallons per year [mgy] of capacity) were idled, and another 10 facilities (representing a combined capacity of 1,301 mgy), were under construction (U.S. EPA, 2010a). The majority of corn starch ethanol facilities are located in the country's major corn-producing states: Iowa (with the largest production capacity and the greatest number of plants) followed by Nebraska, Minnesota, Indiana, and Illinois (U.S. EPA, 2010a).

Conventional ethanol is produced from the fermentation of corn starch. Two methods are currently used:

- **Dry milling,** in which the corn kernel is first ground into a meal, usually without separating out the various component parts of the grain. The meal is then slurried

with water and cooked at high temperatures to form a mash, which then undergoes fermentation. Dry milling is more commonly used than wet milling.
- **Wet milling,** in which the kernels are steeped in water to separate out the germ, fiber, and gluten (fractionation). From this initial separation, co-products such as corn meal, corn gluten meal, and corn gluten feed are recovered. The remaining mash contains the water-soluble starch, which undergoes further processing for biofuel.

In both processes, soluble starch is subsequently converted to a simple sugar (glucose) through saccharification, an enzyme-catalyzed hydrolysis reaction. This is followed by yeast fermentation of the glucose to ethanol. Following fermentation, the mash is distilled to collect the ethanol as a mixture of 95 percent alcohol and 5 percent water. A subsequent dehydration step is required to remove the aqueous portion to yield 99.5 percent pure ethanol. Efforts to improve conversion efficiency have resulted in increased ethanol yields per bushel of corn. A 2008 survey of dry mill corn ethanol plants reported requiring 5.3 percent less corn than in 2001 to produce an equivalent amount of ethanol (Mueller, 2010).

Substantial efforts are under way to develop processes to convert feedstocks containing cellulose into biofuels. These cellulosic feedstocks are primarily composed of cellulose, hemicellulose, and lignin polymers. Currently, two major pathways exist for converting cellulosic feedstocks into biofuel:

- **Biochemical conversion** using a physical and chemical process to liberate tightly bound cellulose and hemicellulose from lignin. The process uses strong acid or enzymes (cellulases) to hydrolyze the cellulose and hemicelluloses to glucose and other simple sugars, followed by microbial fermentation of the sugars into ethanol.
- **Thermochemical conversion** involving gasification or pyrolysis.
  - In the *gasification process*, biomass is heated at high temperatures with a controlled amount of oxygen to decompose the cellulosic material. This yields a mixture composed mainly of carbon monoxide and hydrogen known as syngas.
  - In *pyrolysis*, the biomass is heated in the absence of oxygen at lower temperatures than used in gasification. The product is a liquid bio-oil that can be used subsequently as a feedstock for a petroleum refinery.

Other cellulosic conversion processes are in various stages of development, from concept stage to pilot-scale development to construction of demonstration plants. Although no U.S. commercial-scale plants are operating as of July 2010, several companies are expected to have facilities operating within the next few years.

4.3.1.2. Biodiesel

As of November 2009, there were approximately 191 biodiesel facilities in the United States, with a combined capacity of 2.8 bgy (U.S. EPA, 2010a). Total domestic production of biodiesel in 2009 was 505 mgy—much less than domestic production capacity. The dominant technology used to produce biodiesel involves a transesterification reaction in which triglycerides (fats) from the oil are converted to esters in the presence of an alcohol and a catalyst such as potassium hydroxide. Plant oils (soy, algae, etc.) and other feedstocks (e.g., animal- derived oil such as lard and tallow, recycled oil and grease from restaurants and food processing plants) provide sources of triglycerides for conversion to biodiesel. Free glycerol, or glycerin, is a major co-product in transesterification, constituting an estimated 10 percent of the final product (U.S. EPA, 2010a). Table 3-2 shows the breakdown of feedstocks used to produce biodiesel in the United States in 2009. Vegetable oils, including soybean oil, made up the majority of biodiesel feedstock—nearly 60 percent.

Commercial processes for large-scale algae production and algal oil collection are currently being developed as another plant oil source for biodiesel (U.S. EPA, 2010a). Lipid extraction and drying currently are energy-intensive steps in the algae diesel production process. Other processing techniques are currently being investigated, including enzymatic conversion and catalytic cracking of algal oil, pyrolysis, and gasification of algae. However, lipid extraction via solvents followed by transesterification remains the most commonly used method for algal oil processing (U.S. DOE, 2010). Until commercial facilities using mature technologies go into production, the impacts from algae conversion will be uncertain.

In addition to transesterification, other methods for converting seed oils, algal oils, and animal fats into biofuel have been developed recently using technologies that are already widely employed in petroleum refineries (Huo et al., 2009). Hydrotreating technologies use seed oils or animal fats to produce an isoparaffin-rich diesel substitute referred to as "green diesel" or renewable diesel, which is distinctly different from biodiesel, which is generated using the transesterification process (Huo et al., 2008).

Although the transesterification process can generate much more diesel product than the other processes, as noted above, it requires more energy and chemical inputs (Huo et al., 2008). In some cases, inputs, which are very energy-intensive to produce, must also be taken into consideration in a full life cycle assessment in order to adequately evaluate energy efficiency of each fuel production process. Compared with conventional diesel and biodiesel, renewable diesel fuels have much higher cetane numbers—a measure of diesel fuel quality (Huo et al., 2008).

4.3.2. Air Quality

The net changes in volatile organic compounds (VOCs), carbon monoxide (CO), nitrogen oxides (NO_x), and particulate matter (PM) emissions associated with biofuel production can be attributed to two countervailing effects: (1) emission increases connected with biofuel production and (2) emission decreases associated with reductions in gasoline production and distribution as ethanol displaces gasoline. EPA (2010a) determined that increases in fine particles less than 2.5 micrometers in diameter ($PM_{2.5}$), sulfur oxide (SO_x), and especially NO_x were driven by stationary combustion emissions from the substantial increase in corn and cellulosic ethanol production.

Increasing the production and distribution of ethanol was also found to lead to higher ethanol vapor emissions. To a lesser degree, the production and distribution of greater amounts of ethanol would lead to increases in emissions of formaldehyde and acrolein, as well as very small decreases in benzene, 1,3-butadiene, and naphthalene emissions relative to the total volume of these emissions in the United States. Additional details on EPA's analysis of changes in emissions associated with the revised Renewable Fuels Standard program (RFS2) volumes can be found in www.epa.gov/otaq/fuels/renewablefuels/regulations.

Air pollutant emissions associated with the conversion of biomass to fuel may be mitigated through the use of cleaner fuels during the conversion process and more efficient process and energy generation equipment. The majority of ethanol plants built in recent years, and expected to be built in the near future, use dry mill technology (Wang et al., 2007a). Because they use similar production processes, differences in environmental impacts between plants are primarily due to each plant's choice of fuel. EPA's RFS2 Regulatory Impact Analysis (RIA) assumes a dry mill for the base scenario.

EPA's RFS2 RIA examines the impacts of using energy-saving technologies such as combined heat and power (CHP). CHP is an effective means to reduce air emissions associated with biofuel production (both ethanol and biodiesel). CHP generates electricity by burning natural gas, biogas, or byproducts such as lignin, and then employs a heat recovery unit to capture heat from the exhaust stream as thermal energy. Using energy from the same fuel source significantly reduces the total fuel used by facilities along with the corresponding emissions of carbon dioxide (CO_2) and other pollutants. Fractionation, membrane separation, and raw starch hydrolysis are additional technologies examined in EPA's RFS2 RIA that increase process efficiencies by enabling producers to sell distillers grains (a co-product of the corn-ethanol conversion process) wet rather than dry, thereby reducing GHG emissions and other possible environmental impacts (since drying distillers grains is an energy-intensive process).

4.3.2.1. Ethanol

Ethanol production requires electricity and the use of steam. Electricity is either purchased from the grid or produced on site, and steam is typically produced on site from natural gas. Power and the energy used to fuel boilers are responsible for emissions of VOCs, PM, CO, SO_x and NO_x (U.S. EPA, 2010a; Wang et al., 2007b). For corn-based ethanol, fossil fuels such as natural gas are typically used to produce heat during the conversion process, although a number of corn ethanol facilities are exploring new technologies with the potential to reduce their energy requirements. A 2008 survey of dry corn mill ethanol plants highlighted recent efficiency gains. The survey reported that, compared to 2001, ethanol produced in 2008 required 28.2 percent less thermal energy and 32.1 percent less electricity (Mueller, 2010). A continuation of such efficiency improvements could further reduce the environmental impacts of ethanol production.

A number of processes at ethanol production facilities result in emissions of air toxics. These processes include fermentation, distillation of the resultant mash, and drying of spent wet grain to produce animal feed. Emissions of air toxics vary tremendously from facility to facility due to a variety of factors, and it is difficult to determine how differences in the production processes individually impact emissions (U.S. EPA, 2010a). Ethanol vapor and air toxic emissions associated with biofuel production were projected to increase in EPA's RFS2 RIA, but these increases would be very small compared to current emissions (U.S. EPA, 2010a).

4.3.2.2. Biodiesel

While the production processes for biodiesel and ethanol are fundamentally different, both require thermal and electrical energy for production. The thermal energy required for biodiesel production is usually met using steam generated using a natural gas boiler. In certain situations, the glycerol co-product may also be burned to produce process heat, or a biomass boiler may be used to replace natural gas.

Air quality issues associated with a natural gas-fired biodiesel production process are similar to those for other natural gas applications such as ethanol production, and include emissions of VOCs, PM, CO, SO_x, and NO_x. Glycerol or solid fuel biomass boilers have emissions characteristics similar to those anticipated for cellulosic ethanol plants, including increased particulates and the potential for VOCs, NO_x, and SO_x.

Biodiesel production using a closed hot oil heater system would have none of the air emissions associated with traditional steam production. Air emissions associated with these systems would be associated with the production of the electricity, which would take place outside the biodiesel plant boundary.

Additionally, the extraction of vegetable oil to create biodiesel in large chemical processing plants is typically achieved using hexane, a VOC that EPA has classified as a hazardous air pollutant. Hexane is also commonly used to extract algal oils. Fugitive emissions of hexane may result from increased biodiesel manufacture (Hess et al., 2009a).

4.3.3. Water Quality and Availability

All biofuel facilities use process water to convert biomass to fuel. Water used in the biorefining process is modest in absolute terms compared to the water applied and consumed in growing the plants used to produce biofuel. However, the use of water at production facilities may be locally significant, whereas agricultural water use may be more geographically dispersed. The impacts associated with water use at conversion facilities depend on the location of the facility in relation to water resources. In some regions where water is abundant, increased withdrawals may have little effect. Ground water depletion may result in increased costs to pump water from deeper wells, loss of stream flow, and subsidence of the overlying land (Reilly et al., 2008). Several areas of the country that are already experiencing lowered ground water levels (e.g., the High Plains aquifer, the Lower Mississippi River alluvial aquifer) correspond with regions where increased biofuel production is expected. In addition, minimum in-stream flow for aquatic life can be affected by ground water depletion because ground water discharge into streams is a major source of stream base flow. In some areas, streams have run dry due to ground water depletion, while in other areas, minimum stream flow during the summer has been sustained because of irrigation return flow to streams (Bartolino and Cunningham, 2003). In the case of sole source aquifers, ground water depletion may severely impact drinking water availability, because these areas have no readily available alternative freshwater sources (Levin et al., 2002).

Comprehensive local, state, and regional water planning, as well as state regulatory controls, are critical to ensure that facilities are located in watersheds that can sustain the increased withdrawal without affecting other uses. Siting of biofuel facilities may also be influenced by state laws and regulations designed to avoid or mitigate conflicts among water uses. These vary by state. For example, withdrawals associated with biofuel production facilities may need a state permit to ensure that the proposed withdrawal does not result in unacceptable impacts on other users or on aquatic life. In addition, different states assign water rights in different ways. Some exercise the prior appropriation rule (i.e., water rights are determined by priority of beneficial use, meaning that the first person to use the water can acquire individual rights to the water); some are based on the English law of absolute ownership (i.e., rights to use water are connected to land ownership); some limit withdrawals based on stream flow requirements for aquatic life; and some have a hierarchy to prioritize uses of the water.

Like water quantity impacts, water quality impacts depend on a number of factors including facility location, water source, receiving water, type of feedstock used, biorefinery technology, effluent controls, and water re-use/recycling practices. Water quality impacts are associated with the wastewater discharge from the conversion process. Pollutants of concern discharged from ethanol facilities include biological oxygen demand (BOD), brine, ammonia-nitrogen, and phosphorus. BOD, glycerin, and to a certain extent, total suspended solids (TSS) are primary pollutants of concern found in biodiesel facility effluent. Regulatory controls placed on the quality of biofuel production wastewater discharge can mitigate some water quality and aquatic ecosystem impacts. Discharges to publicly owned wastewater treatment works (POTWs) are subject to general pre-treatment standards (40 CFR 403.5) in the Clean Water Act. Biofuel facilities that discharge their wastewater to POTWs are subject to pre-treatment limitations that are in effect for the receiving POTW. For those facilities that treat and discharge their own wastewater, EPA has enforceable regulations to control production facility effluent discharges of BOD, sediment, and ammonia-nitrogen through the National Pollutant Discharge Elimination System (NPDES) permit program.

Whether effluent is discharged to a POTW or treated on site at the production facility, BOD can lead to methane emissions during the wastewater treatment process. To mitigate the release of methane to atmosphere, facilities can install anaerobic digesters as a treatment step. Anaerobic digesters treat the biosolids contained in wastewater effluent, generating biogas that is approximately 60 to 65 percent methane. This biogas can then either be flared or captured and used as a clean energy source at the biofuel production facility or elsewhere.

Currently there are no effluent limitation guidelines or categorical pretreatment standards that regulate process wastewater discharges from ethanol and biodiesel manufacturing facilities.

4.3.3.1. Ethanol

In 2007–2008, EPA evaluated biodiesel and corn ethanol manufacturing facilities. No major effluent quality issues were found from corn ethanol plants discharging to either surface waters or to wastewater treatment plants.

While some ethanol facilities get their process water from municipal water supplies, most use onsite wells (Wu et al., 2009). However, most untreated ground water sources are generally not suitable for process water because of their mineral content. Ground water high in mineral content is commonly treated by reverse osmosis, which requires energy and concentrates ground water minerals into reject water, with potential water quality impacts upon their release. For every two gallons of pure water produced, about a gallon of brine is discharged as reject water (U.S. EPA, 2010a). Methods to reduce the impact associated with reject water high in mineral concentration include (1) further concentration and disposal or (2) use of in-stream dilution. Some ethanol facilities have constructed long pipelines to access additional water sources to dilute the effluent to levels that meet water quality standards.

Once process water is treated, most is lost as steam during the ethanol production process. Water use varies depending on the age of the facility and the type of milling process. Older generation production facilities use 4 to 6 gallons of process water to produce a gallon of ethanol; newer facilities generally use less than 3. Most of this water savings is gained through improved recycling of water and heat in the process. Dry milling facilities consume on average 3.45 gallons of fresh water per gallon of ethanol produced (Wu, 2008); newer facilities tend to consume about 21 percent less water (Wu et al., 2009). Wet mill facilities consume an average of 3.92 gallons of fresh water per gallon of ethanol produced (Wu,

2008). Most estimates of water consumption in ethanol production are based on the use of clean process water and do not include the water discharged as reject water.

Ethanol plants are designed to recycle water within the plant, and improvements in water use efficiency of ethanol facilities are expected through steam condensate reuse and treated process water recycling (Wu et al., 2009). Continued development of new technologies that improve water efficiency will help mitigate water quantity impacts.

Because no large-scale cellulosic ethanol production facilities are currently operating, water demand for production of cellulosic ethanol is not certain. However, for most cellulosic feedstocks, including agricultural residues like corn stover and dedicated energy crops like switchgrass, water demand is estimated to be between 2 and 10 gallons of water per gallon of ethanol, depending on the conversion technology, with volumes greater than 5 gallons of water per gallon of ethanol cited more often (NRC, 2008; Williams et al., 2009; Wu et al., 2009). Some studies assume water demand for processing woody biomass will be similar to processing cellulosic material from agricultural residues or dedicated energy crops (up to 10 gallons of water per gallon of ethanol) (Evans and Cohen, 2009). Other studies state that new technologies like fast pyrolysis will require less than half that amount of water per gallon of ethanol (Wu et al., 2009). Consumptive use of water is declining as ethanol producers increasingly incorporate recycling and other methods of converting feedstocks to fuels that reduce water use (NRC, 2008; Laser et al., 2009).

Cellulosic ethanol facilities that employ biochemical conversion would be expected to have similar water requirements and brine discharges as the current operating corn ethanol facilities. The additional steps required to separate the lignin from the cellulose could produce wastewater streams high in BOD that would require treatment on site or at wastewater treatment plants.

4.3.3.2. Distillers Grain with Solubles

One important co-product of ethanol production is dried distillers grain with solubles (DDGS). Due to the increase in ethanol production and the prices of corn and soybeans, DDGS has become an increasingly important feed component for confined livestock. About one-third of the corn processed into ethanol is converted into DDGS; therefore, approximately 45 million tons of DDGS will be produced in conjunction with the 15 billion gallons of corn ethanol produced by 2015.

Livestock producers may partially replace corn or other feeds with DDGS for both economic and production reasons. Different livestock species can tolerate varying amounts of DDGS in their diets. Although specific analysis of DDGS can vary among ethanol plants, DDGS are higher in crude protein (nitrogen) and three to four times higher in phosphorus compared to corn (Regassa et al., 2008).

The increase in nitrogen and phosphorus from DDGS in livestock feed has potential implications for water quality and aquatic ecosystems. When nitrogen and phosphorus are fed in excess of animals' needs, excess nutrients are excreted in urine and manure. Livestock manure may be applied to crops, especially corn, as a source of nutrients. When manure is applied at rates above the nutrient needs of the crop or when the crop cannot use the nutrients, the nitrogen and phosphorus can run off to surface waters or leach into ground waters. Excess nutrients from manure nutrients have the same impact on water quality as excess nutrients from other sources.

Livestock producers may limit the potential pollution from manure applications to crops through a variety of techniques. USDA's Natural Resources Conservation Service (NRCS) has developed a standard for a comprehensive nutrient management plan to address the issue of proper use of livestock manure (NRCS, 2009).

4.3.3.3. Biodiesel

Biodiesel facilities use much less water than ethanol facilities to produce biofuel. The primary consumptive water use at biodiesel plants is associated with washing and evaporative processes. Water use is variable, but is usually less than one gallon of water for each gallon of biodiesel produced (U.S. EPA, 2010a); some facilities recycle washwater, which reduces overall water consumption (U.S. EPA, 2010a). However, water use has been reported as high as 3 gallons of water per gallon of biodiesel (Pate et al., 2007). Larger well-designed facilities use water more sparingly, while smaller producers tend to use more water per production volume (U.S. EPA, 2010a). New technologies that improve water efficiency will help mitigate water quantity impacts. Recent plant designs have included either waterless processes or water recycling.

In addition to water use in the washing and evaporation processes, sources of wastewater include steam condensate; process water softening and treatment to eliminate calcium and magnesium salts, iron, and copper; and wastewaters from the glycerin refining process (U.S. EPA, 2008c). In a joint U.S. Department of Energy (DOE)/USDA study, it was estimated that consumptive water use at a biodiesel refinery accounts for approximately one-third of the total water use, or about 0.32 gallons of water per gallon of biodiesel produced (Sheehan et al., 1998b). New technologies have reduced the amount of wastewater generated at facilities. Process wastewater disposal practices include direct discharges (to waters of the United States), indirect discharges (to wastewater treatment plants), septic tanks, land application, and recycling (U.S. EPA, 2008c).

Most biodiesel manufacturing processes result in the generation of process wastewater with free fatty acids (as soap) and glycerin (a major co-product of biodiesel production); however, the quantity of wastewater will be significantly reduced for facilities with waterless processes or water recycling. Despite the existing commercial market for glycerin, the rapid development of the biodiesel industry caused a glut of glycerin production, which resulted in many facilities disposing of glycerin. Glycerin disposal may be regulated under several EPA programs, depending on the practice. Glycerin can be marketed as a feedstock following methanol recovery and additional refining. Significant research on alternative beneficial uses for glycerin is ongoing (U.S. EPA, 2008a). Some potential options for the catalytic and biological conversion of glycerin into value-added products, included bio-based alternatives to petroleum-derived chemicals, have been identified (Johnson and Taconi, 2007).

Other constituents in the wastewater of biodiesel manufacturing include: organic residues such as esters, soaps, inorganic acids and salts, traces of methanol, and residuals from process water softening and treatment (U.S. EPA, 2008c). Solvents used to extract lipids from algae, including hexane, alcohols, and chloroform, could also impact water quality if discharged to surface or ground water. Typical wastewater from biodiesel facilities has high concentrations of conventional pollutants—BOD, TSS, oil, and grease—and also contains a variety of non-conventional pollutants (U.S. EPA, 2008c).

Some biodiesel facilities discharge their wastewater to POTWs for treatment and discharge. In some cases, wastewater with sufficiently high glycerin levels has disrupted wastewater treatment plant function (U.S. EPA, 2010a). There have been several cases of treatment plant upsets due to high BOD loadings from releases of glycerin (U.S. EPA, 2010a). To mitigate wastewater issues, some biodiesel production systems reclaim glycerin from the wastewater. As another option, closed-loop systems in which water and solvents can be recycled and reused can reduce the quantity of water that must be pretreated before discharge.

4.3.4. Impacts from Solid Waste Generation

Biofuels may also lead to significant environmental impacts stemming from solid waste generated by various production processes. EPA defines "solid wastes" as any discarded material, such as spent materials, byproducts, scrap metals, sludge, etc., except for domestic wastewater, nonpoint-source industrial wastewater, and other excluded substances (U.S. EPA, 2010e). Further study is needed to investigate this potential hazard and to examine mitigation strategies.

4.4. Biofuel Distribution

The vast majority of biofuel feedstocks and finished biofuel are currently transported by rail, barge, and tank truck. Ethanol and biodiesel are both generally blended at the end of the distribution chain, just before delivery to retail outlets. Storage of biofuels typically occurs in above-ground tanks at blending terminals, in underground storage tanks (USTs), and at retail outlets (as a petroleum-biofuel blend).

The primary impacts related to transport and storage of biofuels relate to air quality (i.e., emissions from transport vehicles and evaporative emissions) and water quality (i.e., leaks and spills). It should be noted that these impacts are not unique to ethanol and biodiesel, but are associated with the storage, distribution, and transportation system of all fuels.

4.4.1. Air Quality

4.4.1.1. Ethanol

Air pollution emissions associated with distributing fuel come from two sources: (1) evaporative, spillage, and permeation emissions from storage and transfer activities and (2) emissions from vehicles and pipeline pumps used to transport the fuels (see Figure 4-1). Emissions of ethanol occur both during transport from production facilities to bulk terminals, and after blending at bulk terminals.

Although most ethanol facilities are concentrated in the midwestern United States, gasoline consumption is highest along the East and West Coasts. Fleet transport of biofuel, often by barge, rail, and truck, increases emissions of air pollutants such as CO_2, NO_x, and PM due to the combustion of fuels by transport vehicles. EPA's RFS2 RIA found relatively small increases in criteria and air toxics emissions associated with transportation of biofuel feedstocks and fuels (U.S. EPA, 2010a). In addition, transport and handling of biofuel may result in small but significant evaporative emissions of VOCs (Hess et al., 2009a). With the exception of benzene emissions, which were projected to decrease slightly, EPA's RFS2 RIA projected relatively small increases in emissions of air pollutants associated with evaporation (U.S. EPA, 2010a).

Pipeline transport decreases air emissions associated with fleet transport of biofuel because fuel is not combusted in the transport process. However, transport of biofuels by pipeline raises potential technical issues, including internal corrosion and stress corrosion cracking in pipeline walls, and the potential to degrade performance of seals, gaskets, and internal coatings. Additionally, ethanol's solvency and affinity for water can generate concerns about product contamination (U.S. EPA, 2010a). Dedicated ethanol pipelines may alleviate these issues; however, they are costly to construct. Due to the incompatibility issues with the existing petroleum pipeline infrastructure, the growth in ethanol production is expected to increase emissions of criteria and toxic air pollutants from freight transport, while a corresponding decrease in gasoline distribution would decrease emissions related to pipeline pumping (Hess et al., 2009a).

4.4.1.2. Biodiesel

Air pollution emissions from fuel combustion in transport vehicles related to biodiesel feedstocks and fuels are not materially different than those associated with ethanol. Currently, pipeline distribution of biodiesel is still in the experimental phase. Significant evaporative emissions are not expected from storage and transport of biodiesel fuel due to its low volatility (U.S. EPA, 2010a).

4.4.2. Water Quality

Leaks and spills from above-ground, underground, or transport tanks may occur during biofuel transport and storage, potentially contaminating ground water, surface water, or drinking water supplies.

For bulk transport, the major concern is based on an accident scenario in which the transport tank is damaged and a large amount of fuel is spilled. In addition, leaks might occur during transport because of certain fuel-related factors, such as the fuel's corrosivity. Ethanol is slightly acidic and can corrode some active metals; biodiesel is also slightly corrosive. The possibility of leaks during transport is minimized by the selection of appropriate materials and proper design in accordance with the applicable material standards.

Leaks from USTs are also a major concern. Most states report that USTs are a major source of ground water contamination (U.S. EPA, 2000). Releases of biofuels blended with petroleum fuels can migrate to ground and surface water and contaminate drinking water sources. Other health and environmental risks, including the potential for vapor intrusion, are also associated with leaking USTs. Although it is not possible to quantify the risk at this time, EPA is developing modeling software to assess ground water impacts from fuels of varying composition (U.S. EPA, 2010a).

EPA's Office of Underground Storage Tanks is working with other agencies to better understand material compatibility issues associated with UST systems, in order to assess the ability of these systems to handle new fuel blends (U.S. EPA, 2009b). Because most of the current underground storage tank equipment, including approximately 600,000 active USTs, was designed and tested for use with petroleum fuels, many UST system components currently in use may be constructed of materials that are incompatible with ethanol blends greater than 10 percent (U.S. EPA, 2009c) or biodiesel blends greater than 20 percent (NREL, 2009a).

Several measures are already in place to help prevent and mitigate potential water quality impacts. Under the Resource Conservation and Recovery Act (RCRA), owners and operators of regulated UST systems must comply with requirements for financial responsibility, corrosion protection, leak detection, and spill and overfill prevention. Federal regulations require that ethanol and biodiesel storage containers are compatible with the fuel stored. For USTs, leak detection equipment is required and must be functional. Through the Spill Prevention, Control, and Countermeasure (SPCC) rule, EPA has enforceable regulations to control water quality impacts from spills or leaks of biofuel products and byproducts.

Additional details specific to ethanol and biodiesel are discussed below.

4.4.2.1. Ethanol

Ethanol is stored in neat form at the production facility, in denatured form at terminals and blenders, and as E85 (nominally 85 percent ethanol and 15 percent gasoline) and E10 (10 percent ethanol and 90 percent gasoline) mixtures at retail facilities. There is growing availability of blender pumps, primarily in the Midwest, where the consumer can select the desired blend of ethanol in a flex-fuel vehicle. Although there are limited data on the compatibility of storage tanks with ethanol blends greater than 10 percent, studies indicate that mid-level ethanol blends may be more degrading to some materials than the lower ethanol blends (NREL, 2009b).

There are unique fate and transport implications associated with releases of ethanol-gasoline blends compared to releases of gasoline without ethanol. Ethanol is water soluble and can be degraded by microorganisms commonly present in ground water (U.S. EPA, 2009d). In ground water, ethanol's high oxygen demand and biodegradability changes the attenuation of the constituents in ethanol-gasoline blends. This can cause reduced biodegradation of benzene, toluene, and xylene (up to 50 percent for toluene and 95 percent for benzene) (Mackay et al., 2006; U.S. EPA, 2009d). The presence of ethanol can restrict the rate and extent of biodegradation of benzene, which can cause the plumes of benzene to be longer than they would have been in the absence of ethanol (Corseuil et al., 1998; Powers et al., 2001; Ruiz-Aguilar et al., 2002). This could be a significant concern to communities that rely on ground water supplies with the potential to be impacted by leaks or spills (Powers et al., 2001; Ruiz-Aguilar et al., 2002). In surface waters, rapid biodegradation of ethanol can result in depletion of dissolved oxygen with potential mortality to aquatic life (U.S. EPA, 2010a).

There are other potential hazards in addition to those associated with chemical toxicity. Some spills of ethanol-gasoline blends may produce methane concentrations in the soil that pose a risk of explosion (Da Silva and Alvarez, 2002; Powers et al., 2001).

4.4.2.2. Biodiesel

In general, if biodiesel is blended with petroleum diesel, another petroleum product, or a hazardous substance, then state UST regulations may apply to those blends. One-hundred percent biodiesel contains no petroleum-based products or hazardous substances. Therefore, UST regulations generally do not apply to 100 percent biodiesel. However, 100 percent biodiesel also has material compatibility issues with storage and dispensing equipment. According to the National Renewable Energy Laboratory's (NREL's) "Biodiesel Handling and Use Guide" (NREL, 2009a), 100 percent biodiesel is not compatible with some hoses, gaskets, seals, metals, and plastics. These compatibility issues are minimized at lower biodiesel blends; 20 percent biodiesel (B20) and 5 percent biodiesel (B5) are commonly used blends that have not shown significant material compatibility issues with engine or storage tank components. However, it is important that the biodiesel fuels used in these blends meet fuel quality specifications, as outlined in ASTM D6751, the standard specification for biodiesel fuel blend stock.

Biodiesel, like petroleum diesel, is not water soluble. However, when released to the environment, biodiesel degrades approximately four times faster than petroleum diesel. In aquatic environments, biodiesel degrades fairly extensively (Kimble, n.d.). Results of aquatic toxicity testing of biodiesel indicate that it is less toxic than petroleum diesel (Kahn et al., 2007). Biodiesel does have a high oxygen demand in aquatic environments, and can cause fish kills as a result of oxygen depletion (Kimble, n.d.). Water quality impacts associated with spills at biodiesel facilities generally result from discharge of glycerin, rather than biodiesel itself (Kimble, n.d.).

4.5. Biofuel End Use

Most vehicles on the road today can operate on low-level ethanol blends containing up to 10 percent ethanol (E10). Over 90 percent of U.S. gasoline is a low-level ethanol blend such as E10 to boost octane for more complete combustion or to meet air quality requirements (Alternative Fuels and Advanced Vehicles Data Center, 2010). E85 is another form in which ethanol is consumed, but it can only be used in flex-fuel vehicles, which can run on any gasoline ethanol blend up to 85 percent ethanol. Under current market circumstances, greater deployment of flex-fuel vehicles may be needed to meet the Energy Independence and Security Act (EISA) mandated volume standards.

Biodiesel is also commonly used as a blend with petroleum diesel. Because of biodiesel's chemical properties, it is interchangeable with petroleum-based diesel fuel up to a 5 percent blend, as long as it meets the ASTM D6751 biodiesel fuel specification. Biodiesel blends up to 20 percent are also commonly used in the United States, especially for fleet vehicles (NREL, 2009a).

Biodiesel can also be used in its pure form, known as "neat biodiesel" or B100, but most vehicle and engine manufacturers do not recommend its use in non-approved engines and vehicles. Previously, there were some concerns regarding maintenance issues related to engines operated on biodiesel blends greater than 20 percent (B20), because the fuel has been shown to soften and degrade certain types of elastomers and natural rubber compounds over time. This will impair fuel system components such as fuel hoses and fuel pump seals. Such component degradation can lead to leaks, poor performance, and other problems that are likely to result in increased emissions and subsequent environmental impacts. This issue is limited to vehicles from the early 1990s or earlier that used natural or nitrile rubber fuel system components; newer vehicles use biodiesel-resistant fuel system components made of materials such as Teflon®, Viton®, fluorinated plastics, and nylon. Compatibilities between B100 and specific materials have been identified. In general, there are no material compatibility issues with B20 or blends with lower biodiesel fraction, unless the fuel has been oxidized (NREL, 2009a).

Biofuels for jet aircraft require additional refining or need to be blended with typical jet fuels to meet the standards of commercial aviation fuels. There are few long-term studies of biofuel performance on large diesel engines such as stationary power generators, ships, locomotives, and jet engines.

4.5.1. Air Quality

The primary impact associated with biofuel end use is air quality. Section 211(v) of the Clean Air Act (CAA) requires EPA to study the air quality impacts associated with the use of biofuel and biofuel blends. EPA has already adopted mobile source emission control programs that reduce air pollution emissions and improve air quality. If necessary, EPA will issue further regulations to mitigate adverse air quality impacts as a result of increases in biofuels.

4.5.1.1. Ethanol

The following discussion is based on E10, because considerably more information is available about its use. A wide variation in evaporative and tailpipe emissions has been reported due to a range of factors, such as the age of the vehicle, the power output and operating condition of the engine, the fuel characteristics, how the vehicle is operated, and ambient temperatures (Graham et al., 2008; Yanowitz and McCormick, 2009; Ginnebaugh et al., 2010). In 2010, a partial waiver was granted by EPA under Section 211(f)(4) of the CAA that allowed the use of E15 in certain vehicles. Specifically, this waiver allowed the use of E15 in model year 2007 and new light-duty vehicles (i.e., cars, light-duty trucks, and medium-duty passenger vehicles) (U.S. EPA, 2010f). It also denied the use of E15 in model year 2000 and older light-duty vehicles, as well as all heavy-duty gasoline engines and vehicles, highway and off-highway motorcycles, and non-road engines, vehicles, and equipment. Currently, an additional waiver is under consideration by EPA for E15 use in other model year vehicles.

The emission impacts of the 2022 RFS2 volumes in the RFS2 RIA (U.S. EPA, 2010a) were quantified relative to two reference cases: (1) the original RFS program (RFS1) mandate volume of 7.5 billion gallons of renewable fuel (6.7 billion gallons ethanol) and (2) the U.S. Department of Energy (DOE) Annual Energy Outlook (AEO) 2007 projected 2022 volume of 13.6 billion gallons of renewable fuels. In the RFS2 RIA, EPA projected decreases in emissions of CO, benzene, and acrolein in 2022 under the RFS2-mandated volumes of biofuels, while NO_x, hydrocarbons (HC), and the other air toxics, especially ethanol and acetaldehyde, were projected to increase. The inclusion of E85 emissions effects would be expected to yield larger reductions in CO, benzene, and 1,3-butadiene, but more significant increases in ethanol, acetaldehyde, and formaldehyde (U.S. EPA, 2010a).

4.5.1.2. Biodiesel

Air emissions from combustion of some biofuels, such as ethanol, are relatively independent of feedstock or conversion process. However, biodiesel emissions may be highly variable depending on the feedstock type (U.S. EPA, 2002; Lapuerta et al., 2008). With respect to carbon content, plant-based biodiesel is slightly higher percentage-wise than animal-based biodiesel in gallon-per-gallon comparisons. For NO_x, PM, and CO, plant-based biodiesel tends to have higher emissions than animal-based biodiesel for all percent blends (U.S. EPA, 2002). Recent advances in diesel engine emission control technology have resulted in significantly lower levels of air pollutant emissions.

Studies of biodiesel and biodiesel blends show varying results depending on the fuel (i.e., type of biodiesel, biodiesel blend, type of base diesel), the vehicle being tested, and the type of testing. In general, combustion of biodiesel has been shown to decrease PM, CO, and HC emissions, increase NO_x emissions, and increase ozone-forming potential (Gaffney and

Marley, 2009; U.S. EPA, 2002). However, it should be noted that petroleum-based diesel-fueled vehicles are expected to emit significantly lower amounts of SO_2 because of the Heavy-Duty Engine and Vehicle Standards and Highway Diesel Fuel Sulfur Control Requirements (2007 Heavy-Duty Highway Rule) and the availability of low-sulfur diesel fuel in the marketplace, which must be accounted for when considering the emission benefits of low SO_x biodiesel (U.S. EPA, 2001). Blending biodiesel in low percentages will not have much impact on sulfur emissions.

EPA's RFS2 RIA investigated the impacts of 20 percent by volume biodiesel fuels on NO_x, PM, HC, and CO emissions from heavy-duty diesel vehicles, compared to using 100 percent petroleum-based diesel. Average NO_x emissions were found to increase 2.2 percent, while PM, HC, and CO were found to decrease 15.6 percent, 13.8 percent, and 14.1 percent, respectively, for all test cycles run on 20 percent by volume soybean-based biodiesel fuel. Biodiesel results were included in the EPA analysis; however, the biodiesel contribution to overall emissions is quite small (U.S. EPA, 2010a).

5. International Considerations

5.1. Introduction

In the global context, biofuel demands from an increasing number of countries will have direct and indirect impacts, not only on countries that produce biofuels, but also on countries that currently rely on imports of agricultural commodities from biofuel producers (Hertel et al., 2010a, 2010b; Pimentel et al., 2009; Zah and Ruddy, 2009). Section 204 of the Energy Independence and Security Act (EISA) calls for EPA to report to Congress on the environmental impacts outside the United States caused by U.S. biofuel use. Therefore, this section focuses on potential impacts in foreign countries from implementation of the revised Renewable Fuels Standard program (RFS2) standards. Specifically, this section presents the current international production and consumption of two biofuels (ethanol and biodiesel), discusses the composition of future biofuel production, examines current and projected import and export volumes, and discusses the environmental and other impacts of direct and indirect land use changes.

International trade is the primary mechanism through which U.S. biofuel policy will affect foreign nations. Ethanol and, to a much smaller degree, biodiesel, have become global commodities. Both are produced in many countries (Figure 5-1) and both are traded in international markets. Primary producers of ethanol are Brazil, the United States, the European Union, India, and China. Brazil is the only significant exporter of ethanol (See Table 5-1). Based on computer modeling, changes in U.S. production and consumption volumes, such as those in RFS2, are predicted to result in land allocation impacts that have global ramifications through international trade and market price. As a crop price rises, land may be reallocated to grow more of that crop in response to market price; conversely, as a crop price declines, land may tend to be reallocated to grow less of that crop in response to market price. The extent of such conversions will depend on many factors, such as local land use policies and incentives, knowledge of alternatives, access to international markets, and cultural norms. There are differing opinions on the result of possible tradeoffs between land uses and it is not possible at this time to predict with any certainty what type of land use change will result from increased U.S. demand for biofuel and what its environmental consequences will be (Fargione et al., 2008; Goldemberg et al., 2008; Searchinger et al., 2008; Hertel et al., 2010b).

Resulting environmental impacts, both positive and negative, include effects from land use change and impacts on air quality, water quality, and biodiversity. From a U.S. perspective, the severity of these impacts will depend on the volume and location of future imports and exports, both of biofuel and displaced agricultural goods.

In 2008, the United States and Brazil together produced 89 percent of the world's fuel ethanol, with the United States producing around 9 billion gallons (see Table 5-1) (U.S. EIA, n.d.[b]). In 2009, U.S. ethanol production increased to 10.9 billion gallons, and similar increases occurred in most ethanol-producing nations as they attempted to increase the portion of biofuel in their energy mix (U.S. EIA, n.d.[b]). Total world production has nearly doubled from 10.9 billion gallons in 2006 to 20.3 billion gallons in 2009. Figure 5-1 shows the geographical distribution of biofuel production. Patterns of ethanol consumption generally matched those of production, with the largest producers also being the largest consumers (U.S. EIA, n.d.[b]).

On the other hand, the market—and thus the global production—for biodiesel is concentrated in Europe, which represented about 60 percent of world production as of 2009 (U.S. EIA, n.d.[a]). The other 40 percent of global production is largely made up by the United States, Brazil, Argentina, and Thailand, with U.S. production estimated at 505 million gallons for 2009, or about 10 percent of world biodiesel production (U.S. EIA, n.d.[a]). World biodiesel production has been rapidly increasing over the past decade, from 242 million gallons in 2000 to about 4.7 billion gallons in 2009 (U.S. EIA, n.d.[a]). These production increases have been driven by increased consumption targets. For example, Brazil has planned to increase its biodiesel blend from 5 to 10 percent by 2015.

Table 5-1. Top Fuel Ethanol-Producing Countries from 2005 to 2009 (All Figures Are in Millions of Gallons)

Country/Region	2005	2006	2007	2008	2009
United States	3,904	4,884	6,521	9,283	10,938
Brazil	4,237	4,693	5,959	7,148	6,896
European Union	216	427	477	723	951
China	317	369	440	526	567
Canada	67	67	212	250	287
Jamaica	34	80	74	98	106
Thailand	18	34	46	87	106
India	57	63	69	71	89
Colombia	8	71	72	67	80
Australia	6	20	21	38	54
Other	93	216	276	393	274
Total world production	**8,957**	**10,924**	**14,167**	**18,684**	**20,348**

Source: U.S. EIA, n.d.[b].

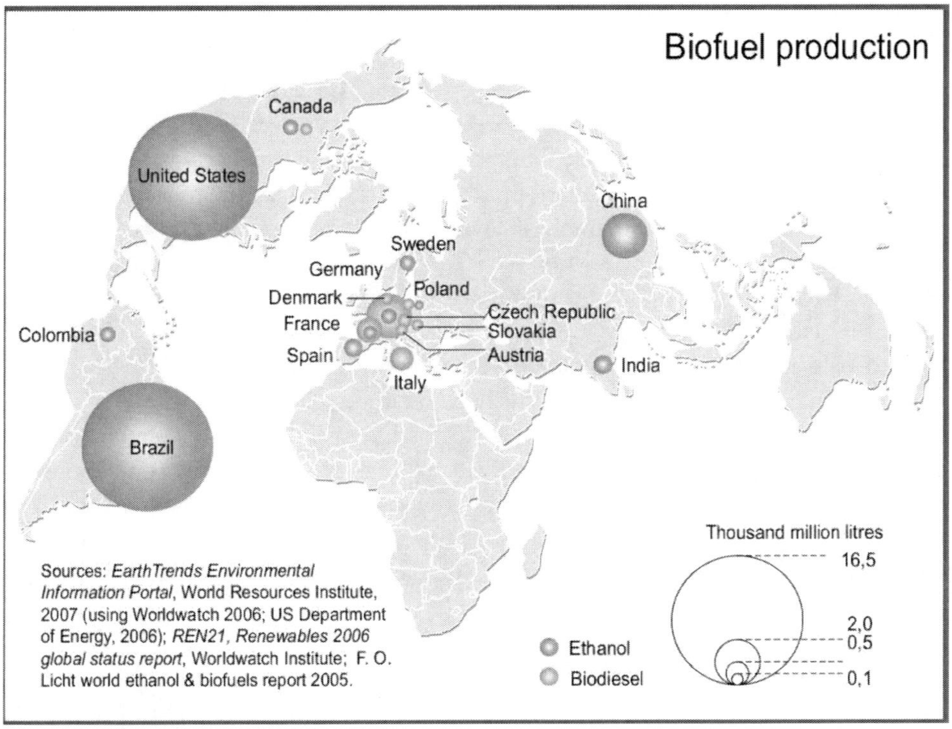

Source: UNEP/GRID-Arendal, 2009.

Figure 5-1. International Production of Biofuels.

5.2. Import/Export Volumes

U.S. biofuel import volumes will depend largely on the relative costs of U.S. biofuel production and imported ethanol. These costs will be determined by domestic production capacity, including the efficiency of the domestic ethanol-producing sector, and the yields attained.

With respect to production capacity, as discussed in Section 2, the renewable fuel volume mandates under EISA require that U.S. biofuel consumption steadily increase to 36 billion gallons by 2022. This biofuel will be composed of both conventional and advanced biofuel (including cellulosic ethanol, algal biodiesel, and other forms of advanced biofuel). Most of the 10.9 billion gallons of conventional ethanol that the United States produced in 2009 came from corn starch. By 2015, the 10.9 billion gallons is expected to increase to the targeted volume of 15 billion gallons provided for in the RFS2 program (as required under EISA) (GAO, 2007; U.S. EPA, 2010a). Future production volumes of advanced biofuel that have not yet been commercially developed are uncertain. In its RFS2 Regulatory Impact Analysis (RIA), EPA estimated that cellulosic technologies could combine to provide an additional 16 billion gallons of ethanol by 2022, with a substantial portion of this, 7.8 billion gallons worth, using corn stover as a cellulosic feedstock source (U.S. EPA, 2010a). In addition, some estimates place U.S. biodiesel production at roughly 1.3 billion gallons by 2019 (FAPRI, 2010b). In 2022, the RFS2 RIA projects that the remaining 4 billion gallons needed to meet the EISA mandate would be composed of a combination of imported sugarcane ethanol from Brazil as well as "other advanced biofuel" (U.S. EPA, 2010a). Figure 5-2 shows the projected import volumes forecasted in the RIA for each year from 2011 to 2022.

Figure 5-2 shows that import volumes are expected to be very low in years preceding 2015, followed by a significant increase in import volumes between 2015 and 2022. This is in part because domestic corn starch ethanol production is expected to increase until it reaches the 2015 peak under RFS2. The total renewable fuel targets may not be reached with domestic production until 2018 or beyond (see Table 2-1). It should also be noted that 2010 import figures have been much lower than those expected when forecasts were made in 2009. Imports of fuel ethanol for the first three-quarters of 2010 have totaled 17 million gallons (USDA, n.d.)—well below EPA's 200 million gallons forecast (U.S. EPA, 2010a). U.S. biofuel imports and exports will also be influenced by trade policy, including tariffs and other incentives in the United States and other countries. Even if the United States succeeds in meeting the RFS2 targets, the United States likely will continue to import and export biofuel as individual producers take advantage of international price differences. Over the past decade (2002 to 2009), U.S. ethanol import quantities varied (see Figure 5-2), mostly due to volatility in the prices of related commodities such as corn, sugar, and other feedstocks, as well as prices of energy commodities such as oil.

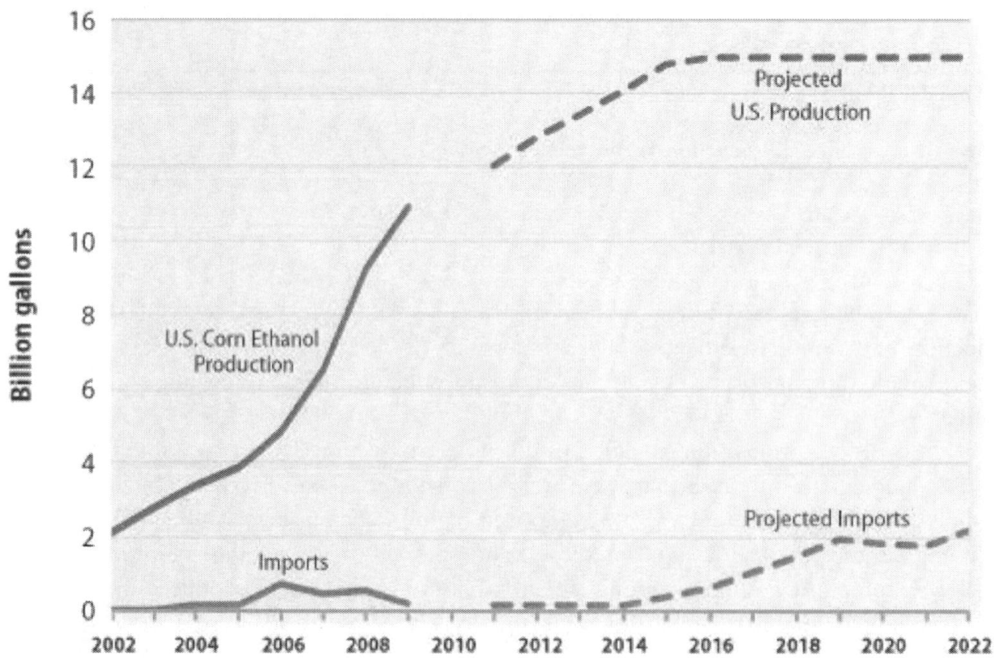

Sources: For 2000 to 2009 production data, U.S. EIA, n.d.[b]. For 2000 to 2009 import data, U.S. EIA, n.d.[c]. For 2011 to 2022 production and import projections, U.S. EPA, 2010a.

Figure 5-2. Annual U.S. Domestic Ethanol Production and Imports Volumes Reported (2002 to 2009) and Projected (2011 to 2022).

The bulk of U.S. ethanol imports are sugarcane-based ethanol from Brazil. In 2008, the United States was the largest importer of Brazilian ethanol, followed by the Netherlands and a number of Caribbean countries (see Table 5-2). However, foreign-produced ethanol is also imported to the United States via these Caribbean countries where the Caribbean Basin Initiative (CBI), a regional trade agreement, enables up to 7 percent of the biofuel consumed in the United States to be imported duty-free from CBI member countries (Yacobucci, 2005; Farinelli et al., 2009). Therefore, most of the Brazilian exports shown as going to CBI member countries such as Costa Rica, Jamaica, El Salvador, and Trinidad and Tobago (see Table 5-2) is eventually re exported to the United States (U.S. EIA, n.d., [c]). Looking closer at the Brazilian export figures in Table 5-2, it is evident that ethanol trade changed somewhat dramatically in 2009, with most destinations experiencing a significant decline in imports. A large part of this decline was due to the drop in U.S. imports caused by a change in energy prices, as well as an increase in sugar prices that made imported Brazilian ethanol less competitive in the U.S. market (Lee and Sumner, 2010). These rising sugar prices, as well as the recent strengthening of Brazil's currency, could significantly hinder Brazil's ability to supply the U.S. market moving forward. While there is currently a tariff in place through December 2011, these factors may limit future imports even were the tariff to expire (USDA, 2010d).

The United States also exports biofuel (including ethanol and biodiesel) to foreign countries. Canada has been the primary recipient of U.S. exports, with Europe becoming a more prevalent destination beginning in 2004 (see Figure 5-3) as its biofuel consumption has increased. U.S. ethanol exports have increased in recent years due to increased production. However, export levels, ranging from about 50 million to 175 million gallons, are no more than 1 percent of domestic production and are far outweighed by imports. Exports are likely to continue to lag behind imports in the near term as consumption rises.

Table 5-2. 2008–2009 Brazilian Ethanol Exports by Country of Destination

Destination Country	Volume (Million Gallons)			
	2008	% of Total	2009	% of Total
Total	1,352.9	100%	870.8	100%
United States	401.6	29.7%	71.9	8.3%
Netherlands	351.9	26.0%	179.2	20.6%
Jamaica	115.3	8.5%	115.6	13.3%
El Salvador	94.1	7.0%	18.8	2.2%
Japan	69.6	5.1%	74.0	8.4%
Trinidad and Tobago	59.3	4.4%	37.0	4.2%
Virgin Islands (U.S.)	49.7	3.7%	3.4	0.4%
Republic of Korea (South Korea)	49.3	3.6%	82.9	9.5%
Costa Rica	28.9	2.1%	26.5	3.0%
Nigeria	25.9	1.9%	30.6	3.5%
United Kingdom	18.4	1.4%	42.7	4.9%

Source: SECEX, n.d.

Note: Percentages do not sum to 100 percent because some destinations are not listed. Original data were converted from liters to gallons.

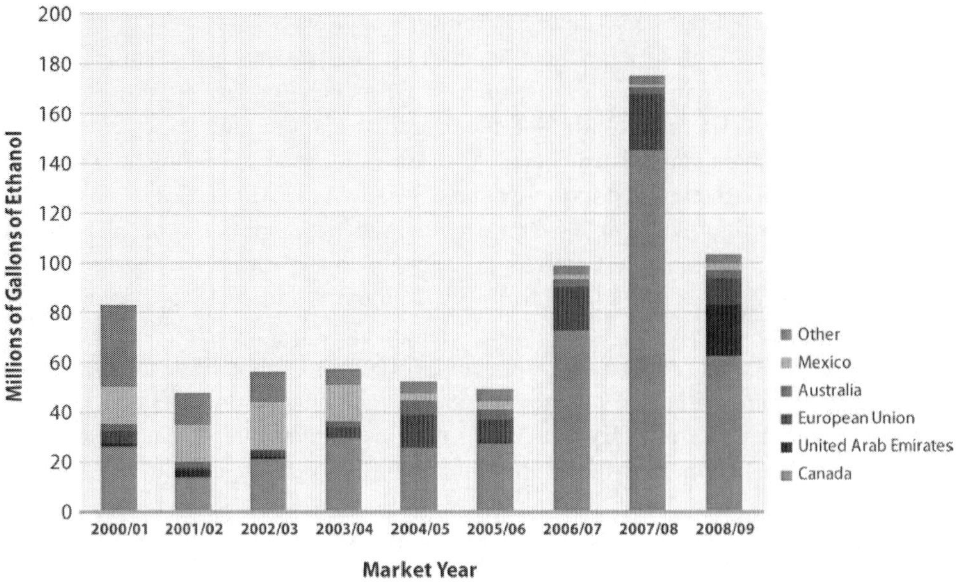

Source: ERS, 2010a.

Note: Original data were converted from liters to gallons.

Figure 5-3. Historic U.S. Ethanol Export Volumes and Destinations.

Table 5-3. 2008 U.S. Biodiesel Balance of Trade

Item	Quantity
U.S. production	774 million gallons
U.S. consumption	412 million gallons
Production – consumption =	362 million gallons
U.S. imports	315 million gallons
U.S. exports	677 million gallons
Exports – imports =	362 million gallons

Source: U.S. EIA, 2009, n.d.[b].

Table 5-3 shows the 2008 U.S. biodiesel trade balance. In 2008, 46.8 percent of domestically produced biodiesel was exported. Biodiesel export volume has increased dramatically in recent years, from about 9 million gallons in 2005 to nearly 677 million gallons in 2008 (U.S. EIA, n.d.[a]). In 2009, biodiesel export volume fell dramatically to only 266 million gallons (USDA, n.d.). Current projections have net U.S. biodiesel exports (i.e., exports minus imports) falling for the next few years and then rising back up to around 100 million gallons by the end of the decade (FAPRI, 2010b).

5.3. Environmental Impacts of Direct and Indirect Land Use Changes

EISA requires that "direct emissions and significant indirect emissions such as significant emissions from land use change" be considered as part of the analysis of environmental impacts stemming from domestic biofuel production and consumption. The issue of land use change inherently includes international considerations, because the demand for biofuel in the United States can influence the international availability of crops such as corn and soybeans for both biofuel and agricultural markets, which in turn can incentivize land use changes in other countries to meet that demand. Land use changes are considered either direct or indirect. In the context of biofuels, *direct land use change* refers to land conversion that is directly related to the biofuel supply chain. An example of direct land use change would be the planting of biofuel feedstock on land, which was previously native forest, to increase the supply of ethanol to export to the United States. *Indirect land use change* refers to land conversion that is a market-oriented response to changes in the supply and demand of goods that arise from increased production of biofuel feedstocks. An example of indirect land use change would be the clearing of foreign land to plant corn in response to an increase in global commodity prices caused by a decrease in U.S. corn exports. Some have argued that these indirect impacts should not be counted as part of the biofuel carbon footprint because they are too difficult to relate back to biofuel production and because of the disparity of highly variable responses based on local policies and conditions. It is instructive, however, to consider the potential impacts if they were to be realized, and this has been done routinely through the use of economic modeling.

In the RFS2 RIA, EPA estimated greenhouse gas (GHG) impacts of direct and indirect land use change using the FAPRI-CARD model.[25] This model predicts world prices by equating excess supply and demand across countries. Changes in world prices determine changes in worldwide commodity production and trade. Under this model, two primary domestic effects directly affect a commodity's worldwide use and trade: changes in U.S. exports and changes in domestic U.S. prices (U.S. EPA, 2010a). Using this model, the RFS2 RIA analysis compares 2022 crop area and production (by crop type and country) predicted to result with and without (i.e., "business as usual") EISA requirements. The results of this analysis are shown in Figures 5-4 through 5-7 and in Table 5-4. In Figures 5-4 through 5-7, each column shows the marginal impact of a scenario that focuses on that particular feedstock in isolation.

The RFS2 RIA forecasts that, by 2022, for every increase of 1,000 gallons of corn starch ethanol production in the United States, corn exports will have decreased by 4 tons. Similarly, for every increase of 1,000 gallons of soybean-based biodiesel produced domestically, soybean oil exports will have decreased by just over 2 tons (see Figure 5-4) (U.S. EPA, 2010a). Thus, as the United States increases domestic production of corn starch ethanol and soybean diesel, exports of corn and soybean for agricultural or other uses are expected to decline, which might result in indirect land use change in the form of land conversion to agriculture in other countries. This result is consistent with the results of a 2009 study, which predicted that due to production increases required by EISA, U.S. coarse grain exports will decrease to all destinations and this could cause dominant export competitors and trading partners, likely in Latin America, China, and the Pacific Rim, to convert more of their lands to make up the difference (Hertel et al., 2010a; Keeney and Hertel, 2009). However, given that RFS2 limits the amount of corn starch ethanol that can be counted toward the mandated volume targets at 15 billion gallons—a level the United States is expected to reach by 2015 or sooner (GAO, 2007; U.S. EPA, 2010a)— indirect land use change impacts resulting from changing trade patterns of corn and other grains may level off at that point. Assuming agricultural yield improvements continue and cellulosic or other production technologies develop to commercialization and replace conventional ethanol production, U.S. biofuel consumption could decrease pressure on conversion of land to agricultural use internationally.

The RFS2 RIA also estimates that the additional biofuel produced to meet the EISA mandates (2.7 billion gallons of corn starch ethanol, 0.5 billion gallons of soy-based biodiesel, 1.6 billion gallons of sugarcane ethanol, and 7.9 billion gallons of switchgrass cellulosic ethanol) compared to "business as usual," will lead to the creation of additional international cropland (approximately 2 million acres of corn, 3.4 million acres of soybeans, 1.1 million acres of sugarcane, and 1.7 million acres of switchgrass) (see Table 5-4) to supply U.S. biofuel imports and respond to the U.S. reductions in exports shown in Figure 5-4 (U.S. EPA, 2010a).

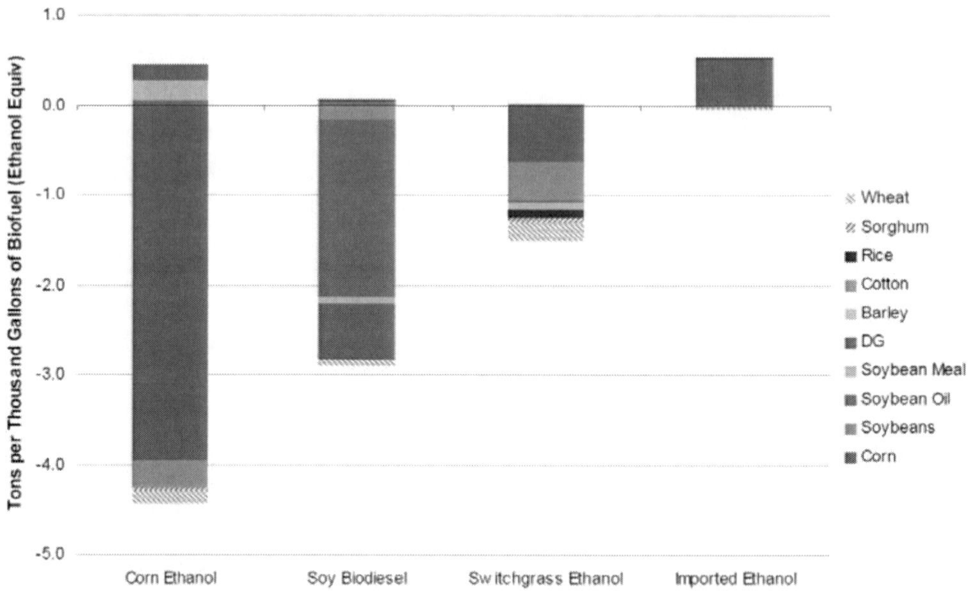

Source: U.S. EPA, 2010a.

Figure 5-4. Change in U.S. Exports by Crop Anticipated to Result from EISA Requirements by 2022.

Table 5-4. Increases in International Crop Area Harvested by Renewable Fuel Anticipated to Result from EISA Requirements by 2022

Feedstock's Marginal Effect Considered	International Crop Area Increase (Thousands of Acres)	Normalized Crop Area Increase (Acres per Billion BTU)
Corn starch ethanol	1,950	9.74
Soy-based biodiesel	1,675	26.32
Sugarcane ethanol	1,063	10.82
Switchgrass ethanol	3,356	5.56

Source: U.S. EPA, 2010a.

Note: Figures converted from hectares to acres. Crop area changes were normalized by dividing by the incremental increase in renewable fuel production in a given scenario and year, on an energy-content basis.

Further, according to the EPA analysis, assuming few or weak local controls that reduce ability or incentive for land conversion, these direct and indirect land use changes will lead to significant GHG emissions (before accounting for GHG savings resulting from petroleum displaced as the biofuel is consumed). Figure 5-5 shows that, based on the model presented in the RFS2 RIA, soy-based biodiesel causes the largest release of GHG emissions (measured in $kgCO_2e/mmBTU$) resulting from international land use change. The RFS2 RIA model results indicate that the majority of emissions resulting from international land use change originate in Brazil in the scenarios for corn ethanol and switchgrass ethanol. This is largely a consequence of projected pasture expansion in Brazil, and especially in the Amazon region where land clearing causes substantial GHG emissions. Of the renewable fuels analyzed, the analysis found that sugarcane ethanol causes the least amount of emissions resulting from land use change. This is due largely to the EPA projection that sugarcane crops would expand onto grasslands in south and southeast Brazil, which results in a net sequestration because sugarcane sequesters more biomass carbon than the grasslands it would replace. Recent data indicate that deforestation rates in Brazil are declining (INPE, n.d.). Given that the largest component of life cycle GHG emissions for corn starch ethanol in the RFS2 RIA results from indirect land use change in Brazil (U.S. EPA, 2010a), understanding the drivers and trends in Brazilian deforestation is important. Various factors influence deforestation rates, including investments in enforcement and monitoring, expanding protected areas, improving land titles for small and medium sized land holders, commitments from agricultural industries, and the establishment of government programs. Changes in land use are being followed closely and more recent analyses may change the outlooks reported here.

The GHG emissions shown in Figure 5-5 can be seen as an international "carbon debt" (Fargione et al., 2008). Clearing forested areas or pasture land for new cropland results in enhanced microbial decomposition of organic carbon and elevated GHG emissions. As described in the RFS2 RIA, the location of land use change is a critical factor determining the GHG impacts of land use change, because these impacts will vary substantially by region (U.S. EPA, 2010a). The conversion of higher carbon-storing types of land such as tropical rainforest will lead to more carbon emissions (U.S. EPA, 2010a).

As noted previously, the results of modeling projected impacts are diverse and it is not possible at this time to predict with any certainty what type of land use change in other countries will result from increased U.S. demand for biofuel or what its environmental consequences will be. (Compare, for example, Fargione et al., 2008; Goldemberg et al., 2008; Hertel et al., 2010b; and Searchinger et al., 2008.) However, if natural ecosystems are converted to cropland, it may take many years for biofuel consumption to "pay down" the carbon debt created from production with GHG savings compared to displaced petroleum. On the other hand, biofuel made from more sustainable grasses or woody crops using higher-yield cellulosic technologies, or from waste biomass or biomass grown on degraded and abandoned agricultural lands, results in much smaller carbon debts and is more likely to lead to overall GHG reductions (Fargione et al., 2008). Figure 5-6 shows forecasted crop area changes by region, with the heaviest impacts occurring in Brazil. It should be noted that the FAPRI-CARD model does not predict what type of cropland will emerge in foreign countries if land use change does occur. This is an important source of uncertainty and GHG and other environmental impacts could vary significantly depending on what crops are grown to offset decreasing U.S. agricultural exports.

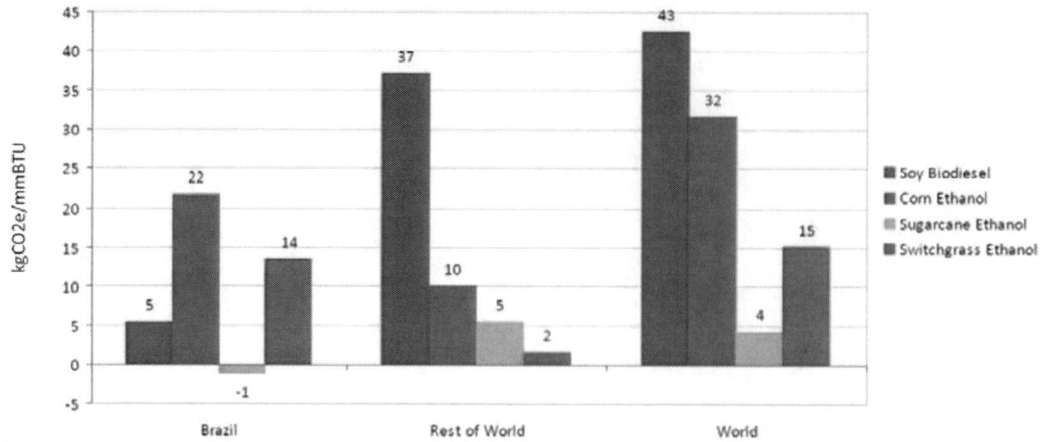

Source: U.S. EPA, 2010a.

Figure 5-5. International Land Use Change GHG Emissions Projected to Result from EISA Requirements by 2022.

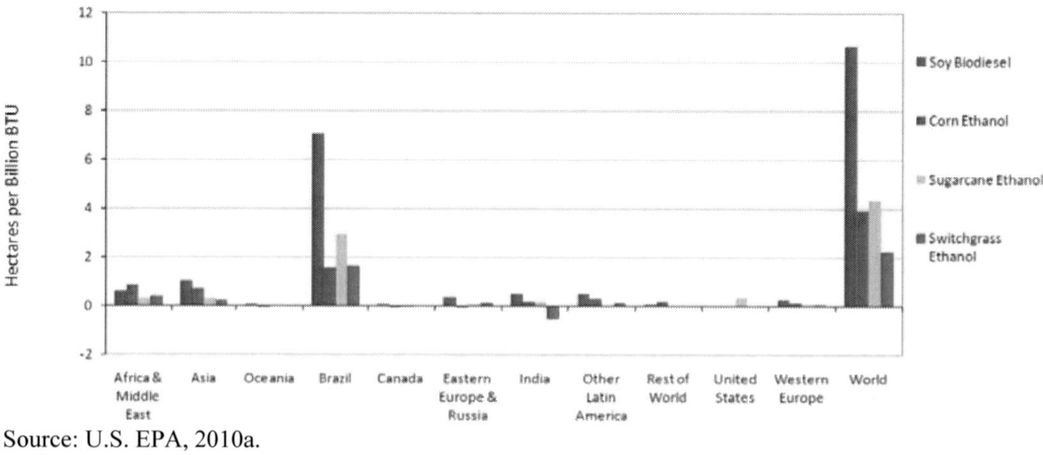

Source: U.S. EPA, 2010a.

Figure 5-6. Harvested Crop Area Changes Projected to Result from EISA Requirements by 2022.

Brazil could continue to be a supplier of U.S. ethanol and will have an important role in international trade of coarse grains and soybeans, so it is informative to consider potential land use changes there. Brazil faces challenges of multiple forms of land use change, both direct and indirect. Land use changes could occur if Brazil increases ethanol production by converting more land from growing other agricultural goods, or from use as pasture land, to grow sugarcane. As pasture lands are converted to sugarcane production, ranchers are pressured to "intensify" livestock production or clear more land (possibly Amazon rainforest or Cerrado woodland) (Bustamante et al., 2009). Figures 5-5 and 5-7 isolate the impacts on Brazil alone. The model results presented in Figure 5-7 appear to be consistent with the prediction that pasture land will decrease in Brazil, while increasing in the rest of the world. However, it is unclear if this would result in rainforest loss or simply mean a greater number of livestock animals per acre. Recent monitoring of the rain forests indicates that deforestation rates in Brazil are declining (INPE, n.d.).

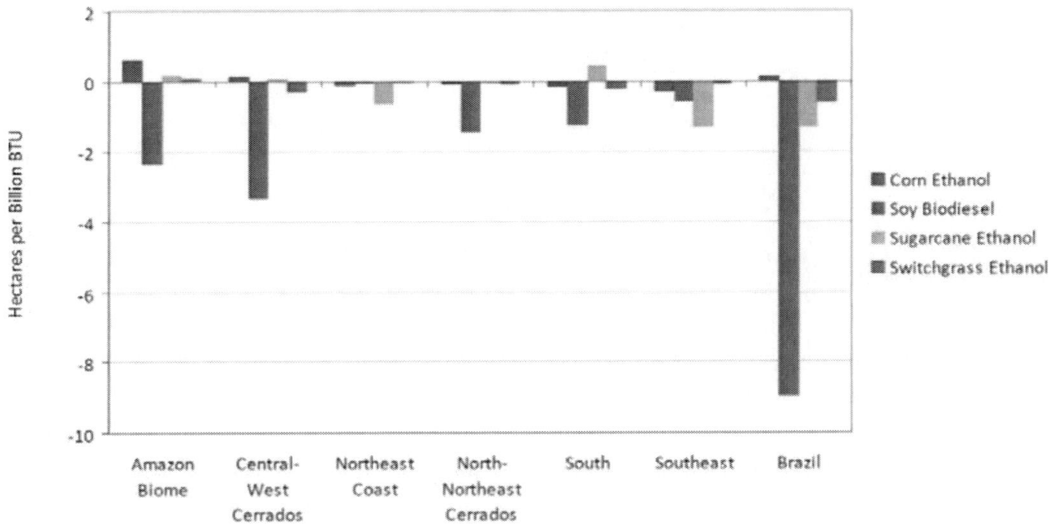

Source: U.S. EPA, 2010a.

Figure 5-7. Modeling of Pasture Area Changes in Brazil Anticipated to Result from EISA Requirements by 2022.

5.4. Other Environmental Impacts

While production of biofuel feedstocks places only one of many demands on water, fertilizer, and other inputs, its impacts will increase as its production increases. Studies have shown that water tables are already declining in the western United States, North India, Pakistan, North China, Mexico, and the Mediterranean (Shah et al., 2007). These trends indicate the vulnerability of various regions to water scarcity issues. The choice of feedstock, cultivation practices, and the location of cultivation will greatly influence how production of biofuel impacts water availability.

Water quality and flooding issues are also relevant. As described in Section 3, U.S. corn production has been a key driver of hypoxia in the Gulf of Mexico. Similar water quality issues could arise or be exacerbated in other countries if feedstock production increases and appropriate management practices are not used. Conversion of land to biofuels feedstock production could have varying impacts, depending on prior ecological function of the converted land and the types of management practices employed. Impacts could include encroachment on wetlands and the discharge of excess nutrients to water resources. For example, Brazilian surface waters suffered from hypoxia during the early stages of their biofuel development when the vinasse, a byproduct of the sugarcane-ethanol production process rich in nitrogen and potassium, was routinely discarded into rivers, lakes, and reservoirs, causing extensive eutrophication (Simpson et al., 2009). Brazilian federal law has prohibited the dumping of vinasse into any water body since 1978. The effluent is now returned to the field as fertilizer, and water quality has improved significantly. However, if other developing countries opt to produce biofuel and do not properly regulate water quality impacts, eutrophication could damage these nations' aquatic ecosystems. Also, if biofuel-related land use change does occur and if it results in deforestation and loss of wetlands, then increased flooding, sedimentation, and lower stream base flows are also likely to occur.

Biofuel production also affects international air quality. While the displacement of petroleum fuels by biofuels does have a positive impact, the air quality issues associated with biofuel feedstock harvesting, refining, and transport could erode these savings if poor management practices are allowed to occur. For instance, the practice of burning sugarcane fields prior to harvesting is a serious air pollution issue in Brazil. It has resulted in large aerosol and trace gas emissions, significant effects on the composition and acidity of rainwater over large areas of southern regions, and elevated ozone levels in those areas affected by the burning. However, harvest burning practices are being phased out in Brazil through state regulations. In 2007, state laws ensured that 40 percent of the sugarcane was harvested without burning in the state of Sao Paulo, and this is forecast to reach 50 percent by 2010 and about 90 percent by 2022 (Goldemberg et al., 2008; U.S. EPA, 2010a). Like many of the effects discussed so far, the severity of air emissions will be highly sensitive to local policies which influence the feedstock selected, location of production, and management practices.

Finally, if increased biofuel consumption in the United States does lead to indirect land use changes and more natural habitat is cleared to create agricultural lands, a loss of biodiversity will occur. Many biofuel production regions coincide with areas with high biodiversity value. For example, many ecosystems in Brazil support high levels of biodiversity. Depending on where biofuel feedstock production occurs, and the manner in which it occurs, impacts on biodiversity could be significant.

5.5. Conclusions about International Impacts

Simulations prepared for the RFS2 indicate that the EISA biofuel targets could alter U.S. and international trade patterns and commodity prices (U.S. EPA, 2010a). The manner in which countries respond to U.S. market conditions, including influences from deforestation, could affect net GHG savings derived from biofuels. As with biofuel production in the United States, these impacts will depend largely on where the crops are grown, forest and agricultural management practices and technologies used, and the efficacy of environmental policies. Global mitigation strategies will have to consider the international implications of biofuel production.

6. Synthesis, Conclusions, and Recommendations

6.1. Introduction

This section presents a qualitative synthesis, major conclusions, and recommendations derived from information assessed in this report. This synthesis is based upon a consensus view of the authors of this report given the broad range of assumptions found in the scientific literature. The synthesis and conclusions illustrate an assessment of the environmental impacts attributable to activities in the biofuels supply chain, from feedstock production to biofuel production, transport and storage. End use impacts and international considerations are also synthesized. The synthesis discusses the range, magnitude, and uncertainty of environmental impacts from the six feedstocks (corn starch, soybean, corn stover, perennial grasses, woody biomass and algae) and three fuels examined (corn ethanol, soybean biodiesel and cellulosic ethanol). This constitutes a set of initial expectations based on available literature through July 2010—it should not be construed as a definitive prediction. Salient points are derived from Sections 3, 4, and 5 across feedstocks and fuels, domestically and internationally, organized by environmental impact category. Finally, the recommendations comprise a set of suggestions concerning environmental assessment, research coordination, impact mitigation, and sustainable biofuels practices. Each of these recommendations will advance approaches for the next Report to Congress, and hopefully promote favorable environmental outcomes as biofuel usage expands in the United States.

6.2. Assessment Scenarios

The published peer-reviewed literature reviewed for this report shows that the production of biofuels can result in a wide range of environmental impacts, including both negative and positive impacts. This range of impacts depends largely on land use changes, management approaches, and regional characteristics such as climate, soil, and ecological factors. In order to effectively synthesize the literature, the authors developed assessment scenarios that represent the range of impacts, as revealed from the literature. The construct of these scenarios is presented below.

6.2.1. Assumptions Underlying the Synthesis of Feedstock Production Impacts

Domestic environmental impacts associated with feedstock production depend upon the conditions under which they are grown. The authors examined available peer-reviewed information to identify reasonable conditions under which a "most negative" and "most positive" environmental impact could arise for each feedstock and impact category. Land-use changes considered were restricted to those allowable under the RFS2 program (U.S. EPA, 2010a). Other conditions considered included: management approaches, regional characteristics, and technologies. For example, it was determined from the literature that the most negative water quality impact from producing corn starch as a feedstock would likely arise if corn grown with conventional tillage and high chemical inputs replaced uncultivated land such as that in the Conservation Reserve Program (CRP). Conversely, the most positive water quality impact from producing corn starch would arise if corn grain was merely diverted from other uses with no increase in corn acreage; utilizing land already in corn production would likely result in a "negligible effect," because sedimentation and nutrient contamination would remain at near current levels. This approach was repeated for each feedstock and impact category and the results are highlighted in associated tables (Tables 6-1, 6-2).

Table 6-1. Assumptions Underlying the Synthesis of Feedstock Production Impacts

Feedstock	Environmental Impact Per Unit Area			
	Most negative	Negligible	Most positive	Most plausible*
Corn Starch	Corn grown with conventional tillage, irrigation, and high chemical inputs replaces uncultivated land such as that in the Conservation Reserve Program (CRP).	Existing corn grown with conservation practices diverted to biofuel supply chain. No change in land use.	Existing corn grown with conservation practices diverted to biofuel supply chain. No change in land use. †Exception for water quantity	Conventionally managed, tilled corn in regions not requiring irrigation replaces conventionally managed, no-till soy or other row crops. Overall trend as reported by USDA is increasing acreage of corn planted since 2005 (see section 3.2.3).
Soybean	Soy grown with conventional tillage, irrigation, and high chemical inputs replaces uncultivated land such as that in the CRP.	Existing soy grown with conservation tillage diverted to biofuel supply chain. No change in land use.	Soy grown with comprehensive conservation practices replaces corn grown with conventional tillage and high chemical inputs.	Existing soy grown with conservation tillage diverted to biofuel supply chain to meet relatively small volumetric RFS2 biodiesel requirements. Overall trend as reported by USDA is relatively stable acreage of soybeans planted since 2005 (excluding 2007, see section 3.2.3).
Corn Stover	High rate of stover removal on highly erodible land requiring additional equipment passes after corn grain harvesting replaces same with no stover removal.	Existing corn with appropriate rate of stover removal to minimize erosion, soil organic matter loss, and fertilizer application given site-specific characteristics replaces same with no stover removal. Single pass harvest with corn.		Stover removal at "logistically removable" rate (see USDA's Billion Ton Study), without considering local characteristics, from conventionally managed, tilled corn in regions not requiring irrigation replaces same with no stover removal. Impacts shown are beyond corn cultivation and from separate pass harvest.
Perennial Grasses	Invasive perennial grasses established with conventional tillage and grown with a short planting interval, high rates of chemical inputs, and irrigation replace uncultivated land such as that in the CRP.	Perennial grasses from currently mowed pasture or other managed grasslands diverted to biofuel supply chain. No change in land use.	Non-invasive perennial grasses established with no till and grown with a long replanting interval, low chemical inputs and no irrigation replace irrigated corn grown with conventional tillage and high chemical inputs.	Switchgrass grown with fertilizer in regions not requiring irrigation replaces CRP and other low management lands. Switchgrass (unlike Giant Miscanthus) cultivated for farm-scale studies on CRP in many areas of US (see section 3.3.3).
Woody Biomass	Invasive short-rotation woody crops (SRWC) with short replanting intervals, high chemical inputs, high isoprene emissions, and no coppicing replace mature, managed, low-isoprene-emitting tree plantations.	Removal of managed forest harvest residues at rates that maintain soil organic matter and minimize erosion replaces residues left on site.	Non-invasive, coppiced SRWC with long replanting intervals, low chemical inputs, and low isoprene emissions replace non-coppiced, managed forests with short replanting intervals and high isoprene emissions. OR Low to moderate rates of forest residue removal or thinning replaces residues left on site.	Removal rate of managed forest harvest residues without considering local characteristics replaces residues left on site. This is the greatest source of woody biomass assumed under the RFS2 RIA (EPA 2010b).

Feedstock	Environmental Impact Per Unit Area			
	Most negative	**Negligible**	**Most positive**	**Most plausible***
Corn Starch	Invasive species of algae grown in open raceway ponds in drier regions with freshwater and high chemical inputs that are not recycled.	Non-invasive algae grown with water where it is abundant, in closed bioreactors; treated effluent is recycled for further use.	Non-invasive algae grown with wastewater in closed bioreactors; treated effluent is recycled for further use.	Algae grown in open ponds on marginal land using nutrient-rich wastewater (see section 3.5.3).

* Sets of assumptions commonly used in the literature.

† Corn replaces CRP with relatively high annual evaporation in non-irrigated areas.

Table 6-2. Assumptions Underlying the Synthesis of Biofuel Production, Transport and Storage Impacts

Environmental Impact Per Unit Volume				
Biofuel	**Most negative**	**Negligible**	**Most positive**	**Most plausible***
Corn Ethanol	Coal powered facility using 3-6 gallons of water per gallon of ethanol; effluent with high biological oxygen demand (BOD); dried distillers grain (DDG) byproduct fed to livestock with inadequate waste management practices; leaking underground storage tanks (USTs).	Given current knowledge, unlikely that corn ethanol facilities will have negligible environmental impacts.	Natural gas powered facility with combined heat and power (CHP) using <3 gallons of water per gallon of ethanol by improving water use efficiency and recycling; effluent effectively treated for BOD; DDG-fed livestock waste used within comprehensive nutrient management plan; USTs do not leak.	Facility uses ~3 gallons of water per gallon of ethanol, utilizes natural gas for heat, produces DDG byproducts fed to livestock with current waste management techniques. Spills and storage tank leaks at currently observed rates
Soybean Biodiesel	Coal powered facility using <1 gallon of water per gallon of biodiesel; effluent with high BOD, total suspended solids (TSS) and glycerin content; USTs leak.	Given current knowledge, biodiesel facilities may approach negligible environmental impacts on water quantity, but not for other impact categories.	Natural gas power facility with CHP using <1 gallon of water per gallon of biodiesel; effluent effectively treated for BOD, TSS and glycerin; USTs do not leak.	Facility produces effluent with current range of pollutants (e.g., glycerin, high BOD), uses <1 gallon of water per gallon of biodiesel, and utilizes natural gas for heat. Spills and storage tank leaks at currently observed rates.
Cellulosic Ethanol	Coal powered facility using 10 gallons of water per gallon of ethanol; effluent with high BOD; USTs leak.	Given current knowledge, unlikely that cellulosic ethanol facilities will have negligible environmental impacts.	Natural gas powered or biomass with CHP using <10 gallons per gallon of ethanol by improving water use efficiency and recycling; effluent treated for BOD; USTs do not leak.	Facility uses >5 gallons of water per gallon of ethanol and utilizes natural gas for heat. Spills and storage tank leaks at same rate as corn ethanol.

* Sets of assumptions commonly used in the literature.

The authors also considered the impacts of the "most plausible" conditions under which a feedstock might be produced. The term "plausible" is used rather than "probable" to make clear that these are not probabilistic predictions. The state of science and advances in technology in this rapidly growing field do not permit a quantitative determination of the likelihood of future practices at this time. "Most plausible" impacts are based on sets of assumptions commonly considered in the literature and include type of land converted, management approaches, regional influences, and technologies, as shown in the last column of Table 6-1. Because corn and soybeans are both currently used in commercial production of biofuels, the impacts depicted can be thought of as plausible now and in the future. On the other hand, impacts attributable to the remaining feedstocks are plausible in the future only since they are not yet in full-scale commercial use.

6.2.2. Assumptions Underlying the Synthesis of Biofuel Production, Transport, and Storage Impacts

In evaluating the potential range of domestic environmental impacts associated with the production, transport, and storage of each of three biofuels, the same iterative process described above was used to determine "most negative" and "most positive". One notable difference is that previous land use is not as essential to this determination. For example, an examination of the literature found that the "most negative" water quality impact from producing corn grain ethanol occurred when biofuel refinery effluent had high biological oxygen demand (BOD); dried distillers grain (DDG) byproduct was fed to livestock with inadequate waste management practices; and the fuel was stored in leaking underground storage tanks (USTs).

Table 6-2 lists key assumptions involving mostly production and storage practices and technologies (because transportation will be similar, though not identical) found in the literature that produce the "most plausible" impacts. Because corn grain ethanol and soybean biodiesel are commercially produced, the impacts depicted can be thought of as plausible now and in the future. On the other hand, impacts attributable to cellulosic ethanol are plausible in the future only since it is not yet in full-scale commercial use. Sets of assumptions underlying the most plausible impacts do not necessarily represent future practices, since technology is rapidly changing.

6.3. Synthesis

EISA Section 204 calls for EPA to report to Congress every three years on the environmental and resource conservation impacts of increased biofuel production and use, including air and water quality, soil quality and conservation, water availability, ecosystem health and biodiversity, invasive species, and international impacts. EPA interpreted the requirements of Section 204 to be those environmental impacts beyond the reductions in greenhouse gas emissions associated with the RFS program and recently analyzed in EPA's RFS2 Regulatory Impact Analysis (U.S. EPA, 2010a). Using the peer-reviewed published literature, this report reviews impacts and mitigation strategies across the entire biofuel supply chain, including feedstock production and logistics, and biofuel production, distribution, and use. This literature demonstrates that production of biofuel feedstocks can result in a wide range of environmental impacts, including negative and positive potential impacts. This range of feedstock impacts depends largely on land use changes, management approaches and regional climate, soil and other ecological characteristics.

6.3.1. Feedstock Production

Synthesis of the feedstock production phase of the biofuel supply chain. Figure 6-1 provides a qualitative synthesis, based on the scientific literature, of the environmental impacts of producing corn, soybeans, corn stover, perennial grasses, woody biomass, and algae. This synthesis is meant to summarize, not substitute, the information contained in this document, providing an illustration of the potential range and the most plausible environmental impacts along the biofuel supply chain. It is based on EPA's review of the scientific literature through July 2010 using a consensus of the authors to assign relative, qualitative values across feedstocks, biofuels, and impact categories.

The impacts presented in Figure 6-1 are on a per unit area basis, rather than another basis of comparison (e.g., per gallon of biofuel) for five reasons: (1) environmental impacts are the focus of this report and are most naturally described on a per unit area basis; (2) the dominant driver of environmental impacts of feedstock production appears to be real changes in land use; (3) a per unit area basis provides a foundation for mapping region-specific impacts; (4) commonly used comparators (e.g., per unit energy) are insufficient for some impacts; and (5) uncertainties in future commercial production were considered too large to develop an alternate consistent unit of comparison. Impacts shown in this figure are relevant for only a unit area of those regions where each feedstock is likely to be grown (see Section 3).

Limitations. No attempt has been made to compare impacts to those of petroleum production, nor do impacts represent possible environmental benefits gained by petroleum displacement. Comparison of petroleum and biofuels with respect to GHGs is presented in the RFS2 Regulatory Impact Analysis (RIA) (U.S. EPA, 2010a); research comparing petroleum and biofuels in all other impact categories is beyond the scope of this report, but is proposed for future reports. In particular, the air quality impacts do not include changes in greenhouse gas (GHG) emissions. Within impact categories, impacts are evaluated strictly relative to each other; no attempt was made to create a common scale to compare the impacts across environmental impacts. For example, the maximum negative impact for water quality is not comparable to the maximum negative impact for air quality. There is more confidence in the synthesis conclusions for corn starch and soybean because we have more experience with them at commercial scales. In general, we have lower confidence in the degree of environmental impacts for the remaining feedstocks.

Figure 6-1 presents an overview of potential impacts by the environmental endpoints called for in EISA Section 204.

Land use. Many of the potential environmental impacts of biofuel production can arise from land use conversion. An increase in cropland extent in response to increasing demand for biofuels has been projected in numerous modeling studies, primarily at the expense of pasture and other less-productive agricultural land. Of the land use conversions allowable under RFS2, the production of row crops such as corn and soybeans on uncultivated land such as that formerly enrolled in the Conservation Reserve Program (CRP) will result in the greatest negative environmental impacts. Since 2006, the year preceding EISA, cropland planted to corn has increased by almost 10 million acres. However, there is little evidence that substantial amounts of land have been converted to cropland to date. Recent crop trends suggest that an annual increase in acres planted to corn has displaced plantings of other row crops. Algae production has the smallest potential land use implications.

Water quality. Impacts on water quality from biofuels in the United States are, and likely will be, primarily driven by chemical inputs at the feedstock production stage. Though there are other impacts, including effluent discharge and other factors associated with processing biomass into biofuel, these will likely be small in comparison and are already regulated. Impacts to date from EISA are considered moderately negative, resulting primarily from an intensification of corn production, which leads to greater erosion and requires more chemical inputs than other feedstocks, especially of nitrogen fertilizer and pesticides. Increased fertilizer runoff contributes to eutrophication, coastal hypoxia, and other areas of concern. Conservation practices, if widely employed, can mitigate these impacts. Cultivation of perennial grasses on land currently used for row crops offers substantial environmental benefits, though such conversion is considered unlikely under current market conditions. These grasses require fewer chemical inputs and have perennial root structures that can lead to lower sediment and nutrient losses to the surrounding environment. Algae grown using wastewater also offer water quality benefits.

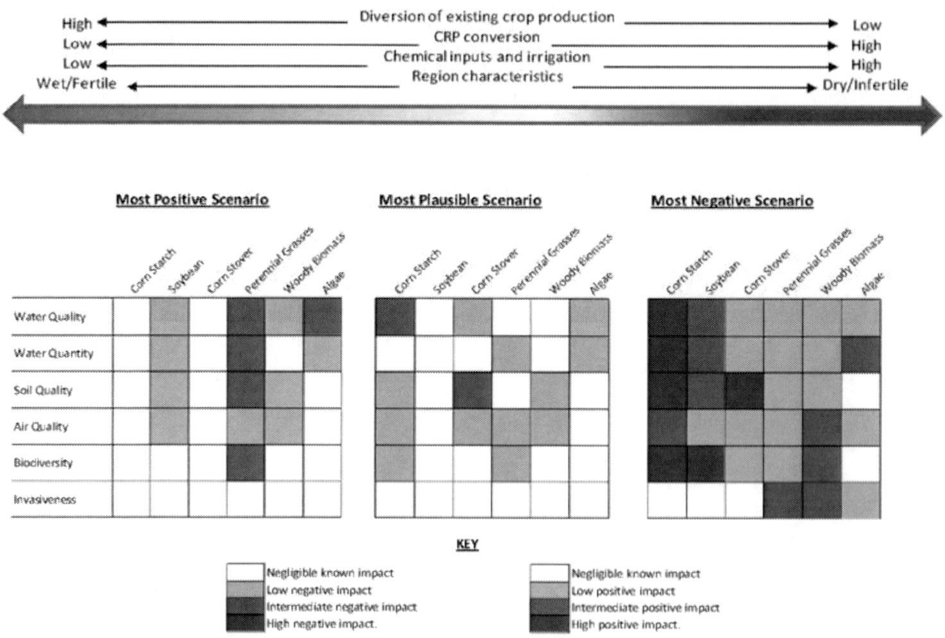

Figure 6-1. Most Positive, Negative, and Plausible Environmental Impacts (on a Per Unit Area Basis) from Feedstock Production.

Water quantity. Water use for feedstock production will likely not change appreciably if production takes place, as the majority does now, in regions where irrigation is not needed. However, water availability for other uses may increase if corn replaces CRP that is covered in relatively high-evapotranspiring perennial grasses. Water use, on the other hand, will increase if feedstock production expands in regions where irrigation is required to achieve profitable yields. Irrigation water use will increase more if row crops are cultivated (with an additional possible requirement if stover is removed from corn fields) than if perennial grasses are cultivated. When used, irrigation can amount to 100 to 1,000 times the volume of water required to convert feedstocks into a given volume of biofuel. Moderate forest thinning and residue removal is unlikely to significantly affect water availability overall. Algae production could consume fresh water, brackish water, saline water or wastewater. The nature of water availability and all its associated impacts on human and ecological communities resulting from feedstock production are difficult to generalize, but impacts are most likely to be adverse in already stressed aquifers or surface watersheds.

Soil quality. Biofuel feedstock production can impact soil quality in a number ways, including through erosion, organic matter content and nutrient losses. High stover removal rates are of particular concern with regard to soil erosion and organic matter. Generally, annual crops, such as soybeans and corn, result in higher erosion rates, lower soil organic matter content, and increased nitrogen and phosphorus losses to waterways compared to perennial feedstocks, such as grasses and woody biomass. However, these impacts may be ameliorated, at least in part, by the use of conservation practices. Perennial feedstocks may not directly replace row crops, and, in such a case, their environmental impacts will be relative to other land uses, such as CRP acreage or abandoned agricultural land. Perennial feedstocks have the potential to improve soil quality on abandoned or idle agricultural land. The opposite is likely if short-rotation woody crops (SRWCs) are planted to replace existing forest land currently managed on longer rotations. Thus, the specific land use conversion will, in large part, determine the soil quality impacts.

Air quality. Combustion of fuels associated with cultivation and harvesting of biofuel feedstocks and airborne particles (dust) generated during tillage and harvesting result in air pollutant emissions, which adversely affect air quality, with effects varying by region. Production of row crops will adversely affect air quality more than non-row crops. Air emissions also result from the production of fertilizers and pesticides used in corn and soybean cultivation, and their application in the field.

Biodiversity. Biofuel feedstock cultivation could significantly affect biodiversity through habitat conversion, especially if CRP lands are put into production. Effects include exposure of flora and fauna to pesticides; sedimentation and eutrophication in water bodies resulting from soil erosion and nutrient runoff, respectively; or water withdrawals resulting in decreased streamflows.

- **Forests.** Changes in existing forests to shorter harvest intervals for SRWCs and residue harvesting can decrease habitat availability and biodiversity, while moderate thinning can increase species diversity and abundances for some species. Use of riparian buffers to reduce erosion and pesticide and fertilizer runoff can increase the availability of forest habitat.
- **Grasslands.** Conversion of grasslands, such as pasture or CRP lands, to row crops negatively impacts grassland-obligate species, while their conversion to perennial grass feedstocks is likely to have fewer impacts. Use of grassland buffers to reduce erosion and pesticide and fertilizer runoff can increase habitat availability.
- **Wetlands.** Some agricultural practices can convert small, unregulated wetlands and increase sediment, nutrient, pesticide, and pathogen runoff into downstream wetlands; while conservation practices, such as constructed or restored wetlands, can

improve habitat availability for some species and improve freshwater habitat conditions.

Invasiveness. Corn and soybeans pose negligible risk of becoming weedy or invasive in the United States. Weed risk assessments predict that in certain regions, switchgrass and some woody crop species or varieties could become invasive in some regions if cultivated without preventative measures, but that the perennial grass Giant Miscanthus poses little risk of becoming invasive. Transport to biofuel production facilities of feedstocks with live seed or vegetative reproduction could facilitate invasion along transportation corridors. The risk of algae escape from production with subsequent establishment is highly uncertain.

6.3.2. Biofuel Production, Transport, and Storage

Key conclusions and synthesis of environmental impacts of biofuel production, transport and storage follow.

Water quality. Pollutants in the wastewater discharged from biofuel production impact water quality. Biological oxygen demand (BOD), brine, ammonia-nitrogen, and phosphorus are primary pollutants of concern from ethanol facilities. BOD, total suspended solids, and glycerin pose the major water quality concerns in biodiesel facility effluent. Actual impacts depend on a range of factors, including the type of feedstock processed, biorefinery technology, effluent controls, and water re-use/recycling practices, as well as the facility location and source and receiving water.

Leaks and spills of biofuel from above-ground, underground, and transport tanks can contaminate ground, surface, and drinking water. A leaking tank can also present other health and environmental risks, including the potential for fire and explosion. Enforcement of existing regulations concerning corrosion protection, leak detection, and spill and overfill prevention will minimize water contamination. Selection and use of appropriate materials and proper design in accordance with the applicable material standards or equipment manufacturer recommendations will help prevent biofuel leaks.

Water quantity. Expansion of biofuel production facilities will increase localized water withdrawals. Volume of withdrawals will depend on the size and water recycling capacity of the facility. On a per volume basis, biofuel production uses 100 to 1,000 times less water than feedstock production. The nature of water availability and associated impacts on human and ecological communities resulting from biofuel production are most likely to be adverse in already stressed aquifers or surface watersheds.

Ecosystem health. Effluent discharges high in nutrients, TSS, and other contaminants decrease aquatic habitat condition and can lead to the loss of sensitive species in rivers and streams. Increased water withdrawals can lead to more frequent low-flow conditions that reduce the availability of aquatic habitat. In areas where low flows and high nutrients, TSS or other contaminants co-occur, aquatic condition will be further reduced.

Air quality. Emissions from biofuel production facilities are generated by a number of processes, such as fermentation and distillation of resulting mash, as well as the stationary combustion equipment used for energy production. Because biofuel production facilities are regulated under the Clean Air Act and subject to state/local permits, enforcement of existing regulations will mitigate air quality impacts. Emissions can be further reduced through use of cleaner fuels (e.g., natural gas instead of coal) and more efficient processes and energy generation equipment. Using energy-saving technologies such as combined heat and power (CHP) is an effective means to reduce air emissions associated with biofuel production (both ethanol and biodiesel).

Air quality will be affected by emissions from biofuel transport via rail, barge, and tank truck and by evaporative, spillage, and permeation emissions from transfer and storage activities. However, the impacts are not expected to be significant.

6.3.3. End-Use

Air Quality: End-use impacts are primarily air-quality impacts. Evaporative and tailpipe emissions from biofuel combustion show great variability due to a range of factors, including the vehicle age, how the vehicle is operated, and ambient temperatures. For ethanol, emissions are expected to be higher for some pollutants (such as nitrogen oxides and hydrocarbons) and lower for others, with large decreases in carbon monoxide emissions in particular. Biodiesel combustion also exhibits a pattern of increases (nitrogen oxide emissions) and decreases (PM, CO, and HC emissions). Emissions from ethanol use are independent of feedstock; in contrast, emissions from biodiesel use differ according to the feedstock. Particulate matter, N_2O, and CO emissions are higher for plant-based biodiesel than for animal-based biodiesel.

In EPA's RFS2 RIA (U.S. EPA, 2010a), these emissions changes were used in air quality models to assess anticipated impacts on ambient concentrations in 2022 as a result of the EISA- mandated biofuel volumes in comparison to two reference scenarios. The effects of ethanol or biodiesel were not separated: rather, the entire landscape of biofuels was assessed collectively. Details of note include findings for ozone and $PM_{2.5}$ levels (two pollutants of ongoing concern because concentrations already exceed National Ambient Air Quality Standards in many areas of the country) and for air toxics.

EISA-mandated biofuels production is expected to increase PM levels in some areas and decrease them in others. The increases are expected as a result of biofuel production and transportation, which is more prevalent in the Midwest. Ozone concentrations over much of the United States are expected to rise: however, ozone air quality improvements are projected in a few highly populated areas that currently have poor air quality (U.S. EPA, 2010a). Ground-level ozone is formed by the reaction of VOCs and NOx in the atmosphere in the presence of heat and sunlight. The projected ozone changes described in the RIA are likely a result of the emissions changes due to the increased volumes of renewable fuels combined with the photochemistry involved, the different background concentrations of VOCs and NOx in different areas of the country, and the different meteorological conditions in different areas of the country.

The RIA's air quality impacts assessment also included compounds that were identified as national- and regional-scale cancer and noncancer risk drivers in past National-Scale Air Toxics Assessments and were also likely to be significantly impacted by the standards. These compounds include benzene, 1,3-butadiene, formaldehyde, acetaldehyde, and acrolein. In addition to these explicit model species, photochemical processes mechanisms model the formation of some of these compounds in the atmosphere from precursor emissions. This aspect of the air quality model requires inventories for a large number of precursor compounds, including compounds such as ethene and methane, and uses atmospheric reaction pathways including that of aldehydes and peroxyacetyl nitrate (PAN). Thus, although numerous other species are not explicitly discussed, their impacts are accounted for in the RIA air quality analysis. Refer to the RIA for additional details and results (U.S. EPA, 2010a).

The RIA found some localized impacts for air toxics, but relatively small changes in national average ambient concentrations. Some urban areas may have small decreases of acetaldehyde and formaldehyde, while some ethanol-producing regions may have small increases (less than 1 percent). Concentrations of 1, 3-butadiene and acrolein are expected to decrease in some southern areas and increase in some northern areas with high altitudes. Small decreases (1 to 10 percent) of benzene are expected.

Finally, the RIA also found the renewable fuel volumes required by RFS2 lead to significant nationwide increases in ambient ethanol concentrations. Increases ranging between 10 to 50 percent are seen across most of the country. The largest increases (more than 100 percent) occur in urban areas with high amounts of nonroad emissions and in rural areas associated with new ethanol plants (U.S. EPA, 2010a).

6.4. Conclusion

Evidence to date from the scientific literature suggests that current environmental impacts from increased biofuels production and use associated with EISA 2007 are negative but limited in magnitude.

- **Environmental impacts along the supply chain are greatest at the feedstock production stage.** Most activities, processes, and products, particularly those occurring after feedstock production, are regulated and subject to limitations.
- **Current environmental impacts are largely the result of corn production.** Corn starch–derived ethanol constituted 95 percent of the biofuel produced in 2009. In general, feedstock demand has been met by diverting existing corn production or by replacing other row crops with corn, resulting in modest additional environmental impacts.

Published scientific literature suggests a potential for both positive and negative environmental effects in the future.

- **Technological advances and market conditions will determine what feedstocks are feasible, and where and how they will be cultivated.**
- **The magnitude of effects will be largely determined by the feedstock(s) selected, land use changes, and cultivation practices.**
- **Overall impacts given most plausible land use changes and production practices will likely be neutral or slightly negative (Figure 6-1).** More adverse or beneficial environmental outcomes are possible.
- **Second-generation feedstocks have a greater potential for positive environmental outcomes relative to first-generation feedstocks (Figure 6-1).**

However, current production levels of second-generation biofuels are negligible and limited by economic and technological barriers.

EISA goals for biofuels production can be achieved with minimal environmental impacts if existing conservation and best management practices are widely employed, concurrent with advances in technologies that facilitate the use of second-generation feedstocks (Figure 6-1).

- **The feedstocks considered in this report all have the potential to support sustainable domestic energy production.** Realizing this potential will require implementation and monitoring of conservation and best management practices, improvements in production efficiency, and implementation of innovative technologies at the commercial scale.
- **International partnerships and federal coordination are needed to accelerate progress towards sustainable and secure energy production.**

6.5. International Considerations

Increases in U.S. biofuel production and consumption volumes will affect many different countries as trade patterns and prices adjust to equate global supply and demand. This will result in environmental impacts, both positive and negative, including effects from land use change and effects on air quality, water quality, and biodiversity. Direct and indirect land use changes could occur internationally as the United States and other biofuel feedstock-producing countries alter their agricultural sectors to allow for greater biofuel production. Many locations where biofuel production is growing are areas of high biodiversity value. For example, Brazil (sugar ethanol) contains ecosystems with high biodiversity. Depending on where biofuel feedstock production occurs, and the manner in which it occurs, impacts to biodiversity could be significant. However, because corn ethanol is limited by the RFS2 and is likely to reach this limit in the next few years, these international impacts projected in the RIA could level off as corn starch ethanol production levels off or is replaced by more advanced technologies.

As with domestic production, the choice of feedstock, how and where it is grown, the resulting land use changes, and how it is produced and transported will have a large effect on how biofuel production and use affects water quality and availability, air quality (e.g., due to emissions from burning crop residue), and biodiversity. The specific impacts will reflect a country's particular circumstances.

6.6. Recommendations

EISA Section 204 specifies that EPA must include recommendations for actions to address any adverse impacts identified in this report. Responding specifically to this request requires a clear understanding of biofuel impacts and their causes and the recognition of the high degree of uncertainty in many areas surrounding the progress of the technologies and implementation of mitigation procedures to ameliorate impacts. For corn starch and soybean production, the impacts are relatively well understood, but more information is needed about the adverse impacts associated with production of other feedstocks and with the production and use of advanced biofuel. This section presents four recommendations to address adverse impacts. Because biofuel impacts cross multiple topics and EPA responsibilities, EPA likely will address these recommendations through continued and **strengthened cooperation with other federal agencies and international partners.**

6.6.1. Comprehensive Environmental Assessment

The biofuel industry is poised for significant expansion in the next few years. A variety of new technologies will likely be implemented and old technologies will be modified to meet the demands of affordable and sustainable alternatives to petroleum fuel. As emphasized by Congress in requiring triennial biofuel impact assessments, it is important to evaluate the environmental implications associated with the ongoing growth of the dynamic biofuel industry. While there is currently available sufficient scientific information to inform environmental decisions, the inherent complexity and uncertainty of environmental impacts across the biofuel supply chain present a challenge to providing definitive assessments and further research is necessary.

RECOMMENDATION: Develop and evaluate environmental life cycle assessments for biofuels. With this report, EPA, USDA, and DOE have begun to develop a framework and partnership that provide an important foundation for future assessments. Future assessments will address advanced biofuel production associated with specific feedstocks and associated byproducts and provide a context for comparison to fossil fuels. As described in Section 7, future assessments should be comprehensive, region-specific and address the major environmental parameters affected by increased biofuel production and use. These assessments will identify gaps and uncertainties in the knowledge base, inform the design and implementation of monitoring strategies and measures for evaluating impacts, provide comprehensive tools for comparing and evaluating development options, and provide the scientific bases for regulatory agencies and the biofuel industry to make environmentally conscious decisions.

6.6.2. Coordinated Research

The expansion of the biofuel industry will be shaped to a large degree by the research behind the technological developments that make biofuel production feasible. It will be important for the scientific infrastructure that supports policy and decision-making to keep pace with industry developments.

RECOMMENDATION: Ensure the success of current and future environmental biofuel research through improved cooperation and sustained support. The Biomass Research and Development Board, co-chaired by DOE and USDA, currently monitors interagency cooperation in biofuel research. The Board recently proposed that an inventory be conducted of federal activities and jurisdictions relevant to environmental, health, and safety issues associated with biofuel production in order to identify issues of concern, research needs, and mitigation options. Efforts to adjust and expand existing research programs to conduct biofuel-relevant research have been initiated. Prioritization and collaboration by the research community will be critical to provide meaningful results in the near term and to meet the wide variety of research needs, including many that have already been identified, that will be important to the industry and to appropriate regulatory oversight.

6.6.3. Mitigation of Impacts from Feedstock Production

As the biofuel industry expands, it will be important to optimize benefits while minimizing adverse impacts. Because many of the known adverse impacts are due to feedstock production, this report has described the potential for mitigation of those impacts through the adoption of conservation systems and practices on farms. USDA has a variety of programs that help agriculture producers implement these conservation systems. As USDA's Conservation Effects Assessment Project (CEAP) report on the upper Mississippi River basin demonstrates, much more needs to be done to control pollution from agriculture, especially from nitrogen. A collaborative effort is needed to develop and foster application of consistent and effective monitoring and mitigation procedures to protect the environment and conserve biodiversity and natural resources as biofuel production expands and advanced biofuels are commercially produced.

RECOMMENDATION: Improve the ability of federal agencies (within their existing authorities) and industry to develop, implement, and monitor best management and conservation practices and policies that will minimize negative environmental impacts and maximize the positive environmental effects of biofuel production and use. This will involve coordination among diverse stakeholders, including state agencies, research scientists, and landowners. These policies and practices should be aligned and assessed within the context of the environmental life cycle assessment and take a multi-factor and multi-scale view of biofuels and their potential environmental effects. Priority areas for development include (1) improved containment processes and technologies that minimize environmental exposure from air emissions and runoff into surface and ground water and (2) methods to monitor, track, and report biofuel environmental impacts.

6.6.4. International Cooperation to Implement Sustainable Biofuel Practices

EISA specifically identifies "significant emissions from land use change" as a potential environmental impact stemming from domestic biofuel production and consumption. This concern is relevant to all countries engaged in biofuel production, but as the United States increases domestic production of corn starch ethanol and soybean diesel, exports of corn and soybean for agricultural or other uses are expected to decline, which may result in indirect land use change in the form of land conversion to agriculture in other countries. Additional biofuel produced to meet the EISA mandates will potentially lead to increases in acreages of international cropland, although these increases may level off after 2015 (see Section 5.3).

RECOMMENDATION: Engage the international community in cooperative efforts to identify and implement sustainable biofuel and land use practices that minimize environmental impact. U.S. and international capacity to minimize the consequences of land use change will depend not only on the willingness of governments and industry to make environmentally sound choices regarding biofuel production, processing, and use, but also on the availability of cost-effective mitigation strategies and sustainable land use practices. The United States can significantly contribute to such an effort by actively engaging the scientific community and biofuel industry to collaboratively develop the body of knowledge needed to support sound environmental decision-making. This effort will be facilitated by a greater understanding and appreciation of how increased biofuel demand may impact the environment internationally, particularly in countries that are most active, or most likely to become active, in biofuel production.

7. ASSESSING ENVIRONMENTAL IMPACTS FROM BIOFUELS: 2013 TO 2022

7.1. Introduction

In requiring EPA to report triennially under EISA Section 204, Congress recognized that the environmental and resource conservation impacts of increased biofuel production and use will be dynamic, changing over time in both nature and scope, based on the amount, type, and location of biofuels produced and used. This first triennial Report to Congress, which reflects the state of scientific knowledge as of July 2010, is a first step toward identifying information that supports future assessment of environmental impacts from increased biofuel production and use. This section outlines an approach EPA may use for its future assessments, beginning with the second (2013) report to Congress. In developing future assessments, EPA will work closely with the U.S. Departments of Agriculture and Energy (USDA and DOE), and other interested federal agencies, such as the Department of Defense, and will seek extensive input from industry and other stakeholders and peer review from the scientific community to create substantive, science-based analyses that facilitate environmental decision-making. Future assessments will benefit from advances in the science of environmental assessment and increased availability of relevant research results on this important topic. EPA anticipates that additional research and analyses will allow for more robust and quantitative assessments of biofuel environmental impacts than are reported here. For example, life cycle assessment (LCA) tools and approaches that are currently used for evaluating "cradle to-grave" resource consumption and waste disposal for specific products can be integrated into risk assessment to form a powerful composite approach for assessing environmental impacts. An approach to more comprehensive environmental analyses that is consistent with the integration of LCA and risk assessment methods has been used in different assessments (Davis and Thomas, 2006; Davis, 2007). This approach would necessitate extending consideration of factors across the entire biofuel life cycle, including current and future feedstock production and biofuel conversion, distribution, and use. The Agency has already applied LCA to assess greenhouse gas (GHG) emissions as part of its revised Renewable Fuel Standard (RFS2) program (U.S. EPA, 2010a) and could adapt this approach to analyze other aspects of biofuel production and use, such as water consumption; evaluation of fossil fuels versus biofuels; net energy balance; production and use scenarios; and market impacts (economics).

7.2. Components of the Second Assessment

This section briefly describes key components that EPA will consider in conducting the next report. A comprehensive environmental assessment framework would facilitate evaluation and quantification of risk and benefits of biofuel production and use. Such a framework would integrate models such as the Soil Water and Assessment Tool (SWAT), the Environmental Policy Integrated Climate Model (EPIC), the Community Multiscale Air Quality Model (CMAQ), and the Daily Century Model (DAYCENT). These modeling efforts would allow greater quantification of the potential impacts of biofuel production and allow for mapping their spatial distribution. For example, the use of EPIC could provide estimates of the soil erosion risk of growing biofuel feedstocks on Conservation Reserve Program (CRP) lands.

A comprehensive framework would also include LCA and environmental risk assessment. The latter could be used to systematically assess environmental risks, both human health risks and ecological risks, for each stage in the life cycle, as well as the potentially cumulative impacts. Conceptual models will illustrate the important factors being considered in each stage of the life cycle and indicate how these factors are interrelated. Where possible, environmental indicators and other metrics will be developed over the next several years to track the impacts of biofuel production and use throughout its life cycle and measure the effectiveness of regulatory and voluntary practices in ameliorating these impacts. A scenario-based approach is currently envisioned to provide a comparative basis for projecting and assessing how biofuel production and use might affect the environment in future years. Finally, the next assessment will include other components, such as a comparison to fossil fuels, net energy balance, and analysis of market impacts, that are important to evaluating biofuel impacts.

7.2.1. Life Cycle Assessments

LCAs have been widely used to assess the potential benefits and potential pitfalls for bio ethanol as a transportation fuel (Von Blottnitz and Curran, 2007; Gnansounou et al., 2009). The majority of such analyses have focused on particular components such as GHG emissions and energy balances (Hill, 2009). Economic models can provide estimates of environmental costs by monetizing ecosystem and human health effects (NRC, 2010b) with varied results depending on the assumptions and input parameters driving the assessments. In some cases, the scientific community seems close to reconciling the various assumptions used by different investigators (Anex and Lifset, 2009). To better address the EISA reporting mandate, however, a broader profile of potential environmental impacts should be considered. This approach has been used in several studies (Von Blottnitz and Curran, 2007) and applied to evaluating trade-offs for fuel options (Davis and Thomas, 2006). As part of the next assessment, EPA anticipates using LCA in a broad context, one that considers a full range of potential environmental effects and their magnitude. A variety of environmental LCA approaches have been developed that would prove useful for such an effort (Puppan, 2002; Ekvall, 2005; Hill et al., 2006; Landis et al., 2007; Duncan et al., 2008).

7.2.2. Environmental Risk Assessment

Environmental risk assessment will be fundamental for systematically evaluating the human and environmental impacts of the activities involved in biofuel production and use. Environmental risk assessment can be used to estimate the risks associated with each stage of the biofuel life cycle, from production of raw materials, through transportation and consumption, to the generation of waste products. Environmental risk assessment is initiated by clearly articulating the problem (i.e., problem formulation); describing the critical sources, stressors, and effects, and the linkages among these factors; quantifying human/ecological exposure and effects; and subsequently characterizing and estimating the risks associated these effects. Environmental risk assessment will identify which stages in the biofuel life cycle contribute the greatest risk so that more informed risk management practices can be developed and implemented for these stages.

7.2.3. Human Health Assessment

Increasing biofuel use presents the potential for distinct health effects separate from the known impacts of fossil fuels. The fate and transport of these new fuel blends in the environment and the subsequent exposures and human health effects have not been fully studied. Drawing definitive conclusions on health impacts is not realistic at this time, given the unknowns surrounding the feedstocks, technologies, and fuel blends that will be used to meet target volumes, and the relatively limited availability of toxicological data to directly evaluate the potential health effects of the various emissions.

Health effects will be assessed in the next report, provided adequate data are available. In examining the health risks and benefits of increased biofuel use, it will be important to understand the unique characteristics of the new fuel blends, how and when releases occur, the fate and transport of these releases, the relevant routes and duration of exposures to humans, and the toxic effects of those exposures. Both individual and population exposures will be important to consider. For example, populations in regions that both produce and use biofuel will experience different exposures than those in regions that only use the fuel. Individuals within the same region may experience different exposures (i.e., occupational, consumer, or public exposures), and vulnerable populations may be at greater risk of adverse effects, depending on their sensitivity.

7.2.4. Conceptual Models

A number of tools are available for use in problem formulation, including conceptual diagrams, which hypothesize relationships between activities and impacts. These diagrams can support multiple purposes, including defining system boundaries; enhancing understanding of the system being analyzed; and supporting communication among assessors, between assessors and stakeholders, and, ultimately, with risk managers.

The information provided in Sections 3, 4, and 5 of this current assessment lays a foundation for constructing initial conceptual models to show relationships among biofuel activities and impacts. Figures 7-1 and 7-2 present generalized conceptual models for feedstock and biofuel production, respectively. Appendix C provides detailed conceptual diagrams for each of the feedstocks and fuels considered in this report. Based on the information gathered during this current assessment, the diagrams show the activities (e.g., crop rotation, water use) associated with the model's domain area and how, through a series of relationships indicated with lines and arrows, these activities are associated with products and impacts. These diagrams are the first step in linking evidence from the literature to show the degree of support for different pathways. They can also lead to mathematically simulating the system and quantifying impacts. Diagrams such as these will be important tools for assessments in EPA's future reports to Congress.

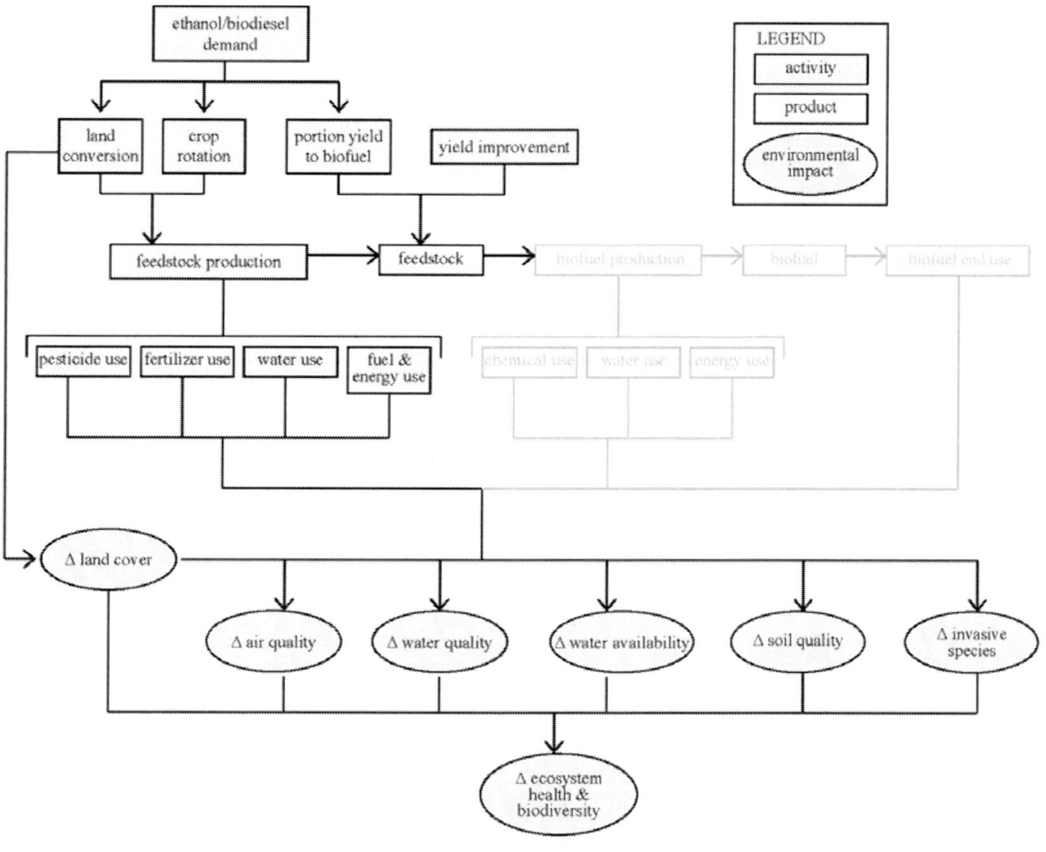

Figure 7-1. Conceptual Diagram of the Potential Environmental Impacts of Biofuel Feedstock Production.

7.2.5. Monitoring, Measures, and Indicators

EPA's ability to accurately assess impacts attributable to biofuels production and use will depend on having timely, relevant, and accurate monitoring information that tracks potential impacts. It will also depend on how effective regulatory and voluntary management practices, risk management practices, and other measures are in protecting the environment.

Current environmental monitoring by various agencies looking at the impacts of traditional land management and energy impacts can provide helpful information, and targeted monitoring for potential biofuel impacts may be needed. Improved monitoring systems will require a collaborative effort across multiple agencies and other organizations. Improved monitoring of indicators and measures are important for a variety of environmental effects, including GHG emissions, human and ecological health, eutrophication, and many other effects. Metrics surrounding these effects will inform decisions at all levels along the biofuel supply chain and well beyond the scope of the individual decision.

7.2.6. Scenarios

EPA's next report to Congress will assess the environmental impacts of all five stages in the biofuel supply chain (see Figure 2-2). One approach may be to create scenarios based on volumetric biofuel requirements for 2022 as presented in the RFS2 (see Table 2-1). Three illustrative scenarios are as follows:

- **Scenario A.** 2022 RFS2-projected feedstock mix produced with comprehensive conservation systems and efficient technologies. Conservation systems include maintenance of crop rotation practices; increased use of conservation tillage, nutrient management, and efficient irrigation systems; crop breeding that results in improved yields and decreased fertilizer, pesticide, and irrigation inputs; minimal expansion of crop land to uncultivated land; and harvest of stover and woody biomass that minimizes soil erosion and nutrient depletion. Efficient technologies include improved fuel conversion processes that require fewer production inputs like energy and fresh water.
- **Scenario B.** 2022 RFS2-projected feedstock mix produced with minimal conservation practice implementation and current technologies. Conventional production practices and non-conservation practices that could be used include decreased crop rotation; minimal use of conservation tillage, nutrient management, and water-saving irrigation; exclusive reliance on increased fertilizer, pesticide, and irrigation inputs to improve crop yields; conversion of CRP and marginal land to crop production that requires fertilizer and irrigation; and harvest of stover and woody biomass that results in erosion and decreased soil nutrients. Current technologies are those now used to convert biomass to fuel with energy and fresh water inputs remaining at current levels.
- **Scenario C.** 2022 conventional feedstock mix (corn starch, corn stover, and soybean) produced with minimal conservation practice implementation. Practices are as in Scenario B, but no perennial grasses, woody biomass, or algae are used as feedstocks to fulfill RFS2 volumetric requirements.

As noted earlier, the landscape of feedstock/biofuel production, conversion, and use is highly dynamic and constantly evolving. Which feedstocks and technologies are used and to what extent they are used will be influenced by technological developments and market forces that are difficult to predict. Development of scenarios for future assessments will need to model or otherwise account for key factors that influence the biofuel market dynamics and associated environmental impacts. These factors include:

- **Regional considerations.** In general, biofuel conversion facilities will tend to be sited at reasonable distances from feedstock production areas, because cost considerations limit the distances over which biofuel feedstocks can be transported. Consequently, environmental impacts of both feedstock production and biofuel conversion will tend to be concentrated in particular regions.
- **Scale and volume of future commercial biofuel operations.** Future development and application of commercially viable biofuel technologies will change the nature of energy feedstocks and conversion processes in use, as well as the scale of their operation. Fossil fuel prices can affect investment in, and adoption of, new biofuel technologies. While the continued use of corn starch for ethanol will likely not change, the future portfolio of feedstocks and biofuels is likely to vary from those

used in 2010. However, which feedstock and biofuel will actually be used and to what extent is highly uncertain and largely dependent on technology advances for the production of second-generation feedstocks.

- **Hybrid processes.** Biofuel conversion processes (e.g., biochemical and thermochemical processes) may evolve in the future to be hybrid processes that would produce not only biofuel but also synthetic chemicals and other industrial co-products. Integrated biorefineries may have the ability to make use of a biofuel-only or a hybrid conversion platform. Each new conversion option will present its own range of potential environmental impacts.
- **Changes in vehicle technologies.** Changes in vehicle technologies, patterns of vehicle sales, and fueling behavior will be needed to accommodate higher ethanol production volumes. Conversely, changes in vehicle technologies driven by other considerations, such as the development of plug-in hybrid electric or all-electric vehicles, could change the demand for liquid biofuels.
- **Changes in agricultural practices due to biofuel production and implications for environmental impacts.** Recent increases in ethanol production have expanded the market demand for corn grain, and farmers have responded to this increased demand by, for example, changing crop rotation practices and/or replacing other row crops with corn. It is not clear what the effects of production shifts, agricultural residue use, and associated farm-level management practice changes will be in the short term.

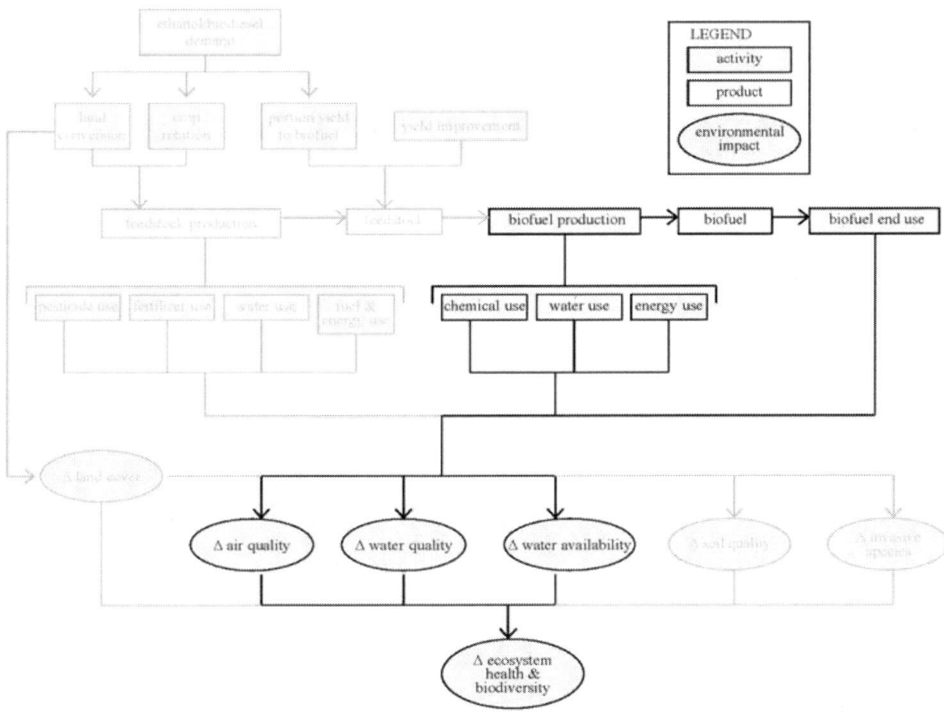

Figure 7-2. Conceptual Diagram of the Potential Environmental Impacts of Biofuel Production and Use.

7.2.7. Other Components

In addition to the above components, EPA plans to include in the next assessment several analyses that provide important perspective for understand and evaluating the impacts of biofuel production and use, as described below.

Comparison of fossil fuel to biofuel. While this current report provides a starting point for comparing the relative impacts associated with a range of different biofuel feedstock and production processes, it is critical to assess biofuel impacts in the larger context of the conventional petroleum fuels that are being displaced under the RFS2 mandates. Ideally, this comparison would cover the full life cycle for each fuel. Such an evaluation would facilitate comprehensive assessment of the relative costs and benefits of RFS2 beyond GHG impacts, and support identification of effective mitigation measures for key impacts. This type of evaluation has been recommended by the National Advisory Council for Environmental Policy and Technology (NACEPT) as a means of conducting integrated environmental decision-making (NACEPT, 2008). Given the limitations of currently available information, a comparative assessment of petroleum fuel and biofuel impacts would be largely qualitative, with significant data gaps and uncertainties. Nevertheless, EPA anticipates that even a qualitative comparative analysis will be an important component of the next assessment.

Net energy balance. Net energy balance (i.e., the amount of energy used to develop biofuels compared to the energy value derived from biofuels) is an important metric that will likely be addressed in the next assessment. It enables comparison of biofuel produced from different feedstocks and via different conversion processes, as well as comparison between biofuel and gasoline. The net energy balance will include consideration of energy embedded in co-products of the fuel conversion process. For example, increases in corn ethanol production will increase the amount of co-products used in animal feed, which in turn displaces whole corn and soybean meal used for the same purpose—the "displaced" energy is credited to the ethanol system and offsets some of the energy required for production (Hammerschlag, 2006; Liska et al., 2008).

Market impacts. Biofuels displace fossil energy resources, but also consume petroleum products, natural gas, electricity (much of which comes from nonrenewable energy sources), and even coal at different points along their supply chain. Consequently, changes in fossil fuel prices will impact the economics of biofuel production in unpredictable ways. The next report will likely address market impacts and incorporate modeling of coupled energy systems and agricultural markets.

APPENDIX A. GLOSSARY AND ACRONYMS

advanced biofuel: A renewable fuel, other than ethanol derived from corn starch, that has life cycle greenhouse gas emissions that are at least 50 percent less than life cycle GHG emissions from petroleum fuel. Cellulosic biofuels must achieve a 60 percent reduction in GHG to get credit for being "advanced."

agricultural residue: Plant parts, primarily stalks and leaves that are not removed from fields used for agriculture during harvesting of the primary food or fiber product. Examples include corn stover (stalks, leaves, husks, and cobs), wheat straw, and rice straw.

algae: Plant-like organisms (usually photosynthetic and aquatic) that do not have true roots, stems, leaves, or vascular tissue, and that have simple reproductive structures. Algae are distributed worldwide in the sea, in fresh water, and in wastewater. Most are microscopic, but some are quite large (e.g., some marine seaweeds that can exceed 50 meters in length).

B20: A fuel mixture that includes 20 percent biodiesel and 80 percent conventional diesel and other additives. Similar mixtures, such as B5 or B10, also exist and contain 5 and 10 percent biodiesel, respectively.

B100: Pure (i.e., 100 percent) biodiesel, also known as "neat biodiesel."

best management practices (BMPs): The techniques, methods, processes, and activities that are commonly accepted and used to facilitate compliance with applicable requirements, and that provide an effective and practicable means of avoiding or reducing the potential environmental impacts.

biodiesel (also known as "biomass-based diesel"): A renewable fuel produced through transesterification of organically derived oils and fats. May be used as a replacement for or component of diesel fuel. According to 40 CFR 80.1401, biodiesel means "a mono-alkyl ester that meets ASTM D6751 ('Standard Specification for Biodiesel Fuel Blend Stock (B100) for Middle Distillate Fuels')."

biodiversity: The variety and variability among living organisms and the ecological complexes in which they occur. Biodiversity can be defined as the number and relative frequency of different items, from complete ecosystems to the biochemical structures that are the molecular basis of heredity. Thus, the term encompasses ecosystems, species, and genes.

biofuel: Any fuel made from organic materials or their processing and conversion derivatives.

biofuel blend: Fuel mixtures that include a blend of renewable biofuel and petroleum-based fuel. This is opposed to "neat form" biofuel that is pure, 100 percent renewable biofuel.

biofuel distribution: Transportation of biofuel to blending terminals and retail outlets by a variety of means, including rail, barge, tankers, and trucks. This almost always includes periods of storage.

biofuel end use: Combustion of biofuel in vehicles and various types of engines, usually as a blend with gasoline or diesel, or in some cases in neat form.

biofuel life cycle: All the consecutive and interlinked stages of biofuel production and use, from feedstock generation to biofuel production, distribution, and end use by the consumer.

biofuel production: The process or processes involved in converting a feedstock into a consumer-ready biofuel.

biofuel supply chain: The five main stages involved in the life cycle of a biofuel: feedstock production, feedstock logistics, fuel production, fuel distribution, and fuel use.

biogenic: Produced by living organisms or a biological process.

biomass: Any plant-derived organic matter (e.g., agricultural crops and crop wastes; wood and wood wastes and residues; aquatic plants; perennial grasses).

biomass-based diesel: See "biodiesel" above. Biomass-based diesel includes non-co-processed renewable diesel, which does not use the transesterification technology.

According to 40 CFR 80.1401, biomass-based diesel is "a renewable fuel that has lifecycle greenhouse gas emissions that are at least 50 percent less than baseline lifecycle greenhouse gas emissions and meets all of the following requirements:

- Is a transportation fuel, transportation fuel additive, heating oil, or jet fuel;
- Meets the definition of either biodiesel or non-ester renewable diesel; and
- Registered as a motor vehicle fuel or fuel additive under 40 CFR part 79, if the fuel or fuel additive is intended for use in a motor vehicle.

Renewable fuel that is coprocessed with petroleum is not biomass-based diesel."

cellulosic biofuel: A renewable fuel derived from lignocellulose (a plant biomass composed of cellulose, hemicellulose, and lignin that is a main component of nearly every plant, tree, and bush in meadows, forests, and fields). Lignocellulose is converted to cellulosic biofuel by producing sugars from the residual material, mostly lignin, and then fermenting, distilling, and dehydrating this sugar solution. According to 40 CFR 80.1401, cellulosic biofuel is "renewable fuel derived from any cellulose, hemicelluloses, or lignin that has lifecycle greenhouse gas emissions that are at least 60 percent less than the baseline lifecycle greenhouse gas emissions."

Conservation Reserve Program (CRP): A U.S. Department of Agriculture program that provides technical and financial assistance to eligible farmers and ranchers to address soil, water, and related natural resource concerns on their lands in an environmentally beneficial and cost- effective manner. It encourages farmers to convert highly erodible cropland or other environmentally sensitive acreage to vegetative cover, such as tame or native grasses, wildlife plantings, trees, filter strips, or riparian buffers. Farmers receive an annual rental payment for the term of the multi-year contract.

conservation tillage: Any cultivation system that leaves at least one-third of the land surface covered with residue after planting in order to reduce soil erosion and conserve soil productivity. One example would be "no-till," where fields are not tilled at all and crops are planted directly into the existing residue. Other variations include "strip-till" or "ridge-till," which remove some, but not all, of the residue from the harvested area.

conventional biofuel: In the context of this report, "conventional biofuel" refers to ethanol derived from corn starch that does *not* lead to at least a 50 percent reduction in greenhouse gas emissions compared to petroleum. EISA 2007 requires conventional biofuel to achieve a 20 percent reduction in life cycle GHG emissions compared to gasoline.

corn stover: The stalks, leaves, husks, and cobs that are *not* removed from the fields when corn is harvested.

crop yield: The quantity of grains or dry matter produced from a particular area of land. (In this report, crop yield is most often measured in corn or soybean bushels per acre.)

direct land use change: In the context of biofuel, "direct land use change" refers to land conversion that is directly related and easily attributable to the biofuel supply chain. For example, conversion by a U.S. farmer of grasslands to corn for ethanol would be considered a direct land use change.

double cropping: The process of planting two different crops (not including cover crops) on the same piece of land over the course of a growing season.

dry milling: A process for producing conventional corn starch ethanol in which the kernels are ground into a fine powder and processed without fractionating the grain into its component parts. Most ethanol comes from dry milling.

E10: A fuel mixture of 10 percent ethanol and 90 percent gasoline based on volume.

E85: A fuel mixture of 85 percent ethanol and 15 percent gasoline based on volume.

ecosystem health: The ability of an ecosystem to maintain its metabolic activity level and internal structure and organization, and to resist external stress over time and space scales relevant to the ecosystem.

effluent: Liquid or gas discharged in the course of industrial processing activities, usually containing residues from those processes.

Energy Independence and Security Act (EISA): Signed into law as Public Law 110-140 on December 19, 2007, this legislation established energy management goals and requirements while also amending portions of the National Energy Conservation Policy Act. EISA's stated goals are to move the U.S. toward greater energy independence and security; increase production of clean renewable fuels; protect consumers; increase the efficiency of products, buildings, and vehicles; promote research on and deploy greenhouse gas capture and storage options; and improve the energy performance of the federal government.

environmental life cycle assessment: In the context of this report, an environmental life cycle assessment is an assessment in which the LCA methodology (see "life cycle assessment") is applied to address the full range of potential environmental impacts over all environmental media.

ethanol (also known as "bioethanol"): A colorless, flammable liquid produced by fermentation of sugars. Ethanol is generally blended with gasoline and used as a fuel oxygenate.

eutrophication: Nutrient enrichment of aquatic ecosystems, in which excessive nutrient levels cause accelerated algal growth, which in turn can reduce light penetration and oxygen levels in water necessary for healthy aquatic ecosystems. Eutrophication can cause serious deterioration of both coastal and inland water resources and can lead to hypoxia.

feedstock: In the context of biofuel, "feedstock" refers to a biomass-based material that is converted for use as a fuel or energy product.

feedstock logistics: All activities associated with handling, storing, and transporting feedstocks after harvest to the point where the feedstocks are converted to biofuel.

feedstock production: All activities associated with cultivation and harvest of biofuel feedstock.

filter strip: A strip or area of herbaceous vegetation that may reduce nutrient loading, soil erosion, and pesticide contamination by removing soil particles and contaminants from overland water flow.

forest residue: Includes tops, limbs, and other woody material *not* removed in forest harvesting operations in commercial hardwood and softwood stands.

forest thinning: Removal of trees from overgrown forests to reduce forest fire risk or increase forest productivity. These trees are typically too small or damaged to be sold as round wood but can be used as biofuel feedstock.

genetically engineered feedstock: Plants, trees, and other organisms that have been modified by the application of recombinant DNA technology and produce the biomass-based material converted for use as a fuel or energy product.

greenhouse gases (GHGs): Gases that trap the heat of the sun in the Earth's atmosphere, producing the greenhouse effect. Greenhouse gases include water vapor, carbon dioxide, hydrofluorocarbons, methane, nitrous oxide, perfluorocarbons, and sulfur hexafluoride.

hemicellulose: any of various plant polysaccharides less complex than cellulose and easily hydrolysable to monosaccharides (simple sugars) and other products.

hybrid: A plant species created from the offspring of genetically different parents, both within and between species. Hybrids combine the characteristics of the parents or exhibit new ones.

hypoxia: The state of an aquatic ecosystem characterized by low dissolved oxygen levels (less than 2 to 3 parts per million) due to accelerated algal growth and reduced light penetration because of excessive nutrient levels (eutrophication). Low dissolved oxygen can reduce fish populations and species diversity in the affected area.

indirect land use change: In the context of biofuel, "indirect land use change" refers to land conversion that occurs as a market response to changes in the supply and demand of *goods other than biofuel* (e.g., food commodities) that result from changes in biofuel demand. For example, clearing of foreign land to plant corn as a food crop in response to reduced U.S. corn exports caused by increased use of U.S. corn to produce ethanol is considered to be an indirect land use change.

integrated pest management (IPM): An environmentally sensitive approach to pest management that uses current, comprehensive information on the life cycles of pests and their interaction with the environment to manage pest damage by the most economical means, and with the least possible hazard to people, property, and the environment.

invasive plant (also called an invasive or a noxious plant): A novel species or genotype whose introduction does or is likely to cause economic or environmental harm or harm to human health.

land cover: Vegetation, habitat, or other material covering a land surface.

land use: The human use of land involving the management and modification of natural environment or wilderness into human-dominated environments such as fields, pastures, and settlements.

legumes: Plants belonging to the pea family that typically host symbiotic nitrogen-fixing bacteria.

life cycle assessment: A comprehensive systems approach for measuring the inputs, outputs, and potential environmental impacts of a product or service over its life cycle, including resource extraction/generation, manufacturing/production, use, and end-of-life management.

life cycle greenhouse gas emissions: The aggregate quantity of greenhouse gas emissions (including direct emissions and significant indirect emissions such as significant emissions from land use changes), as determined by the EPA Administrator, related to the full fuel life cycle, where the mass values for all greenhouse gases are adjusted to account for their relative global warming potential. (See above for definition of "biofuel life cycle.")

milling residues (primary and secondary): Wood and bark residues produced in processing (or milling) logs into lumber, plywood, paper, furniture, or other wood-based products.

mitigation: In the context of the environment, action to reduce adverse environmental impacts.

neat biofuel: See "B100."

net energy balance: In the context of biofuel, refers to the energy content in the resulting biofuel minus the total amount of energy used over the production and distribution process.

nitrogen fixation: The transformation of atmospheric nitrogen into nitrogen compounds that growing plants can use. Nitrogen-fixing species, such as soybeans, can accomplish this process directly.

nutrient loading: A process in which compounds from waste and fertilizers, such as nitrogen and phosphorus, enter a body of water. This can happen, for example, when sewage is managed poorly, when animal waste enters ground water, or when fertilizers from residential and agricultural runoff wash into a stream, river, or lake.

oxygenated fuels: Fuels, typically gasoline, that have been blended with alcohols or ethers that contain oxygen in order to reduce carbon monoxide and other emissions.

ozone: A form of oxygen consisting of three oxygen atoms. In the stratosphere (7 to 10 miles or more above the Earth's surface), ozone is a natural form of oxygen that shields the Earth from ultraviolet radiation. In the troposphere (the layer extending up 7 to 10 miles from the Earth's surface), ozone is a widespread pollutant and major component of photochemical smog.

perennial grass: A species of grass that lives more than two years and typically has low nutrient demand and diverse geographical growing range, and offers important soil and water conservation benefits.

photobioreactor: A vessel or closed-cycle recirculation system containing some sort of biological process that incorporates some type of light source. Often used to grow small phototrophic organisms such as cyanobacteria, moss plants, or algae for biodiesel production.

renewable biomass: As defined by the 2007 Energy Independence and Security Act, renewable biomass is any of the following:

- Planted crops and crop residue from agricultural land cleared before December 19, 2007, and actively managed or fallow on that date.

- Planted trees and tree residue from tree plantations cleared before December 19, 2007, and actively managed on that date.
- Animal waste material and byproducts.
- Slash and pre-commercial thinnings from non-federal forestlands that are neither old-growth nor listed as critically imperiled or rare by a State Natural Heritage program.
- Biomass cleared from the vicinity of buildings and other areas at risk of wildfire.
- Algae.
- Separated yard waste and food waste.

renewable fuel: A fuel produced from renewable biomass that is used to replace or reduce the use of fossil fuel.

Renewable Fuels Standard (RFS) program: An EPA program created under the Energy Policy Act of 2005 that established the first renewable fuel volume mandate in the United States. The original RFS program (RFS1) required 7.5 billion gallons of renewable fuel to be blended into gasoline by 2012. (See below for RFS2.)

RFS2: The Renewable Fuels Standard program as revised in response to requirements of the 2007 Energy Independence and Security Act. RFS2 increased the volume of renewable fuel required to be blended into transportation fuel to 36 billion gallons per year by 2022.

RFS2 Regulatory Impact Analysis (RIA): EPA's analysis of the impacts of the increase in production, distribution, and use of the renewable fuels need to meet the RFS2 volumes established by Congress in the 2007 Energy Independence and Security Act.

riparian forest buffer: An area of trees and shrubs adjacent to streams, lakes, ponds, and wetlands that may reduce nutrient loading, soil erosion, and pesticide contamination by removing soil particles and contaminants from overland water flow.

row crop: A crop planted in rows wide enough to allow cultivators between the rows. Examples include corn, soybeans, peanuts, potatoes, sorghum, sugar beets, sunflowers, tobacco, vegetables, and cotton.

sedimentation: Soil particles, clay, sand, or other materials settle out of a fluid suspension into the bottom of a body of water.

short-rotation woody crop (SRWC): Fast-growing tree species grown on plantations and harvested in cycles shorter than is typical of conventional wood products, generally between three and 15 years. Examples include hybrid poplars (*Populus* spp.), willow (*Salix* spp.), and eucalyptus.

soil erosion: The movement and loss of soil by the action of wind or water or a combination thereof.

soil organic matter: Decomposed plant and animal material fully incorporated into the soil.

soil quality: The capacity of a specific kind of soil to function, within natural or managed ecosystem boundaries, to sustain plant and animal productivity, maintain or enhance water and air quality, and support human health and habitation.

sugarcane bagasse: The fibrous material that remains after sugar is pressed from sugarcane.

sweet sorghum pulp: The bagasse or dry refuse left after the juice is extracted from sweet sorghum stalks during the production of ethanol and other sweet sorghum products. The pulp is usually treated as farm waste in plantations that grow sweet sorghum for biofuel production.

transesterification: In the context of biofuel, the chemical process that reacts an alcohol with triglycerides in vegetable oils and animal fats to produce biodiesel and glycerin.

turbidity: A cloudy condition in water due to suspended silt or organic matter.

vegetative reproduction: A form of asexual reproduction in plants by which new individuals arise without the production of seeds or spores. It can occur naturally or be induced by horticulturists.

water availability: In the context of this report, water availability refers to the amount of water that can be appropriated from surface water sources (e.g., rivers, streams, lakes) or ground water sources (e.g., aquifers) for consumptive uses.

water quality: Water quality is a measure of the suitability of water for a particular use based on selected physical, chemical, and biological characteristics. It is most frequently measured by characteristics of the water such as temperature, dissolved oxygen, and pollutant levels, which are compared to numeric standards and guidelines to determine if the water is suitable for a particular use.

wet milling: In the context of biofuel, a process for producing conventional corn starch ethanol in which the corn is soaked in water or dilute acid to separate the grain into its component parts (e.g., starch, protein, germ, oil, kernel fibers) before converting the starch to sugars that are then fermented to ethanol.

woody biomass: Tree biomass thinned from dense stands or cultivated from fast-growing plantations. This also includes small-diameter and low-value wood residue, such as tree limbs, tops, needles, and bark, which are often byproducts of forest management activities.

APPENDIX B. SUMMARY OF EPA STATUTORY AUTHORITIES HAVING POTENTIAL IMPACT ON THE PRODUCTION AND USE OF BIOFUELS

Table B-1. Summary of Selected Statutory Authorities Having Potential Impact on the Production and Use of Biofuels

Summary of Statute/Program	Stage of Life Cycle		
	Feedstock Production and Transport	Biofuel Production, Transport, and Storage	Use of Biofuel
Clean Air Act (CAA) (http://www.epa.gov/air/caa/)			
The CAA defines EPA's responsibilities for protecting and improving air quality and stratospheric ozone. It requires EPA to set national ambient air quality standards (NAAQS) for widespread pollutants from numerous and diverse sources considered harmful to public health and the environment. EPA and states must develop regulations to achieve and maintain the NAAQS and to control other pollutants.	Vehicles used for the transportation of feedstock may be subject to an inspection and maintenance program for tailpipe emissions and vehicle emission standards for air quality.	• A biofuel plant will need to obtain an air operating permit for day-to-day facility operations. Based on potential-to-emit, a facility may be required to obtain a Title V Air Operating Permit. Operating permits will be issued containing emission limits, monitoring, and recordkeeping requirements. • Pre-construction permits will be required for initial construction and for changes made to the plant. There are two types of major pre-construction permits under the New Source Review (NSR) Program: Prevention of Significant Deterioration permits and Nonattainment NSR permits. A minor pre-construction permit would be required if major NSR is not required. • A vehicle used to transport biofuels may be subject to an inspection and maintenance program.	The CAA regulates the amount of ethanol mixed in gasoline as part of the reformulated gasoline program.
The goal of the CWA is to restore and maintain the chemical, physical, and biological integrity of the nation's waters. Entities that discharge to waters of the United States through point sources (i.e., pipes, ditches, concentrated animal feeding operations), must obtain a National Pollutant Discharge Elimination System (NPDES) permit. These entities include many municipal, industrial, and construction-related sources of stormwater. States develop water quality standards (WQS) that define the goals for a water body by designating its uses, setting criteria to protect those uses, and establishing	Agricultural stormwater and irrigation returns flows are exempted from NPDES permit requirements. Under Section 319, EPA provides grants to states to address nonpoint sources of pollution.	A biofuel production facility typically uses water for cooling and also for washing the biofuel product to remove impurities. The wastewater is discharged either directly to a water body or indirectly to a municipal wastewater treatment plant. Both are point-source discharges, regardless whether the facility uses a septic tank or treatment prior to discharge. Any discharge into a water body by a point source must have an NPDES permit prior to discharge. Permits may be required for discharge to a municipal wastewater treatment system, which could include pre-treatment requirements. Land application of wastewater may be covered by an NPDES permit if it is determined that pollutants run off the application site to a waterway in a discernible channel or pipe. To minimize the impact of site runoff on water quality, a NPDES stormwater permit must be obtained for discharges to waters of the United States from any construction activity that disturbs 1 acre or more of land (including smaller sites that are part of a larger common plan of development).	Management of emergency response oil discharges must be reported to the National Response Center if they are in a quantity that "may be harmful."

Table B-1. (Continued)

Summary of Statute/Program	Stage of Life Cycle		
	Feedstock Production and Transport	Biofuel Production, Transport, and Storage	Use of Biofuel
provisions to protect that water body. The CWA requires states to identify waters not meeting WQS and to develop Total Maximum Daily Loads (TMDLs) for those waters. TMDLs identify point- and nonpoint-source loads that can be discharged to a water body and still meet WQS.			
CWA: Section 404 Wetlands Program (www.epa.gov/owow/wetlands/			
Section 404 addresses the discharges of dredged or fill material into waters of the United States, including wetlands. Permits are required for activities such as expanded water resource projects (including dams, impoundments, and levees) and altering or dredging a water of the United States.	Most ongoing agricultural maintenance practices are exempt from Section 404.	Generally, Section 404 requires a permit before these materials may be placed in a U.S. water, such as a wetland, stream, river, slough, lake, bay, etc., during construction activities.	
Comprehensive Environmental Response, Compensation and Liability Act (CERCLA) (http://www.epa.gov/lawsregs/laws/			
CERCLA provides a federal "Superfund" to clean up uncontrolled or abandoned hazardous waste sites as well as accidents, spills, and other emergency releases of pollutants and contaminants into the environment. Through CERCLA, EPA was given authority to ensure responsible parties' cooperation in site cleanup. CERCLA also regulates the property transfer of these sites.		Requirements under CERCLA that may apply include: • Reporting requirements for hazardous substances. • Implementation and periodic revision of the National Contingency Plan. • Management by emergency response authorities and responses to discharges of biofuels.	
Emergency Planning and Community Right-to-Know ACT (EPCRA) (http://www.epa.gov/oecaagct/lcra.html)			
The objective of the EPCRA is to: (1) allow state and local planning for chemical emergencies, (2) provide for notification of emergency releases of chemicals, and (3) address communities' right- to-know about toxic and hazardous chemicals.		Section 302 requires any facility with regulated chemicals (extremely hazardous substances) above threshold planning quantities to notify the state emergency response commission (SERC) and the local emergency planning committee (LEPC). Section 304 requires the facility to report a release of an extremely hazardous substance. Section 311 requires the facility to have	Electric utilities are subject to EPCRA Section 313, "Toxics Release Inventory Reporting."

Summary of Statute/Program	Stage of Life Cycle		
	Feedstock Production and Transport	Biofuel Production, Transport, and Storage	Use of Biofuel
		material safety data sheets (MSDSs) on site for hazardous chemicals, as defined by the Occupational Safety and Health Act, that exceed certain quantities and to submit copies to its SERC, LEPC, and local fire department. Section 312 establishes reporting for any hazardous chemical or extremely hazardous chemical that is stored at a facility in excess of the designated threshold planning quantity. These reports are also known as the Tier II hazardous chemical inventory form. Section 313 requires owners or operators of certain facilities that manufacture, process, or otherwise use any listed toxic chemicals, or chemical categories, in excess of threshold quantities to report annually to EPA and to the state in which such facilities are located.	
Federal Insecticide, Fungicide, and Rodenticide Act (FIFRA) (http://www.epa.gov/oecaagct/lfra.html)			
The objective of FIFRA is to provide federal control of pesticide distribution, sale, and use.	EPA reviews and registers pesticides for specified uses and can cancel the registration if information shows continued use would pose unreasonable risk. Consideration is given to worker exposure ecological exposure and food-chain imports.		
Hazardous Materials Transportation Act (Regulations Codified 49 CFR) (http://www.phmsa.dot.gov/hazmat/regs and http://www.fmcsa.dot.gov/safety-security/hazmat/security-plan-guide.htm)			
The Department of Transportation regulations require procedures to be put in place ensuring the safe transport of hazardous materials. Also, regulation HM-232 requires companies to complete a written security assessment and to develop a security plan based on the assessment.		Requirements are in place for shippers and carriers of hazardous materials to prepare shipments for transport, placard containers for easy identification of hazards, and ensure the safe loading, unloading, and transport of materials. HM-232 requires companies to complete a written security assessment and to develop a security plan based on the assessment.	

Table B-1. (Continued)

Summary of Statute/Program	Stage of Life Cycle		
	Feedstock Production and Transport	Biofuel Production, Transport, and Storage	Use of Biofuel
National Environmental Policy Act (NEPA) (http://www.epa.gov/compliance/			
NEPA requires federal agencies to integrate environmental values into their decision-making processes by considering the environmental impacts of their proposed actions and reasonable alternatives to those actions. To meet NEPA requirements in certain circumstances, federal agencies prepare a detailed statement known as an Environmental Impact Statement (EIS).		If federal money is being used to partially or entirely finance the construction of a biofuel plant or any associated facility, such as an access road or water supply facility, then construction of the plant may be subject to NEPA. NEPA requires federal agencies to incorporate environmental considerations in their planning and decision-making and to prepare a detailed statement assessing the environmental impact of activities and alternatives that significantly affect the environment.	
Oil Pollution Act (OPA) of 1990 (http://www.epa.gov/lawsregs/laws/			
The OPA of 1990 streamlined and strengthened EPA's ability to prevent and respond to catastrophic oil spills. A trust fund financed by a tax on oil is available to clean up spills when the responsible party is incapable or unwilling to do so. The OPA requires oil storage facilities and vessels to submit to the federal government plans detailing how they will respond to large discharges.		Provides that the responsible party for a vessel or facility from which oil is discharged, or which poses a substantial threat of a discharge, is liable for: (1) certain specified damages resulting from the discharged oil; and (2) removal costs incurred in a manner consistent with the National Contingency Plan. Provides for spill contingency plans and mandates development of response plans for worst case discharge; provides requirements for spill removal equipment. Oil Spill Plans must be in place before operation at facilities that could spill oil to navigable waters.	

Summary of Statute/Program	Stage of Life Cycle		
	Feedstock Production and Transport	**Biofuel Production, Transport, and Storage**	**Use of Biofuel**
Renewable Fuel Standard (RFS) (http://www.epa.gov/otaq/fuels/renewablefuels/index.htm)			
The RFS program was created under the Energy Policy Act (EPAct) of 2005, and established the first renewable fuel volume mandate in the United States. As required under EPAct, the original RFS program (RFS1) required 7.5 billion gallons of renewable fuel to be blended into gasoline by 2012. Under the Energy Independence and Security Act (EISA) of 2007, the RFS program was expanded. EISA also required EPA to apply life cycle greenhouse gas (GHG) performance threshold standards. The GHG requirement is that the life cycle GHG emissions of a qualifying renewable fuel must be less than the life cycle GHG emissions of the 2005 baseline average gasoline or diesel fuel that it replaces. Four different levels of reductions are required for the four different renewable fuel standards: Renewable Fuel (20 percent); Advanced Biofuel (50 percent); Biomass-based Diesel (50 percent); and Cellulosic Biofuel (60 percent).		If a facility produces 10,000 gallons or more of renewable fuel per year, it may participate in the RFS program, though it is not required to do so. A facility that chooses to participate in the RFS program must satisfy the following criteria: • Register • Generate renewable identification • Transfer renewable identification numbers with fuel • Provide product transfer documents • Follow blending requirements • Follow exporting requirements • Follow non-road use of fuel • Attest engagements • Keep records for five years • Report quarterly	
Resource Conservation and Recovery Act (RCRA) (http://www.epa.gov/lawsregs/laws/			
RCRA gives EPA the authority to control hazardous waste generation, transportation, treatment, storage, and disposal of hazardous waste. Any facility that handles hazardous waste must obtain an operating permit from the state agency or EPA. RCRA regulates USTs.		Regulatory issues related to solid and hazardous waste generated by biofuel production include: • New regulations on storage and transport of fuel related to expanded use of biofuels. • New concerns related to assessing compatibility of fuel storage systems, managing water in storage tanks, protecting against corrosiveness and conductivity, managing methane formation, and detecting, preventing and responding to storage tank and pipe leaks and spills. • Management of emergency response authorities and responses to biofuel spills.	UST leak detection and prevention are required.

Table B-1. (Continued)

Summary of Statute/Program	Stage of Life Cycle		
	Feedstock Production and Transport	Biofuel Production, Transport, and Storage	Use of Biofuel
Safe Drinking Water Act (SDWA) (http://www.epa.gov/ogwdw/sdwa/)			
SDWA is the federal law that protects the safety of water distributed by public water systems. Under SDWA, EPA has National Primary Drinking Water Regulations for more than 90 contaminants and rules regarding monitoring of treated drinking water as well as reporting and public notification.	There are a number of threats to drinking water: anthropogenic chemicals including pesticides and improperly disposed chemicals, animal wastes, and naturally occurring substances. A primary impact to drinking water is nitrate pollution from row crops.	Wastewater from biofuel production facilities or corn starch ethanol facilities and leaking biofuel storage tanks can contaminate surface and ground drinking water resources, requiring treatment under SDWA.	
Safe Drinking Water Act: Underground Injection Control (UIC) Program (http://www.epa.gov/safewater/uic/)			
The UIC program protects underground sources of drinking water by regulating the construction, operation, permitting, and closure of injection wells that place fluids underground for storage or disposal.	Agriculture drainage wells are Class V UIC wells. They are primarily regulated under state law.	A biofuels plant is subject to the requirements of the UIC Program if any of the following apply: • It is disposing of stormwater, cooling water, or industrial or other fluids into the subsurface via an injection well. • It has an onsite sanitary waste disposal system (e.g., aseptic system) that serves or has the capacity to serve 20 or more persons. • It has an onsite sanitary waste disposal system that is receiving other than a solely sanitary waste stream regardless of its capacity. • It is undergoing a remediation process where fluids are being introduced into the subsurface via an injection well to facilitate or enhance the cleanup.	
Spill Prevention, Control and Countermeasure (SPCC) and Facility Response Plans (FRP) (http://www.epa.gov/oem/content/spcc/index.htm)			
The SPCC rule includes requirements for oil spill prevention, preparedness, and response to prevent oil discharges to navigable waters and adjoining shorelines. The rule requires specific facilities to prepare, amend, and implement SPCC Plans. The SPCC rule is part of the Oil Pollution Prevention regulation, which also includes the FRP rule.	The SPCC program requires certain farms (e.g., those that store oil and could reasonably be expected to discharge oil to waters of the United States) to prepare and implement SPCC Plans.	A biofuel facility is subject to this regulation if the following apply: • It is non-transportation-related. • It has a total above-ground oil storage capacity greater than 1,320 gallons or a completely buried oil storage capacity greater than 42,000 gallons. • There is a reasonable expectation of an oil discharge into or upon navigable waters of the United States or adjoining shorelines. • Secondary containment cannot be provided for all regulated oil storage tanks.	

Summary of Statute/Program	Stage of Life Cycle		
	Feedstock Production and Transport	Biofuel Production, Transport, and Storage	Use of Biofuel
Toxic Substances Control Act (TSCA) (http://www.epa.gov/lawsregs/laws/			
TSCA gives EPA broad authority to identify and control chemical substances that may pose a threat to human health or the environment. EPA's Office of Pollution Prevention and Toxics operates both the New Chemicals Program and the Biotechnology Program under Section 5 of TSCA. Both programs were established to help manage the potential risk from chemical substances and genetically engineered (intergeneric) microorganisms new to the marketplace or applied in significant new uses. Additional sections of TSCA give EPA the broad authority to issue toxicity testing orders or to regulate the use of any existing chemicals that pose unreasonable risk.	Notification and review of new intergeneric genetically engineered microbes (e.g., bacteria, fungi and algae) used to produce biofuel feedstocks.	Mandatory notification and approval for new chemicals and new biological products, before manufacture and commercial use. New uses of chemicals are subject to review for potential environmental hazards under the Significant New Use Notification process. As a result of the review process, health and environmental effects testing of existing or new chemicals that pose unreasonable risk may be required. EPA may also restrict use and handling of chemicals or biological products as a result of their review.	

APPENDIX C. CONCEPTUAL MODELS

As described in this report, the activities associated with cultivation of biofuel feedstocks and their conversion to fuel result in a complex set of in ter-related environmental impacts. Conceptual models provide a useful tool to describe, understand, and communicate the complex pathways by which these activities lead to impacts. As noted in Section 7, EPA anticipates developing and using conceptual models as an important tool for the assessment in its next Report to Congress.

The conceptual models presented in this appendix lay a foundation for this future effort. Figures C-1 to C-7 present conceptual models for feedstock cultivation and harvest. Figures C-8 and C-9 present models for biofuel production and distribution. (Note that models are not included for end use of biofuel.) These early renditions graphically present the environmental effects most commonly identified in current peer-reviewed literature; while comprehensive, they do not attempt to include all possible effects.

TERMS AND ABBREVIATIONS USED IN THE CONCEPTUAL MODELS

From the Legend

- *biotic response*: Response of living parts of terrestrial or aquatic ecosystems, in terms of either number of species or numbers of individuals of a particular species
- *ecosystem service*: Direct or indirect contribution of the environment to human well-being
- *environmental parameter*: A measureable attribute of the environment

From the Diagrams

- *aquatic life use support*: A beneficial use designation in which the water body provides suitable habitat for survival and reproduction of desirable fish, shellfish, and other aquatic organisms (this is a synthetic quality made up of many different environmental parameters)
- *BOD*: biological oxygen demand
- *contamination*: Release of nutrients or pesticides used in feedstock production to waterways or bodies of water
- *PM*: particulate matter
- *T & E species*: threatened and endangered species
- *VOC*: volatile organic compound

Feedstock Production

Figures C-1 to C-7 present seven models for six feedstocks covered in this report: corn starch, soybeans, corn stover, perennial grass, woody biomass (short-rotation woody crops and forest thinning/residue removal), and algae production.

Different pathways are introduced at the tops of several of these feedstock models. These pathways were selected because (1) they will likely be pursued in combination in order to grow enough feedstock to meet RFS2 2022 biofuel requirements (see Section 2 for a description of requirements) and (2) they result in different environmental impacts.

Arrows in the impact boxes (below the initial row of activities) depict whether the impacts are negative or positive. The number(s) by each arrow designate the pathway to which the arrows refer. A few pathways can have both negative and positive impacts (e.g., corn starch cultivation could result in increased or decreased use of ground and surface water). Dotted borders denote impacts that have a relatively large degree of uncertainty due to a lack of information. Dotted boxes without arrows depict highly uncertain impacts that nonetheless are described in the literature.

Fuel Production and Distribution

Figures C-8 and C-9 present conceptual models for production and distribution of the two biofuels covered in this report: ethanol and biodiesel.

Ethanol Production

Figure C-8 shows the activities and impacts associated with production and distribution of ethanol from both starch (i.e., corn grain) and cellulosic feedstocks, including corn stover, perennial grasses, and woody biomass. A single model is provided for these four types of feedstocks because their impacts and associated uncertainty are largely similar, with a few exceptions (e.g., water use will likely be slightly higher for cellulose conversion).

As depicted in the upper left of Figure C-8, conversion of starch to ethanol consists of several sequential steps, including milling, hydrolysis, and fermentation. There currently are two distinct alternatives for converting cellulosic feedstock into ethanol: (1) biochemical conversion (which is preceded by a catalysis step to separate cellulose and hemicellulose from their tightly bound state with lignin), and (2) thermochemical conversion. These alternatives involve slightly different chemical processes and byproducts. As with Figures C-1 to C-7, a dotted border is used to denote impacts with relatively large uncertainty due to a lack of information.

Biodiesel Production

Figure C-9 shows the activities and impacts associated with production of biodiesel from soybeans and algae. Several techniques may be used to convert plant oils into biodiesel, including hydrogenation, catalytic cracking, and transesterification. All these processes produce biodiesel, with glycerin as a byproduct.

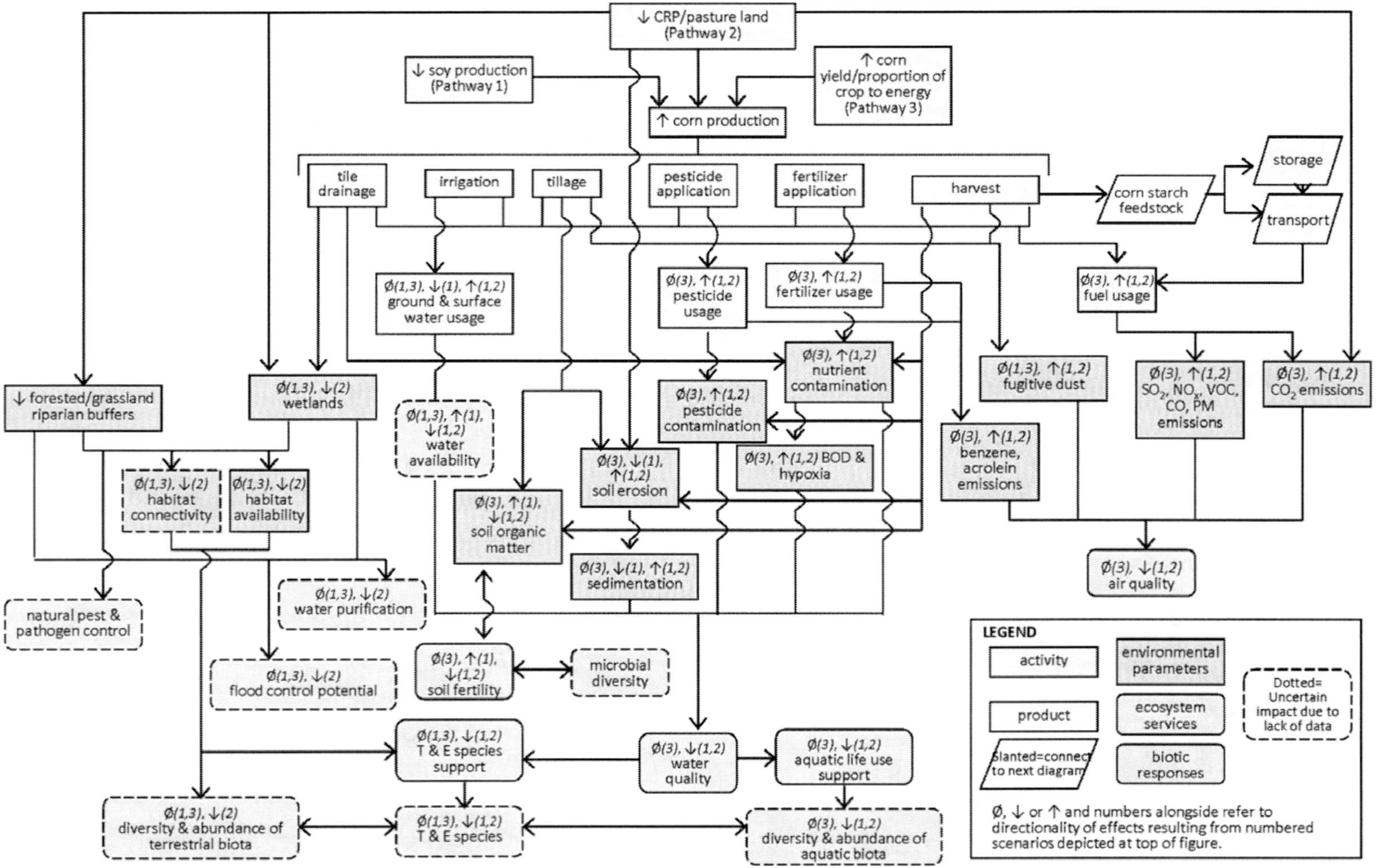

Figure C-1. Pathways for Potential Environmental Impacts of Corn Starch Feedstock Cultivation.

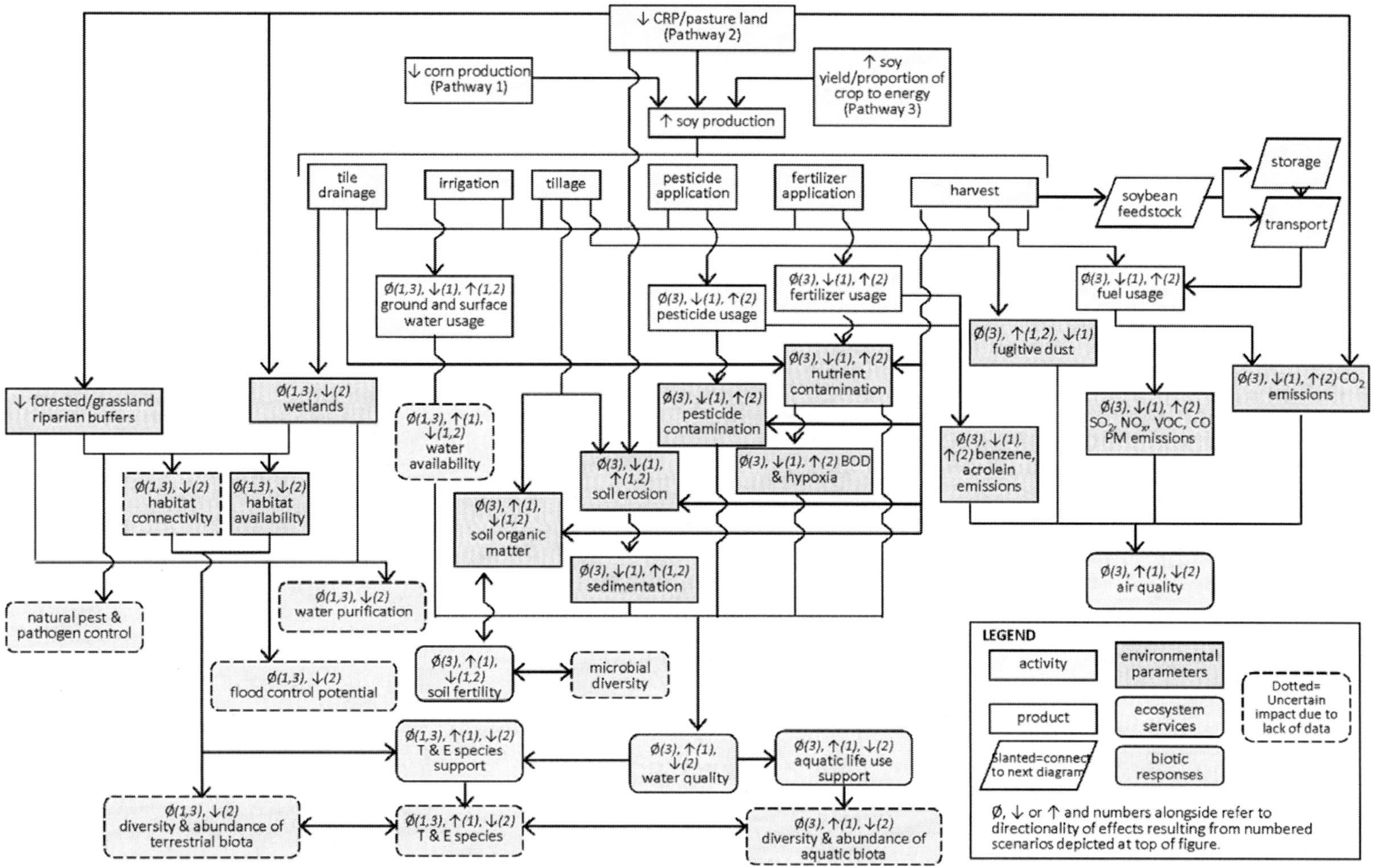

Figure C-2. Pathways for Potential Environmental Impacts of Soybean Feedstock Cultivation.

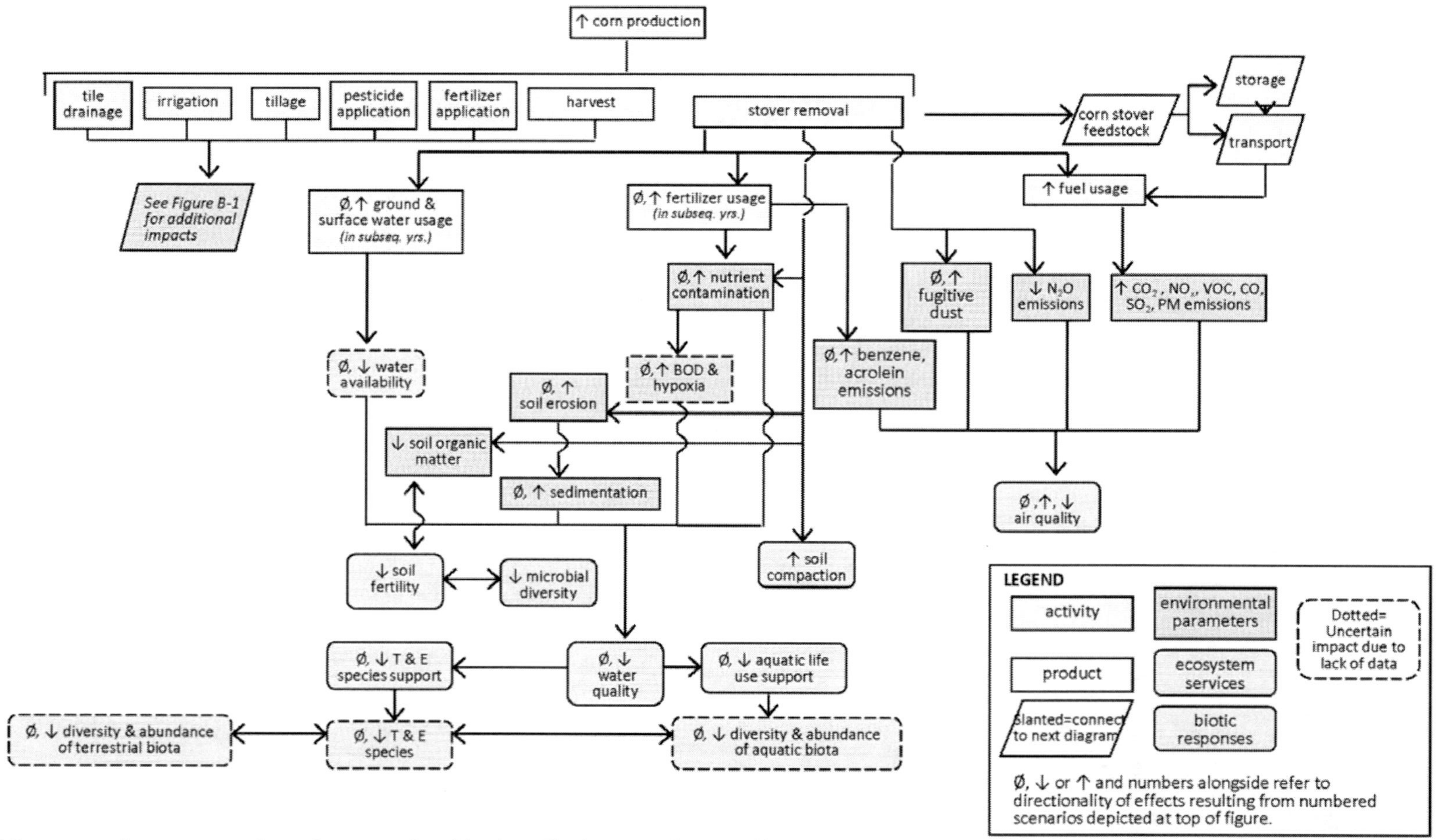

*Corn stover is a waste product of corn starch cultivation. The impacts of corn cultivation are shown in Figure C-1. Figure C-3 highlights the environmental impacts of stover removal *above and beyond* those impacts attributable to corn grain production.

Figure C-3. Pathways for Potential Environmental Impacts of Corn Stover Feedstock Cultivation.*

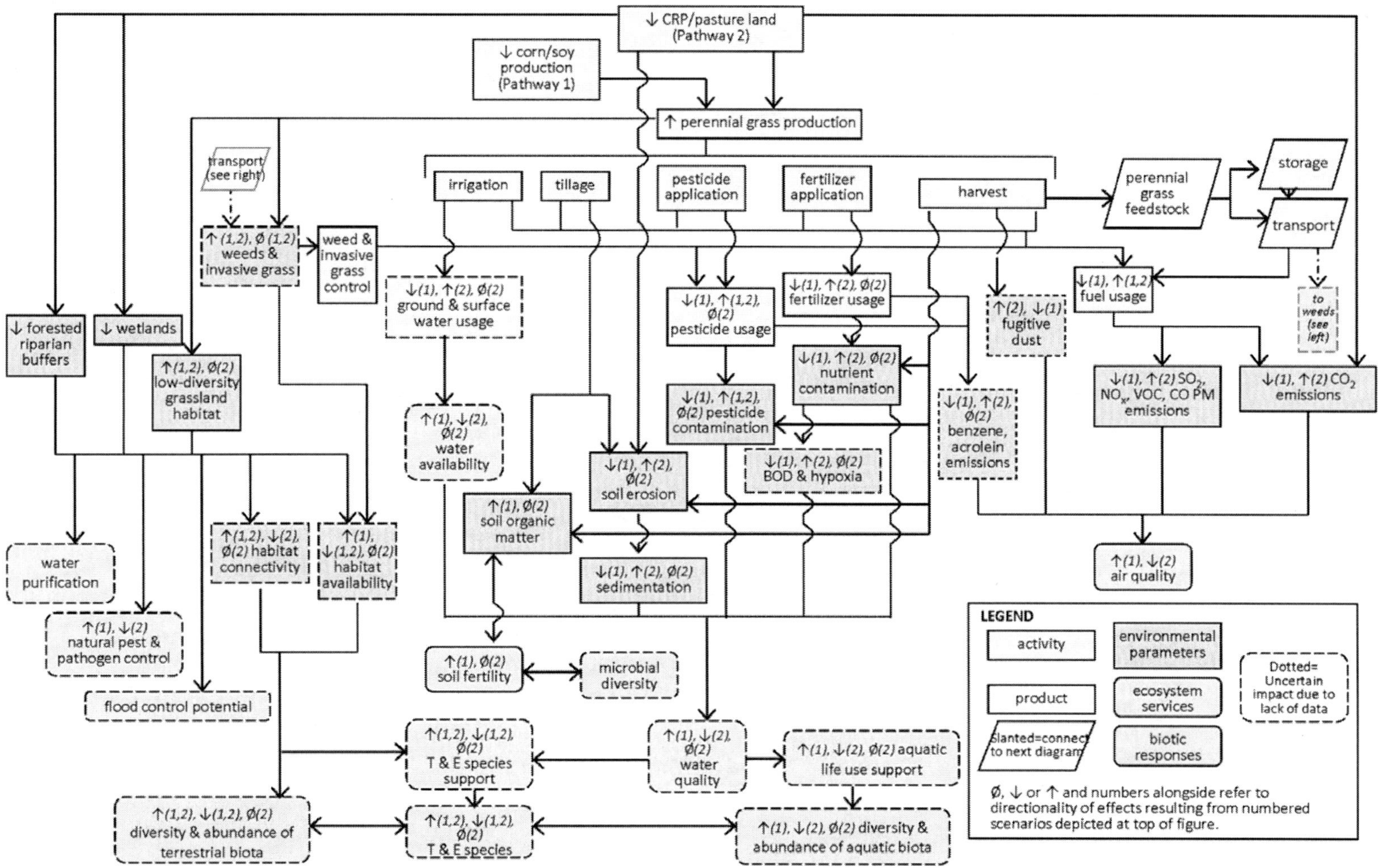

Figure C-4. Pathways for Potential Environmental Impacts of Perennial Grass Feedstock Cultivation.

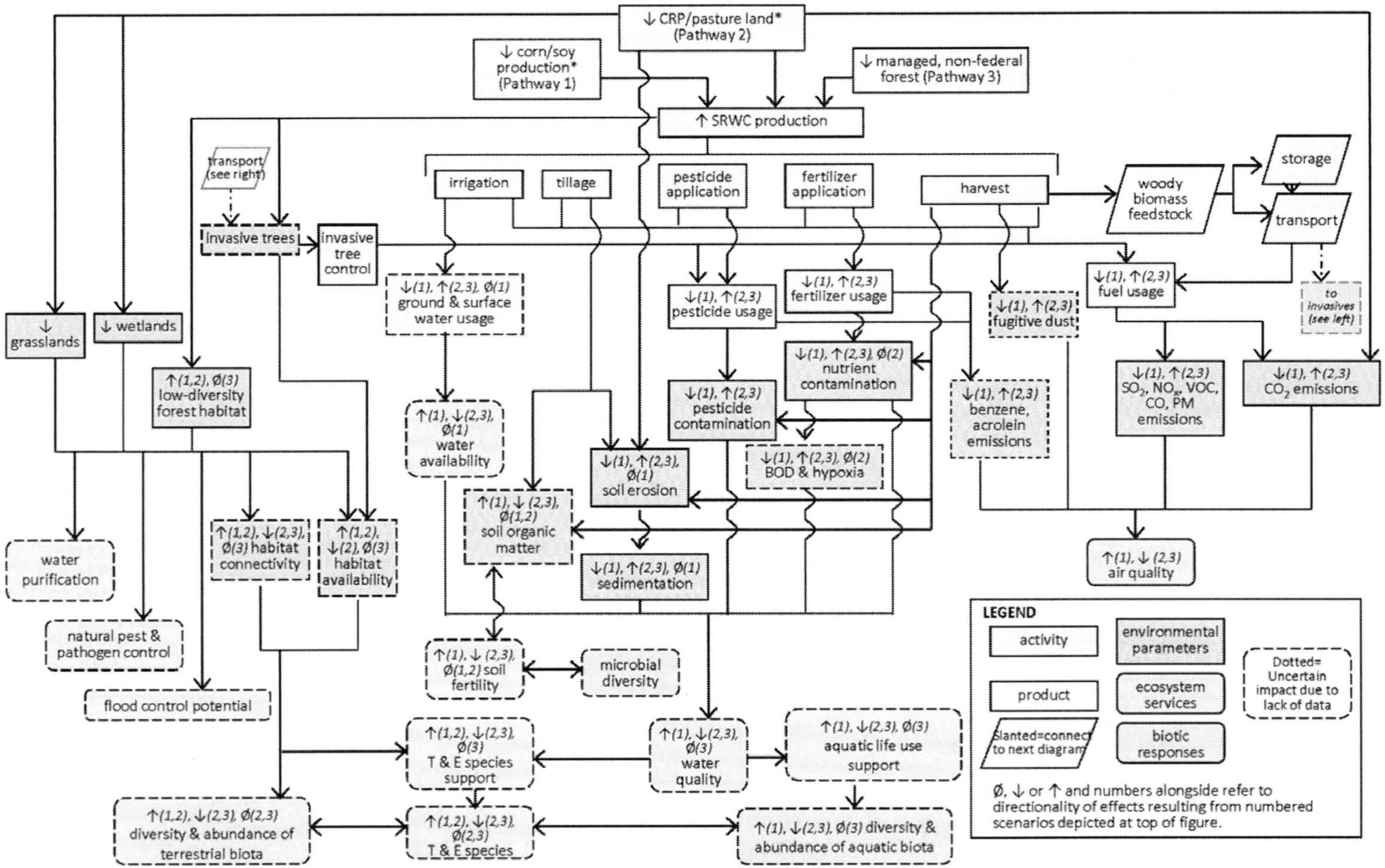

* These particular land use changes may not currently be allowable under RFS2.

Figure C-5. Pathways for Potential Environmental Impacts of Short-Rotation Woody Crop Feedstock Cultivation.

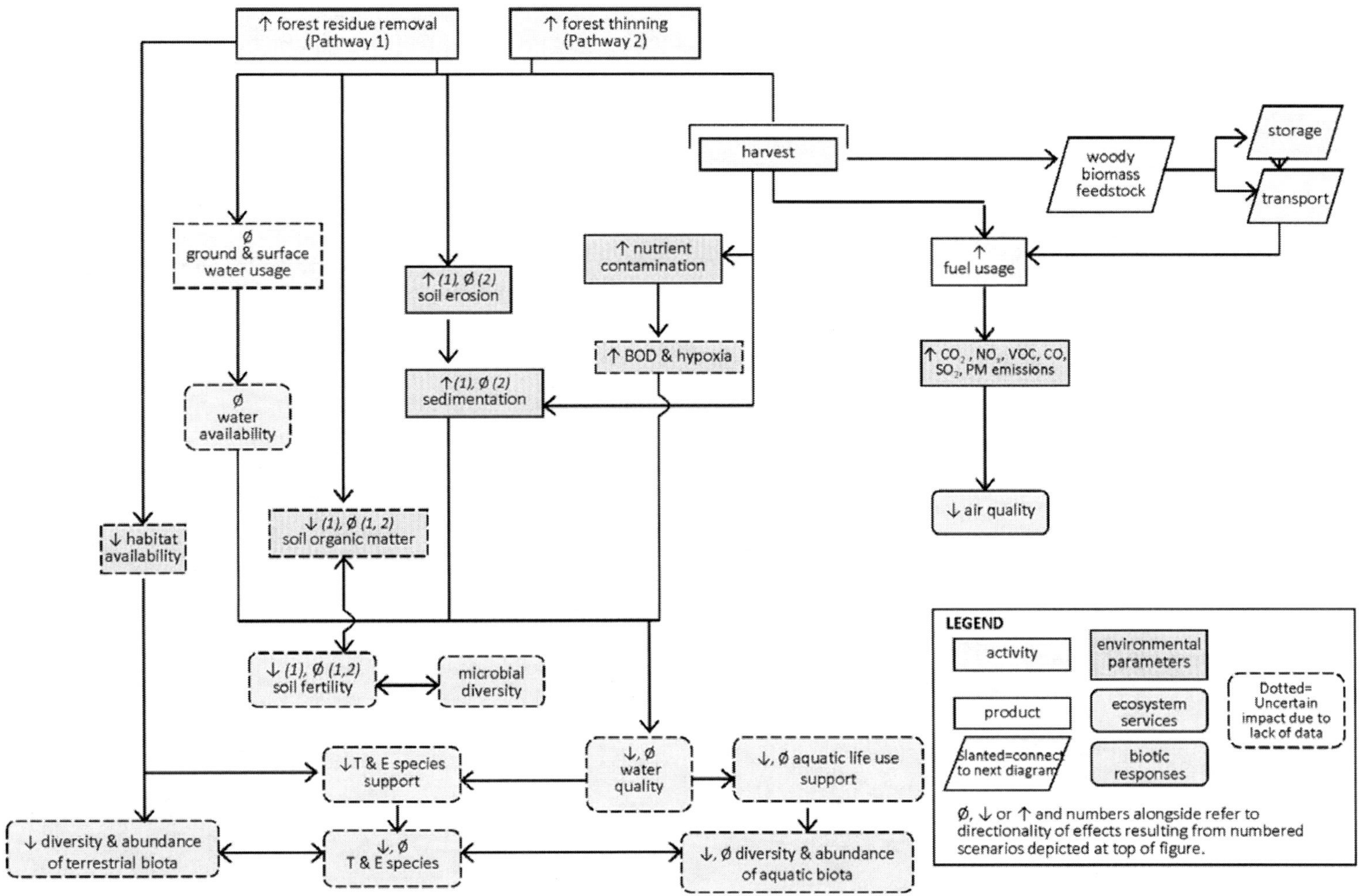

Figure C-6. Pathways for Potential Environmental Impacts of Forest Thinning and Residue Removal.

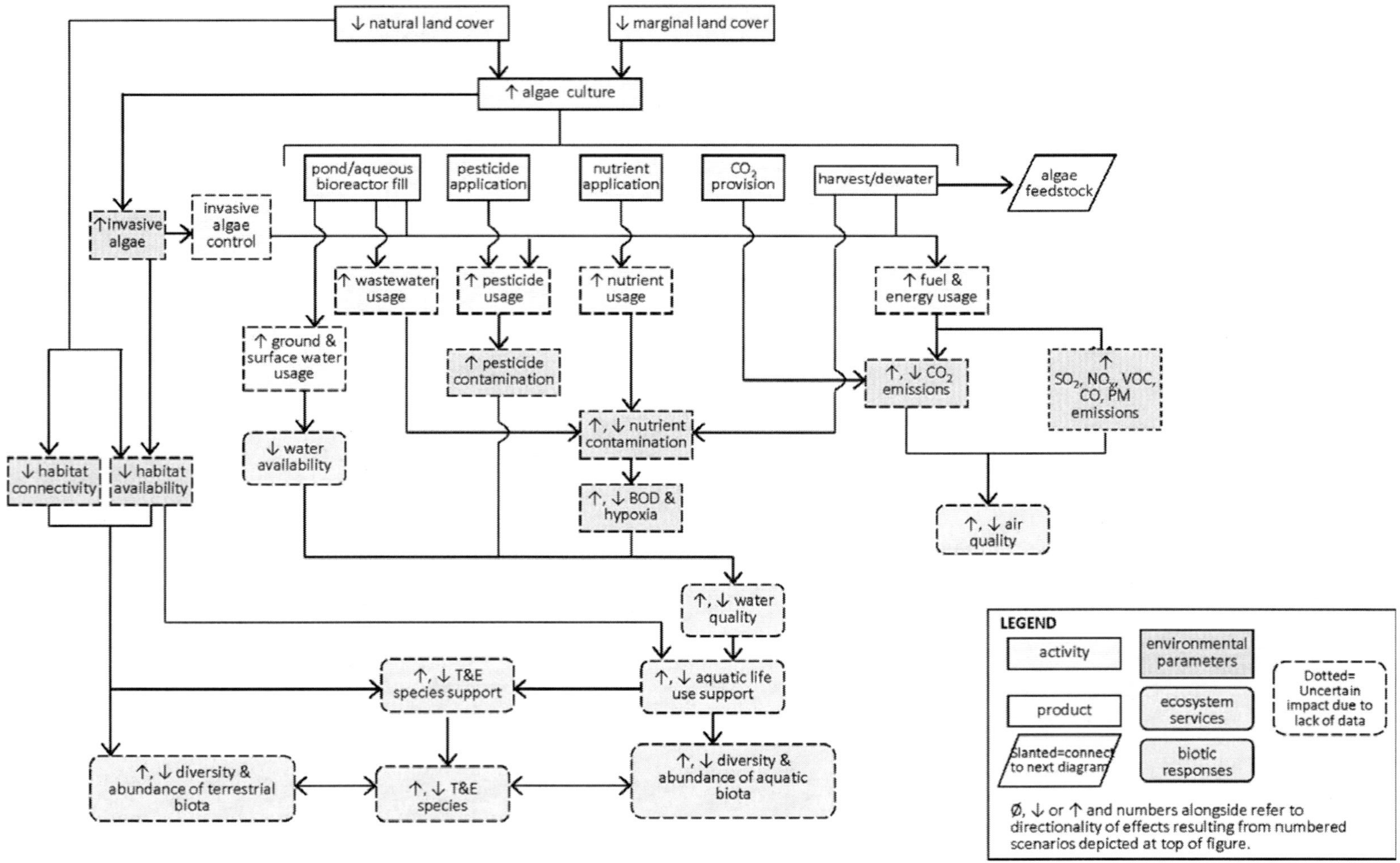

Figure C-7. Potential Environmental Impacts of Algae Feedstock Production.

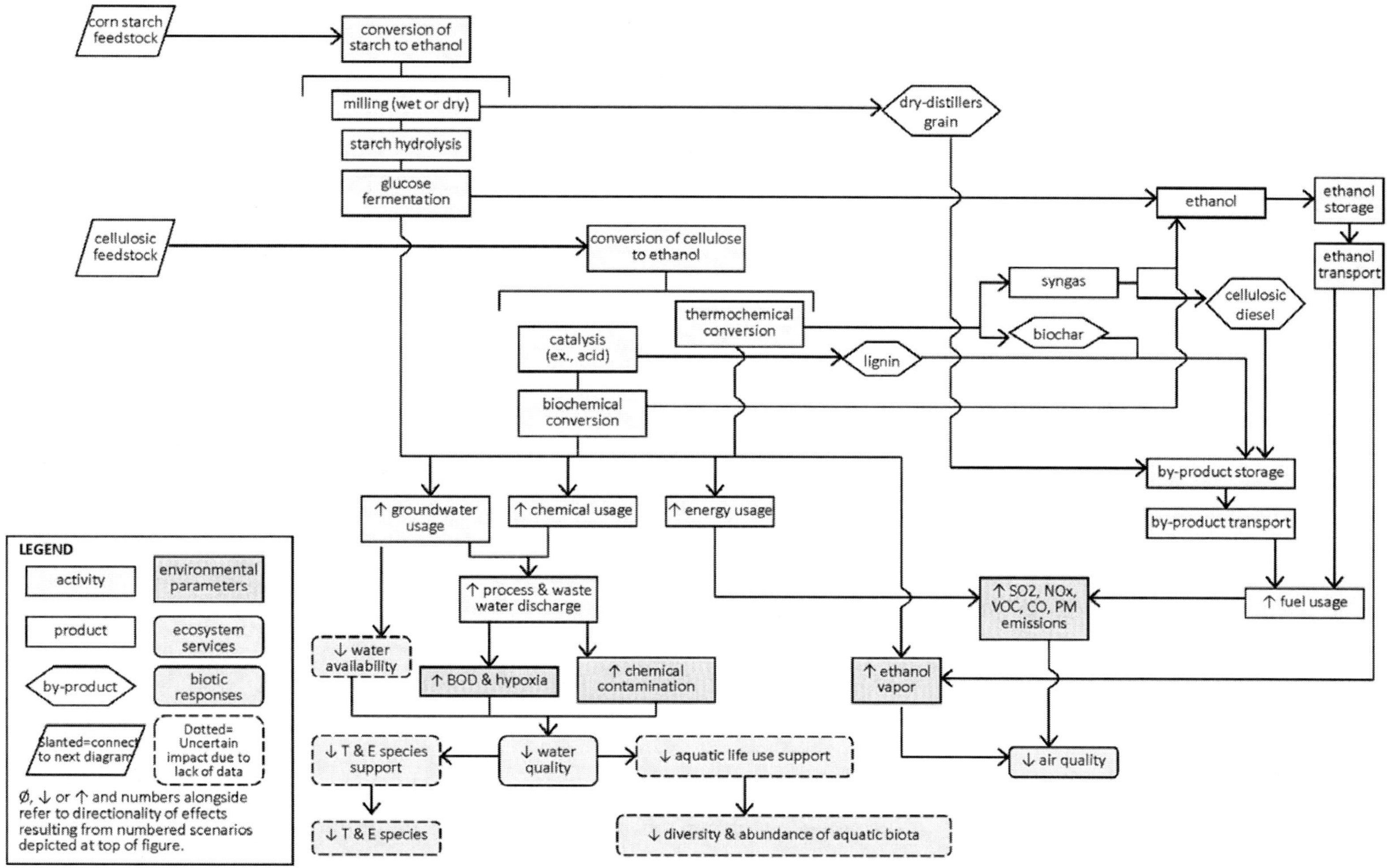

Figure C-8. Potential Environmental Impacts of Producing and Distributing Conventional and Cellulosic Ethanol (Impacts of Fuel Use Not Included).

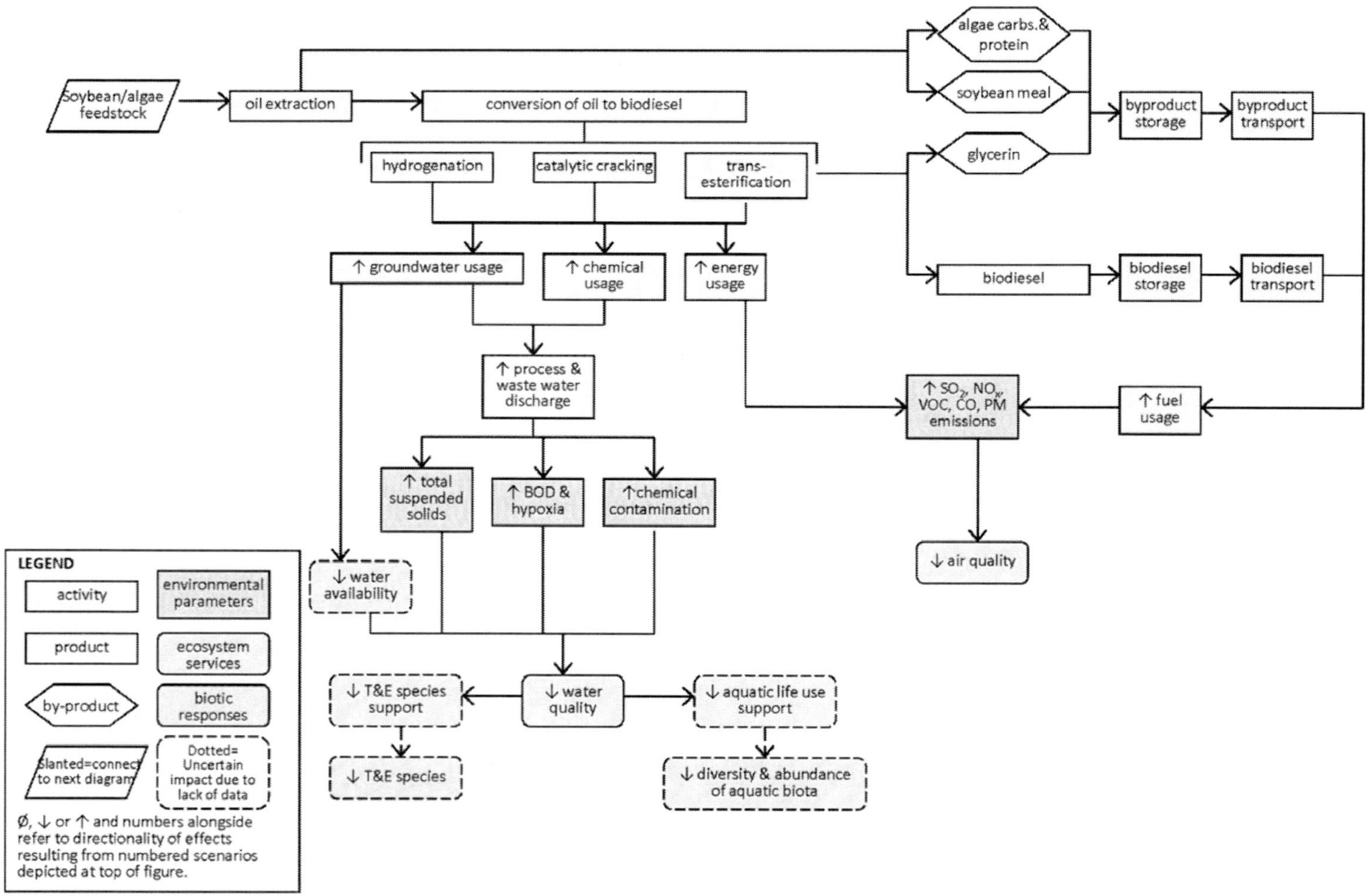

Figure C-9. Potential Environmental Impacts of Producing and Distributing Biodiesel (Impacts of Fuel Use Not Included).

References

Adegbidi, HG; Volk, TA; White, EH; Abrahamson, LP; Briggs, RD; Bickelhaupt, DH. 2001. Biomass and nutrient removal by willow clones in experimental bioenergy plantations in New York State. *Biomass and Bioenergy*, 20(6), 399-411.

Adler, PR; Sanderson, MA; Weimer, PJ; Vogel, KP. 2009. Plant species composition and biofuel yields of conservation grasslands. *Ecological Applications*, 19, 2202-2209.

Advanced Biofuels Task Force. 2008. Advanced Biofuels Task Force report. *Commonwealth of Massachusetts*. Available at: http://www.mass.gov/Eoeea/docs/eea/biofuels/biofuels_complete.pdf.

Adviento-Borbe, MAA; Haddix, ML; Binder, DL; Walters, DT; Dobermann, A. 2007. Soil greenhouse gas fluxes and global warming potential in four high-yielding maize systems. *Global Change Biology*, 13(9), 1972-1988.

Alexander, RB; Smith, RA; Schwarz, GE; Boyer, EW; Nolan, JV; Brakebill, JW. 2008. Differences in phosphorus and nitrogen delivery to the Gulf of Mexico from the Mississippi River Basin. *Environmental Science and Technology*, 42(3), 822-830.

Alley, WM; Healy, RW; LaBaugh, JW; Reilly, TE. 2002. Flow and storage in ground water systems. *Science*, 296, 1985-1990.

Alternative Fuels and Advanced Vehicles Data Center. 2010. E10 and other low-level ethanol blends. U.S. Department of Energy. Available at, http://www.afdc.energy.gov/afdc/ethanol/blends_e10.html.

Anderson-Teixeira, KJ; Davis, SC; Masters, MD; Delucia, EH. 2009. Changes in soil organic carbon under biofuel crops. Global Change Biology *Bioenergy*, 1(1), 75-96.

Andow, DA; Zwalen, C. 2006. Assessing environmental risks of transgenic plants. *Ecology Letters*, 9, 196-214.

Andrews, SS. 2006. Crop Residue Removal for Biomass Energy Production: Effects on Soils and Recommendations. USDA-Natural Resource Conservation Service. Available at: http://soils.usda.gov/sqi/management/files/agforum_residue_white_paper.pdf

Anex, R; Lifset, R. 2009. Post script to the corn ethanol debate, Reaching consensus. *Journal of Industrial Ecology*, 13(6), 996-998.

Angermeier, PL; Karr, JR. 1984. Relationships between woody debris and fish habitat in a small warmwater stream. *Transaction of the American Fisheries Society*, 113(6), 716-726.

Antizar-Ladislao, B; Turrion-Gomez, J. 2008. Second-generation biofuels and local bioenergy systems. *Biofuels, Bioproducts and Biorefining*, 2, 455-469.

Aqua Terra Consultants. 2010. Phase III/IV modeling water quality impacts of corn production for ethanol in the upper Mississippi River basin. U.S. Environmental Protection Agency. EPA-HQ-OAR-2005-0161-2978. Available at: http://www.regulations.gov/#!documentDetail;D=EPA-HQ-OAR-2005-0161-2978.

Aronsson, PG; Bergstrom, LF; Elowson, SNE. 2000. Long-term influence of intensively cultured short-rotation willow coppice on nitrogen concentrations in ground water. *Journal of Environmental Management*, 58(2), 135-145.

ARS (United States Department of Agriculture, Agricultural Research Service). 2003. *Agricultural phosphorus and eutrophication*. Second edition. ARS-149.

ARS (United States Department of Agriculture, Agricultural Research Service). 2006.

Accomplishment report (2002–2006). *National Program 207, Integrated Agricultural Systems*. Available at, http://www.ars.usda.gov/research/programs/programs.htm?np_code=207&docid=17659.

ARS (United States Department of Agriculture, Agricultural Research Service). n.d. FY-2005 annual report Manure and Byproduct Utilization National Program 206. Available at, http://www.ars.usda.gov/research/programs/programs.htm?np_code=206&docid=13337.

Artuzi, JP; Contiero, RL. 2006. Herbicidas aplicados na soja e produtividade do milho em sucessão. *Pesquisa Agropecuária Brasileira*, 41, 1119-1123.

Auer, C. 2008. Ecological risk assessment and regulation for genetically-modified ornamental plants. *Critical Reviews in Plant Sciences*, 27, 255-271.

Aust, WM; Blinn, CR. 2004. Forestry Best Management Practices for timber harvesting and site preparation in the eastern United States, An overview of water quality and productivity research during the past 20 years (1982-2002). *Water, Air, and Soil Pollution, Focus*, 4(1), 5-36.

Aust, WM; Lea, R; Gregory, JD. 1991. Removal of floodwater sediments by a clearcut tupelo-cypress wetland. *Journal of the American Water Resources Association*, 27(1), 111-116.

Baeumler, R; Zech, W. 1998. Soil solution chemistry and impact of forest thinning in mountain forests in the Bavarian Alps. *Forest Ecology and Management*, 108(3), 231-238.

Baker, A; Zahniser, S. 2006. Ethanol Reshapes the Corn Market. USDA ERS. *Amber Waves*. Available at, http://www.ers.usda.gov/AmberWaves/April06/Features/Ethanol.htm.

Baker, JM; Ochsner, TE; Venterea, RT; Griffis, TJ. 2007. Tillage and soil carbon sequestration—what do we really know? *Agriculture Ecosystems and Environment*, 118(1-4), 1-5.

Baliga, R; Powers, S. 2010. Sustainable algae biodiesel production in cold climates. *International Journal of Chemical Engineering*, 2010. Article ID 270532. Available at: http://downloads.hindawi.com/journals/ijce/2010/102179.pdf.

Barney, JN; DiTomaso, JM. 2008. Nonnative species and bioenergy: Are we cultivating the next invader? *Bioscience*, 58(1), 1-7.

Bartolino, JR; Cunningham, WL. 2003. *Ground-water depletion across the nation*. USGS Fact Sheet 103-03. U.S. Geological Survey.

Bates, SL; Zhao, JZ; Roush, RT; Shelton, AM. 2005. Insect resistance management in GM crops, past, present and future. *Nat Biotechnol*, 23(1), 57-62.

Bäumler, R; Zech, W. 1999. Effects of forest thinning on the streamwater chemistry of two forest watersheds in the Bavarian Alps. *Forest Ecology and Management*, 116(1-3), 119 128.

Beach, RH; McCarl, BA. 2010. U.S. Agricultural and Forestry Impacts of the Energy Independence and Security Act: FASOM Results and Model Description. RTI. Project number 0210826.003.

Beale, CV; Long, SP. 1997. Seasonal dynamics of nutrient accumulation and partitioning in the perennial C-4-grasses Miscanthus x giganteus and Spartina cynosuroides. *Biomass and Bioenergy*, 12(6), 419-428.

Beale, CV; Morison, JIL; Long, SP. 1999. Water use efficiency of C4 perennial grasses in a temperate climate. *Agricultural and Forest Meteorology*, 96, 103-115.

Beaty, ER; Engel, JL; Powell, JD. 1978. Tiller development and growth in switchgrass. *Journal of Range Management*, 31(5), 361-365.

Bell, GP. 1997. *Ecology and management of Arundo donax, and approaches to riparian habitat restoration in Southern California*. In: Brock, JH; Wade, M; Pysek, P; Green, D (eds). Plant invasions: Studies from North America and Europe. Backhuys Publishers. 103-113.

Bellamy, PE; Croxton, PJ; Heard, MS; Hinsley, SA; Hulmes, L; Hulmes, S; Nuttall, P; Pywell, RF; Rothery, P. 2009. The impact of growing Miscanthus for biomass on farmland bird populations. *Biomass and Bioenergy*, 33, 191-199.

Berndes, G. 2002. Bioenergy and water—the implications of large-scale bioenergy production for water use and supply. *Global Environmental Change*, 12, 253-271.

Bianchi, FJJA; Booij, CJH; Tscharntke, T. 2006. Sustainable pest regulation in agricultural landscapes, A review on landscape composition, biodiversity and natural pest control. *Proceedings of the Royal Society B*, 273, 1715-1727.

Binkley, D; Brown, TC. 1993. Forest practices as nonpoint sources of pollution in North America. *Water Resources Bulletin*, 29(5), 729-740.

Biomass Research and Development Institute (BRDI). 2008. *Increasing Feedstock Production for Biofuels: Economic Drivers, Environmental Implications, and the Role of Research.* Available at: http://www.usbiomassboard.gov/pdfs/increasing_feedstock_ revised.pdf.

Blanco-Canqui, H. 2010. Energy crops and their implications on soil and environment. *Agronomy Journal*, 102(2), 403-419.

Blanco-Canqui, H; Lal, R. 2008. No-tillage and soil-profile carbon sequestration, An on-farm assessment. *Soil Science Society of America Journal*, 72(3), 693-701.

Blanco-Canqui, H; Lal, R. 2009a. Corn stover removal for expanded uses reduces soil fertility and structural stability. *Soil Science Society of America Journal*, 73(2), 418-426.

Blanco-Canqui, H; Lal, R. 2009b. Crop residue removal impacts on soil productivity and environmental quality. *Critical Reviews in Plant Sciences*, 28(3), 139-163.

Blanco-Canqui, H; Gantzer, CJ; Anderson, SH; Alberts, EE. 2004. Grass barriers for reduced concentrated flow induced soil and nutrient loss. *Soil Science Society of America Journal*, 68(6), 1963-1972.

Blann, KL; Anderson, JL; Sands, GR; Vondracek, B. 2009. Effects of agricultural drainage on aquatic ecosystems: A review. *Critical Reviews in Environmental Science and Technology*, 38(11), 909-1001.

Bracmort, K; Gorte RW. 2010. Biomass: Comparison of definitions in legislation. Congressional Research Service, R40529.

Bradshaw, JD; Prasifka, JR; Steffey, KL; Gray, ME. 2010. First report of field populations of two potential aphid pests of the bioenergy crop Miscanthus giganteus. *The Florida Entomologist*, 93, 135-137.

Brady, SJ. 2007. *Effects of cropland conservation practices on fish and wildlife habitat*. In: Fish and Wildlife Response to Farm Bill Conservation Practices. J.B. Haufler, ed. The Wildlife Society, Technical Review 07-1, 9-23.

Bransby, DI; McLaughlin, SB; Parrish, DJ. 1998. A review of carbon and nitrogen balances in switchgrass grown for energy. *Biomass and Bioenergy* 14(4), 379-384.

Breitburg, DL; Hondorp, DW; Davias, LA; Diaz, RJ. 2009. Hypoxia, Nitrogen, and Fisheries: Integrating Effects Across Local and Global Landscapes. *Annual Review of Marine Science*, 1, 329-349.

Brinmer, TA; Gallivan, GJ; Stephenson, GR. 2005. Influence of herbicide-resistant canola on the environmental impact of week management. *Pest Management Science*, 61, 47-52.

Brookes, G; Barfoot P. 2006. Global impact of biotech crops: Socio-economic and environmental effects in the first ten years of commercial use. *AgBioForum*, 9(3), 139–151.

Brookes, G; Barfoot P. 2008. Global impact of biotech crops: Socio-economic and environmental effects, 1996-2006. *AgBioForum*, 11(1), 21-38.

Brookes, G; Barfoot, P. 2009. Global impact of biotech crops: Income and production effects, 1996-2007. *AgBioForum*, 12(2), 184-208.

Brookes, G; Barfoot, P. 2010. Global impact of biotech crops: Environmental effects, 1996 2008. *AgBioForum*, 13(1), 76-94.

Brooks, JP; Adeli, A; Read, JJ; McLaughlIn: MR. 2009. Rainfall simulation in greenhouse microcosms to assess bacterial-associated runoff from land-applied poultry litter. *Journal of Environmental Quality*, 38(1), 218-229.

Brown, AE; Zhang, L; McMahon, TA; Western, AW; Vertessy, RA. 2005 A review of paired catchment studies for determining changes in water yield resulting from alterations in vegetation. *Journal of Hydrology*, 310, 28-61.

Buddenhagen, CE; Chimera, C; Clifford, P. 2009. Assessing biofuel crop invasiveness: A case study. *PLoS ONE*, 4(4), e5261.

Bullard, MJ; Christian, DG; Knight, JD; Lainsbury, MA; Parker, SR (eds.). 2001. Biomass and energy crops II, University of York, York, UK. 18-21 December 2001. *Aspects of Applied Biology*, 65, 1-374.

Bustamante, MMC; Melillo, J; Connor, DJ; Hardy, Y; LambIn: E; Lotze-Campen, H; Ravindranath, NH; Searchinger, T; Tschirley, J; Watson, H. 2009. What are the final land limits. In: Howarth, RW; Bringezu, S (eds.). Biofuels: Environmental consequences and interactions with changing land use. *Proceedings of the Scientific Committee on Problems of the Environment (SCOPE) International Biofuels Project Rapid Assessment*. pp. 271-291. Available at: http://cip.cornell.edu/DPubS?service=UI&version=1.0&verb= Display & handle=scope.

Campbell, JE; Lobell, DB; Genova, RC; Field, CB. 2008. The global potential of bioenergy on abandoned agriculture lands. *Environmental Science and Technology*, 42, 5791-5794.

Cardinale, BJ; Srivastava, DS; Duffy, JE; Wright, JP; Downing, AL; Sankaran, M; Jouseau, C. 2006. Effects of biodiversity on the functioning of trophic groups and ecosystems. *Nature*, 443(7114), 989-992.

Cardinale, BJ; Wright, JP; Cadotte, MW; Carroll, IT; Hector, A; Srivastava, DS; Loreau, M; Weis, JJ. 2007. Impacts of plant diversity on biomass production increase through time because of species complementarity. Proceedings of the National Academy of Sciences of the United States of America, 104, 18123-18128.

Carpenter, SR; Caraco, NF; Correll, DL; Howarth, RW; Sharpley, AN; Smith, VH. 1998. Nonpoint pollution of surface waters with phosphorus and nitrogen. *Ecological Applications*, 8(3), 559-568.

Carroll, A; Somerville, C. 2009. Cellulosic biofuels. *Annual Review of Plant Biology*, 60, 165-182.

Cassel, DK; Raczkowski, CW; Denton, HP. 1995. Tillage effects on corn production and soil physical conditions. *Soil Science Society of America Journal*, 59(5), 1436-1443.

CENR (Committee on Environment and Natural Resources). 2010. *Scientific assessment of hypoxia in U.S. coastal waters*. Interagency Working Group on Harmful Algal Blooms, Hypoxia, and Human Health of the Joint Subcommittee on Ocean Science and Technology. Available at: http://www.whitehouse.gov/sites/default/files/microsites/ostp/hypoxia-report.pdf.

ChapotIn: SM; Wolt, JD. 2007. Genetically modified crops for the bioeconomy, meeting public and regulatory expectations. *Transgenic Research*, 16, 675-688.

Chen, W; Wei, X. 2008. Assessing the relations between aquatic habitat indicators and forest harvesting at watershed scale in the interior of British Columbia. *Forest Ecology and Management*, 256(1-2), 152-160.

Chester, M; MartIn: E. 2009. Cellulosic ethanol from municipal solid waste, A case study of the economic, energy, and greenhouse gas impacts in California. *Environmental Science and Technology*, 43(14), 5183-5189.

Chester, M; PlevIn: R; Rajagopal, D; Kammen, D. 2007. *Biopower and waste conversion technologies for Santa Barbara County, California*. Report for the Community Environmental Council. Available at: http://www.ce.berkeley.edu/~chester/library/santa_barba_biopower_potential.pdf.

Chisti, Y. 2007. Biodiesel from microalgae. *Biotechnology Advances*, 25, 294-306.

Chiu, Y-W; Walseth, B; Suh, S. 2009. Water embodied in bioethanol in the United States. *Environmental Science and Technology*, 43(8), 2688-2692.

Chizinski, CJ; Vondracek, B; Blinn, CR; Newman, RM; Atuke, DM; Fredricks, K; Hemstad, NA; Merten, E; Schlesser, N. 2010. The influence of partial timber harvesting in riparian buffers on macroinvertebrate and fish communities in small streams in Minnesota, USA. *Forest Ecology and Management*, 259, 1946-1958.

Christian, DG; Riche, AB; Yates, NE. 2001. The yield, composition and production costs for seven varieties of switchgrass and one panic grass grown as a biofuel. *Aspects of Applied Biology*, 65, 199-204.

Christian, DP; Hoffman, W; Hanowski, JM; Niemi, GJ; Beyea, J. 1998. Bird and mammal diversity on woody biomass plantations in North America. *Biomass and Bioenergy*, 14, 395-402.

Clarens, AF; Resurreccion, EP; White, MA; Colosi, LM. 2010. Environmental life cycle comparison of algae to other bioenergy feedstocks. *Environmental Science and Technology*, 44(5), 1813-1819.

Clifton-Brown, JC; Breuer, J; Jones, MB. 2007. Carbon mitigation by the energy crop, Miscanthus. *Global Change Biology*, 13(11), 2296-2307.

Conner, AJ; Glare, TR; Nap, JP. 2003. The release of genetically modified crops into the environment. Part II. Overview of ecological risk assessment. *The Plant Journal*, (33), 19-46.

Corseuil, HX; Hunt, CS; Ferreira dos Santos, RC; Alvarez, PJJ. 1998. The influence of the gasoline oxygenate ethanol on aerobic and anaerobic BTX degradation. *Water Resources Research*, 33(7), 2056–2072.

Craig, W; Tepfer, M; Degrassi, G; Ripandelli, D. 2008. An overview of general features of risk assessments of genetically modified crops. *Euphytica*, 164(3), 853-880.

Cram, DS; Baker, TT; Fernald, AG; Madrid, A; Rummer, B. 2007. Mechanical thinning impacts on runoff, infiltration, and sediment yield following fuel reduction treatments in southwestern dry mixed conifer forest. *Journal of Soil and Water Conservation*, 62(5), 359-366.

Crosti, R; Cascone C; Cipollaro S. 2010. Use of a weed risk assessment for the Mediterranean region of Central Italy to prevent loss of functionality and biodiversity in agro ecosystems. *Biological Invasions*, 12(6), 1387-3547.

Cruse, RM; Herndl, CG. 2009. Balancing corn stover harvest for biofuels with soil and water conservation. *Journal of Soil and Water Conservation*, 64, 286-291.

Daehler, CC; Carino, DA. 2000. Predicting invasive plants, Prospects for general screening system based on current regional models. *Biological Invasions*, 2, 93-102.

Dahl, T.E. 2000. Status and trends of wetlands in the conterminous United States 1986 to 1997. U.S. Department of the Interior, Fish and Wildlife Service, Washington, D.C. 82 Available at: http://library.fws.gov/Pubs9/wetlands86-97_lowres.pdf.

Dale, VH; Kline, KL; Wiens, J; Fargione, J. 2010. Biofuels: Implications for land use and biodiversity. Ecological Society of America. Available at: http://esa.org/biofuelsreports/files/ESA%20Biofuels% 20Report_VH%20Dale%20et%20a l.pdf.

Darnall, DW; Greene, B; Hosea, H; McPherson, RA; Henzl, M. 1986. Recovery of heavy metals by immobilized algae. In: Thompson, R (ed). Trace metal removal from aqueous solution. Special Publication No. 61. *Royal Society of Chemistry*. 1-24.

Da Silva, MLB; Alvarez, PJJ. 2002. Effects of ethanol versus MTBE on benzene, toluene, ethylbenzene, and xylene natural attenuation in aquifer columns. *Journal of Environmental Engineering*, 128, 862-867.

Das, P; Kim, K-H; Sa, J-H; Bae, WS; Kim, J-C; Jeon, EC. 2008. Emissions of ammonia and nitric oxide from an agricultural site following application of different synthetic fertilizers and manures. *Geosciences Journal*, 12(2), 177-190.

Das, MK; Taliaferro, CM. 2009. Genetic variability and interrelationships of seed yield and yield components in switchgrass. *Euphytica*, 167(1), 95-105.

Davidson, EA; Ackerman, IL. 1993. Changes in soil carbon inventories following cultivation of previously untilled soils. *Biogeochemistry*, 20(3), 161-193.

Davis, JM. 2007. How to assess the risks of nanotechnology, Learning from past experience. *Journal of Nanoscience and Nanotechnology*, 7, 402-409.

Davis, JM; Thomas, VM. 2006. Systematic approach to evaluating trade-offs among fuel options: The lessons of MTBE. *Annals of the New York Academy of Sciences*, 1076, 498-515.

Davis, SC; Anderson-Teixeira, KJ, DeLucia, EH. 2009. Life-cycle analysis and the ecology of biofuels. *Trends in Plant Science*, 14(3).

Davis, SC; Parton, WJ; Dohleman, FG; Smith, CM; Del Grosso, S; Kent, AD; Delucia, EH. 2010. Comparative biogeochemical cycles of bioenergy crops reveal nitrogen-fixation and low greenhouse gas emissions in a Miscanthus x giganteus agro-ecosystem. *Ecosystems*, 13(1), 144-156.

Day, JD; Edwards, AP; Rodgers, GA. 1991. Development of an industrial-scale process for the heterotrophic production of a micro-algal mollusc feed. *Bioresource Technology*, 38(2-3), 245-249.

de la Torre Ugarte, DG; Walsch, ME; Shapouri, H; Slinsky, SP. 2003. *The economic impacts of bioenergy crop production on U.S. agriculture*. Agricultural Economic Report Number

816. United States Department of Agriculture. Available at: http://www.usda.gov/oce/reports/energy/AER816Bi.pdf.

Demura, T; Ye, ZH. 2010. Regulation of plant biomass production. *Current Opinion in Plant Biology*, 13(3), 299-304.

Dhondt, AA; Wrege, PH; Cerretani, J; Sydenstricker, KV. 2007. Avian species richness and reproduction in short rotation coppice habitats in central and western New York. *Bird Study*, 54, 12-22.

Dinnes, DL; Karlen, DL; Jaynes, DB; Kaspar, TC; Hatfield, JL; Colvin, TS; Cambardella, CA. 2002. Nitrogen management strategies to reduce nitrate leaching in tile-drained midwestern soils. *Agronomy Journal*, 94(1), 153-171.

Dismukes, GC; Carrieri, D; Bennette, N; Ananyev, GM; Posewitz, MC. 2008. Aquatic phototrophs, efficient alternatives to land-based crops for biofuels. *Current Opinion in Biotechnology*, 19, 235–240.

DiTomaso, JM; Reaser, JK; Dionigi, CP; Doering, OC; Chilton, E; Schardt, JD; Barney, JN. 2010. Biofuel vs. bioinvasion, Seeding policy priorities. *Environmental Science and Technology*, 44, 6906-6910.

Dohleman, FG; Long, SP. 2009. More productive than maize in the Midwest, How does Miscanthus do it? *Plant Physiology*, 150(4), 2104-2115.

Dominguez-Faus, R; Powers, S; Burken, J; Alvarez, P. 2009. The water footprint of biofuels, A drink or drive issue? Environmental Science and Technology 43(9), 3005-3010.

Donner, SD; Kucharik, CJ. 2008. Corn-based ethanol production compromises goal of reducing nitrogen export by Mississippi River. *Proceedings of the National Academy of Sciences*, 105, 4513-4518. Available at: http://www.pnas.org/content/105/11/4513.full.pdf.

Donner, SD; Kucharik, CJ; Foley, JA. 2004. Impact of changing land use practices on nitrate export by the Mississippi River. *Global Biogeochemical Cycles*, 18, GB1028.

Dosskey, MG. 2001. Toward quantifying water pollution abatement in response to installing buffers on crop land. *Environmental Management*, 28(5), 577-598.

Drinkwater, LE; Wagoner, P; Sarrantonio, M. 1998. Legume-based cropping systems have reduced carbon and nitrogen losses. *Nature*, 396(6708), 262-265.

Duncan, M; Lippiatt, BC; Haq, Z; Wang, M; Conway, RK. 2008. Metrics to support informed decision making for consumers of biobased products. USDA/OCE Agriculture Information Bulletin 803. Available at: http://www.usda.gov/oce/energy/Metrics 1.pdf.

Edeso, JM; Merino, A; Gonzalez, MJ; Marauri, P. 1999. Soil erosion under different harvesting managements in steep forestlands from northern Spain. *Land Degradation and Development*, 10(1), 79-88.

Eghball, B; Gilley, JE; Kramer, LA; Moorman, TB. 2000. Narrow grass hedge effects on phosphorus and nitrogen in runoff following manure and fertilizer application. *Journal of Soil and Water Conservation*, 55, 172-176.

Ekvall, T. 2005. SETAC summaries. *Journal of Cleaner Production*, 13, 1351-1358.

Ellstrand, NC; Prentice, HC; Hancock, JF. 1999. Gene flow and introgression from domesticated plants into their wild relatives. *Annual Review of Ecology and Systematics*, 30, 539-563.

ERS (United States Department of Agriculture, Economic Research Service). 2008. *2008 farm bill side-by-side*. Available at: http://www.ers.usda.gov/FarmBill/2008/Titles/TitleIIConservation.htm#conservation.

ERS (United States Department of Agriculture, Economic Research Service). 2010a. NEW feed grains data: Yearbook tables: U.S. exports of ethyl alcohol by selected destinations. Available at: http://www.ers.usda.gov/Data/feedgrains/Table.asp?t=32.

ERS (United States Department of Agriculture, Economic Research Service). 2010b. Feed grains data yearbook tables: Table 1. Corn, sorghum, barley, and oats: Planted acreage, harvested acreage, production, yield, and farm price. Available at: http://www.ers.usda.gov/data/feedgrains/Table.asp?t=01.

ERS (United States Department of Agriculture, Economic Research Service). 2010c. Feed grains data yearbook tables: Table 4. Corn: Supply and disappearance. Available at: http://www.ers.usda.gov/data/feedgrains/Table.asp?t=4.

ERS (United States Department of Agriculture, Economic Research Service). 2010d. Feed grains data yearbook tables: Table 31. Corn, Food, seed, and industrial use. Available at: http://www.ers.usda.gov/data/feedgrains/Table.asp?t=31.

Evans, JM; Cohen, MJ. 2009. Regional water resource implications of bioethanol production in the southeastern United States. *Global Change Biology*, 15, 2261-2273.

Ewers, RM; Didham, RK. 2006. Confounding factors in the detection of species responses to habitat fragmentation. *Biological Reviews*, 81, 117-142.

Fagan, WF; Cantrell, RS; Cosner, C. 1999. How habitat edges change species interactions. *American Naturalist*, 153, 165-182.

FAPRI (Food and Agricultural Policy Research Institute). 2010a. *U.S. and world agricultural outlook*. Available at: http://www.fapri.iastate.edu/outlook/2010/.

FAPRI (Food and Agricultural Policy Research Institute). 2010b. *World biofuels*. In: U.S. and world agricultural outlook. Available at: http://www.fapri.iastate.edu/outlook/2010/text/15Biofuels.pdf.

Fargione, J; Hill, J; Tilman, D; Polasky, S; Hawthorne, P. 2008. Land clearing and the biofuel carbon debt. *Science*, 319, 1235-1238.

Fargione, JE; Cooper, TR; Flaspohler, DJ; Hill, J; Lehman, C; McCoy, T; McLeod, S; Nelson, EJ; Oberhauser, KS; Tilman, D. 2009. Bioenergy and wildlife: Threats and opportunities for grassland conservation. *BioScience*, 59(9), 767-777.

Faria, MS; Ré, A; Malcato, J; Silva, PCLD; Pestana, J; Agra, AR; Nogueira, AJA; Soares, AMVM. 2006. Biological and functional responses of *in situ* bioassays with *Chironomus riparius* larvae to assess river water quality and contamination. *Science of the Total Environment*, 371, 125-137.

Farinelli, B; Carter, CA; LIn: C-YC; Sumner, DA. 2009. Import demand for Brazilian ethanol: A cross-country analysis. *Journal of Cleaner Production*, 17, S9-S17.

Farrell, AE; PlevIn: RJ; Turner, BT; Jones, AD; O'Hare, M; Kammen, DM. 2006. Ethanol can contribute to energy and environmental goals. *Science*, 311(5760), 506-508.

Federer, CA; Hornbeck, JW; Tritton, LM; MartIn: CW; Pierce, RS; Smith, CT. 1989. Long-term depletion of calcium and other nutrients in Eastern US forests. *Environmental Management*, 13(5), 593-601.

Fenn, ME; Poth, MA; Aber, JD; Baron, JS; Bormann, BT; Johnson, DW; Lemly, AD; McNulty, SG; Ryan, DE; Stottlemyer, R. 1998. Nitrogen excess in North American ecosystems; Predisposing factors, ecosystem responses, and management strategies. *Ecological Applications*, 8(3), 706-733.

FIFRA Scientific Advisory Board. 2007. A set of scientific issues being considered by the Environmental Protection Agency regarding: the potential for atrazine to affect amphibian gonadal development. SAP minutes 2007-08. Available at http://www.epa.gov/oscpmont/sap/meetings/2007/october/finalminutes.pdf.

Filippi-Codaccioni, O; Devictor, V; Bas, Y; Julliard, R. 2010. Toward more concern for specialization and less for species diversity in conserving farmland biodiversity. *Biological Conservation*, 143, 1493-1500.

Firbank, L. 2007. Assessing the ecological impacts of bioenergy projects. Bioenergy Research doi: 10.1007/s12155-007-9000-8. Available at: https://files.pbworks.com/download/HjG9dZBkTr/npnet/12639072/Firbank%20%282008%29%20Assessing%20ecological%20impacts%20bioenergy%20projects.pdf.

Fletcher Jr., RJ; Robertson, BA; Evans, J; Doran, PJ; Alavalapati, JRR; Schemske, DW. 2010. Biodiversity conservation in the era of biofuels: risks and opportunities. *Frontiers in Ecology and the Environment*, 9(3), 161-168.

Flynn, KJ; Greenwell, HC; Lovitt, RW; Shields, RJ. 2010. Selection for fitness at the individual or population level: Modeling effects of genetic modification in microalgae on productivity and environmental safety. *Journal of Theoretical Biology*, 263(3), 269-280.

Follett, RF; Varvel, GE; Kimble, JM; Vogel, KP. 2009. No-till corn after bromegrass: Effect on soil carbon and soil aggregates. *Agronomy Journal*, 101(2), 261-268.

Fornara, DA; Tilman, D. 2009. Ecological mechanisms associated with the positive diversity-productivity relationship in an N-limited grassland. *Ecology*, 90(2), 408-418.

Foust, TD; Aden, A; Dutta, A; Phillips, S. 2009. An economic and environmental comparison of a biochemical and a thermochemical lignocellulosic ethanol conversion processes. *Cellulose*, 16(4), 547-565.

Franklln: JF; Spies, TA; Van Pelt, R; Carey, AB; Thornburgh, DA; Berg, DR; Lindenmayer, DA; Harmon, ME; Keeton, WS; Shaw, DC; Bible, K; Chen, J. 2002. Disturbances and structural development of natural forest ecosystems with silvicultural implications, using Douglas-fir forests as an example. *Forest Ecology and Management*, 155, 399-423.

Franzluebbers, AJ; Wright, SF; Stuedemann, JA. 2000. Soil aggregation and glomalin under pastures in the Southern Piedmont USA. *Soil Science Society of America Journal*, 64(3), 1018-1026.

Fridley, JD; Stachowicz, JJ; Naeem, S; Sax, DF; Seabloom, EW; Smith, MD; Stohlgren, TJ; Tilman, D; Von Holle, B. 2007. The invasion paradox: Reconciling pattern and process in species invasions. *Ecology*, 88(1), 3-17.

FSA (United States Department of Agriculture, Farm Service Agency). 2009. *Conservation Reserve Program: Summary and enrollment statistics*, FY 2008. Available at: http://www. fsa.usda.gov/ Internet/FSA_File/annualsummary2008.pdf.

FSA (United State Department of Agriculture, Farm Service Agency). 2010. *Conservation Reserve Program: Summary and enrollment statistics, FY2009*. Available at: http://www.fsa.usda.gov/Internet/FSA_File/fyannual2009.pdf

Gaffney, JS; Marley, NA. 2009. The impacts of combustion emissions on air quality and climate—from coal to biofuels and beyond. *Atmospheric Environment*, 43, 23-36.

GAO (Government Accountability Office). 2007. Biofuels: DOE lacks a strategic approach to coordinate increasing production with infrastructure development and vehicle needs. GAO-07-713. Available at: http://www.gao.gov/new.items/d07713.pdf.

Gardiner, MA; Tuell, JK; Isaacs, R; Gibbs, J; Ascher, JS; Landis, DA. 2010. Implications of three biofuel crops for beneficial arthropods in agricultural landscapes. *Bioenergy Research*, 3, 6-19.

Garrett, KA; Dendy, SP; Power, AG; Blaisdell, GK; Alexander, HM; McCarron, JK. 2004. Barley yellow dwarf disease in natural populations of dominant tallgrass prairie species in Kansas. *Plant Disease*, 88, 574.

Garten, CT; Wullschleger, SD. 2000. Soil carbon dynamics beneath switchgrass as indicated by stable isotope analysis. *Journal of Environmental Quality*, 29(2), 645-653.

George, RR; Farris, AL; Schwartz, CC; Humburg, DD; Coffey, JC. 1979. Native prairie grass pastures as nest cover for upland birds. *Wildlife Society Bulletin*, 7, 4-9.

Georgescu, M; Lobell, DB; Field, CB. 2009. Potential impact of US biofuels on regional climate. *Geophysical Research Letters*, 36(L21806).

Gerbens-Leenes, PW; Hoekstra, AY; van der Meer, T. 2009. The water footprint from biomass: A quantitative assessment and consequences of an increasing share of bio-energy in energy supply. *Ecological Economics*, 68, 1052-1060.

Ginnebaugh, D; Liang, J; Jacobson, M. 2010. Examining the temperature dependence of ethanol (E85) versus gasoline emissions on air pollution with a largely-explicit chemical mechanism. *Atmospheric Environment*, 44, 1192-1199.

Glaser, JA; Matten, SR. 2003. Sustainability of insect resistance management strategies for transgenic Bt corn. *Biotechnol Adv*, 22(1-2), 45-69.

Gleason, RA; Euliss Jr., NH; Hubbard, DE; Duffy, WG. 2003. Effects of sediment load on emergence of aquatic invertebrates and plants from wetland soil egg and seed banks. *Wetlands*, 23(1), 26-34.

Gnansounou, E; Dauriat, A; Villegas, J; Panichelli, L. 2009. Life cycle assessment of biofuels, Energy and greenhouse gas balances. *Bioresource Technology*, 100(21), 4919-4930.

Goldemberg, J; Teixeira Coelho, S; Guardabassi, P. 2008. The sustainability of ethanol production from sugarcane. *Energy Policy*, 36, 2086-2097.

Goodlass, G; Green, M; Hilton, B; McDonough, S. 2007. Nitrate leaching from short-rotation coppice. *Soil Use and Management*, 23(2), 178-184.

Goolsby, DA; BattaglIn: WA; Lawrence, GB; Artz, RS; Aulenbach, BT; Hooper, RP; Keeney, DR; Stensland, GJ. 1999. Flux and sources of nutrients in the Mississippi-Atchafalaya River Basin: Topic 3 report for the integrated assessment of hypoxia in the Gulf of Mexico. Decision Analysis Series No. 17. National Oceanic and Atmospheric Administration.

Gordon, DR; Onderdonk, DA; Fox, AM; Stocker, RK; Gantz, C. 2008. Predicting Invasive Plants in Florida Using the Australian Weed Risk Assessment. *Invasive Plant Science and Management*, 1(2), 178-195.

Gorte, RW. 2009. *Wildfire fuels and fuel reduction*. Congressional Research Service Report R40811. Available at: http://ncseonline.org/NLE/CRSreports/09Sept/R40811.pdf.

Grady, KC; Hart, SC. 2006. Influences of thinning, prescribed burning, and wildfire on soil processes and properties in southwestern ponderosa pine forests: A retrospective study. *Forest Ecology and Management*, 234(1-3), 123-135.

Graham, RL; Nelson, R; Sheehan, J; Perlack, RD; Wright, LL. 2007. Current and potential US corn stover supplies. *Agronomy Journal*, 99(1), 1-11.

Graham, LA; Belisle, SL; Baas, C-L. 2008. Emissions from light duty gasoline vehicles operating on low blend ethanol gasoline and E85. *Atmospheric Environment*, 42, 4498- 4516.

Gressel, J. 2008. Transgenics are imperative for biofuel crops. *Plant Science,* 174, 246-263.

Grigal, DF; Berguson, WE. 1998. Soil carbon changes associated with short-rotation systems. *Biomass and Bioenergy*, 14(4), 371-377.

Groom, MJ; Gray, EM; Townsend, PA. 2008. Biofuels and biodiversity: Principles for creating better policies for biofuel production. *Conservation Biology*, 22, 602-609.

Guan, TY; Holley, RA. 2003. Pathogen survival in swine manure environments and transmission of human enteric illness: A review. *Journal of Environmental Quality*, 32(2), 383-392.

Hammerschlag, R. 2006. Ethanol's energy return on investment: A survey of the literature 1990 present. *Environmental Science and Technology*, 40(6), 1744-1750.

Hansen, EA. 1988. Irrigating short rotation intensive culture hybrid poplars. *Biomass*, 16, 237-250.

Heaton, EA; Clifton-Brown, J; Voigt, T; Jones, MB; Long, SP. 2004a. Miscanthus for renewable energy generation: European Union experience and projections for Illinois. *Mitigation and Adaptation Strategies for Global Change*, 9, 433-451.

Heaton, E; Voigt, T; Long, SP. 2004b. A quantitative review comparing the yields of two candidate C-4 perennial biomass crops in relation to nitrogen, temperature and water. *Biomass and Bioenergy*, 27, 21-30.

Heaton, EA; Dohleman, FG; Long, SP. 2008. Meeting US biofuel goals with less land: The potential of Miscanthus. *Global Change Biology*, 14, 2000-2014.

Heaton, EA; Dohleman, FG; Long, SP. 2009. Seasonal nitrogen dynamics of Miscanthus x giganteus and Panicum virgatum. *Global Change Biology Bioenergy*, 1(4), 297-307.

Hector, A; Bagchi, R. 2007. Biodiversity and ecosystem multifunctionality. *Nature*, 448, 188-U6.

Herkert, JR. 2007. Evidence for a recent Henslow's sparrow population increase in Illinois. *Journal of Wildlife Management*, 71(4), 1229-1233.

Herron, PM; Martine, CT; Latimer, AM; Leict-Young, SA. 2007. Invasive plants and their ecological strategies: prediction and explanation of woody plant invasion in New England. *Diversity and Distributions*, 13(5), 633-644.

Hertel, TW; Tyner, WE; Birur, DK. 2010a. The global impacts of biofuel mandates. *The Energy Journal*, 31(1), 75-100.

Hertel, TW; Golub, AA; Jones, AD; O'Hare, M; Plevln: RJ; Kammen, DM. 2010b. Effects of US maize ethanol on global land use and greenhouse gas emission: Estimating market-mediated responses. *Bioscience*, 60, 223-231.

Hess, JR; Kenney, KL; Ovard, LP; Searcy, EM; Wright, CT. 2009a. Uniform-format bioenergy feedstock supply system design report series: Commodity-scale production of an infrastructure-compatible bulk solid from herbaceous lignocellulosic biomass. Idaho National Lab INL/Ext-09017527.

Hess, P; Johnston, M; Brown-Steiner, B; Holloway, T; de Andrade, J; Artaxo, P. 2009b. Air quality issues associated with biofuel production and use. Section 10. In: Howarth, RW; Bringezu, S (eds.). Biofuels: Environmental consequences and interactions with changing land use. Proceedings of the Scientific Committee on Problems of the Environment (SCOPE) International Biofuels Project Rapid Assessment. 169-194. Available at: http://cip.cornell.edu/DPubS?service=UI&version=1.0&verb=Display&handle=scope.

Hickman, G; VanLoocke, A; Dohleman, F; Bernacchi, C. 2010. A Comparison of canopy evapotranspiration for maize and two perennial grasses identified as potential bioenergy crops. *Global Change Biology*, 2(4), 157-168.

Hill, J. 2007. Environmental costs and benefits of transportation biofuel production from food- and lignocelluloses-based energy crops, a review. *Agronomy for Sustainable Development*, 27(1), 1-12.

Hill, J. 2009. Environmental costs and benefits of transportation biofuel production from food-and lignocellulose-based energy crops: A review. In: Lichtfouse, E; Navarrete, M; Debaeke, P; Souchère, V; Alberola, C (eds.). Sustainable agriculture. Springer New York. 125-139.

Hill, J; Nelson, E; Tilman, D; Polasky, S; Tiffany, D. 2006. Environmental, economic, and energetic costs and benefits of biodiesel and ethanol biofuels. *Proceedings of the National Academy of Sciences*, 103(30), 11206-11210.

Hill, J; Polasky, S; Nelson, E; Tilman, D; Huo, H; Ludwig, L; Neumann, J; Zheng, H; Bonta, D. 2009. Climate change and health costs of air emissions from biofuels and gasoline. *Proceedings of the National Academy of Sciences*, 106(6), 2077-2082. Available at: http://www.pnas.org/content/early/2009/02/02/0812835106.abstract.

Hodkinson, TR; Chase, MW; Lledo, MD; SalamIn: N; Renvoize, SA. 2002. Phylogenetics of Miscanthus, Saccharum and related genera (Saccharinae, Andropogoneae, Poaceae) based on DNA sequences from ITS nuclear ribosomal DNA and plastid trnL intron and trnL-F intergenic spacers. *Journal of Plant Research*, 115, 381-392.

Hoffmann, JP. 1998. Wastewater treatment with suspended and nonsuspended algae. *Journal of Phycology*, 34(5), 757-763.

Homyack, JA; Harrison, DJ; Krohn, WB. 2005. Long-term effects of precommercial thinning on small mammals in northern Maine. *Forest Ecology and Management*, 205, 43-57.

Hooper, DU; ChapIn: FS; Ewel, JJ; Hector, A; Inchausti, P; Lavorel, S; Lawton, JH; Lodge, DM; Loreau, M; Naeem, S; Schmid, B; Setälä, H; Symstad, AJ; Vandermeer, J; Wardle, DA. 2005. Effects of biodiversity on ecosystem functioning: A consensus of current knowledge. *Ecological Monographs*, 75(1), 3-35.

Huesemann, M; Benemann J. 2009. Biofuels from microalgae: Review of products, processes and potential, with special focus on Dunaliella sp. Section 19. In: Ben-Amotz, A; Polle, JEW; Subba Rao, DW (eds.). The alga Duneliella, Biodiversity, physiology, genomics and biotechnology. Science Publishers.

Huntington, TG. 1995. Carbon sequestration in an aggrading forest ecosystem in the southeastern USA. *Soil Science Society of America Journal*, 59(5), 1459-1467.

Huo, H; Wang, M; Bloyd, C; Putsche, V. 2008. *Life-cycle assessment of energy and greenhouse gas effects of soybean-derived biodiesel and renewable fuels.* Available at: http://www.transportation.anl.gov/pdfs/AF/467.pdf.

Huo, H; Wang, M; Bloyd, C; Putsche, V. 2009. Life-cycle assessment of energy use and greenhouse gas emission of soybean-derived biodiesel and renewable fuels. *Environmental Science and Technology*, 43, 750-756.

INPE (Instituto Nacional de Pesquisas Espaciais) n.d. Monitora mento da floresta Amazonica *Brasileira por satelite*. Available at: http://www.obt.inpe.br/prodes/.

Iowa State University. 2008. General guide for crop nutrient and limestone recommendations in Iowa. Iowa State University Extension Publication PM 1688.

Isebrands, JG; Guenther, AB; Harley, P; Helmig, D; Klinger, L; Vierling, L; Zimmerman, P; Geron, C. 1999. Volatile organic compound emission rates from mixed deciduous and coniferous forest in Northern WisconsIn: USA. *Atmospheric Environment*, 33, 2527- 2536.

ISO (International Organization for Standardization). 2006. ISO 14040: 2006: Environmental management—life cycle assessment—principles and framework.

Jackson, RB; Jobbagy, EG; Avissar, R; Roy, SB; Barrett, DJ; Cook, CW; Farley, KA; le Maitre, DC; McCarl, BA; Murray, BC. 2005. Trading water for carbon with biological carbon sequestration. *Science*, 310, 1944-1947.

Jakob, K; Zhou, FS; Paterson, AH. 2009. Genetic improvement of C4 grasses as cellulosic biofuel feedstocks. *In Vitro Cellular and Developmental Biology–Plant*, 45(3), 291-305.

James, LK; Swinton, SM; Thelen, KD. 2010. Profitability analysis of cellulosic energy crops compared with corn. *Agronomy Journal*, 102(2), 675-687.

Jandl, R; Lindner, M; Vesterdal, L; Bauwens, B; Baritz, R; Hagedorn, F; Johnson, DW; Minkkinen, K; Byrne, KA. 2007. How strongly can forest management influence soil carbon sequestration? *Geoderma*, 137(3-4), 253-268.

Janowiak, MK; Webster, CR. 2010. Promoting ecological sustainability in woody biomass harvesting. *Journal of Forestry*, (108), 16-23.

Janzen, HH; BeauchemIn: KA; Bruinsma, Y; Campbell, CA; Desjardins, RL; Ellert, BH; and Smith, E G. (2003). The fate of nitrogen in agroecosystems: An illustration using Canadian estimates. *Nutr. Cycl. Agroecosyst*. 67, 85-102.

Jarecki, MK; ParkIn: TB; Chan, ASK; Hatfield, JL; Jones, R. 2008. Greenhouse gas emissions from two soils receiving nitrogen fertilizer and swine manure slurry. *Journal of Environmental Quality*, 37(4), 1432-1438.

Jeanneret, P; Schüpbach, B; Luka, H. 2003. Quantifying the impact of landscape and habitat features on biodiversity in cultivated landscapes. *Agriculture, Ecosystems and Environment*, 98, 311-320.

Jiang, Y; Swinton, SM. 2009. Market interactions, farmer choices, and the sustainability of growing advanced biofuels: A missing perspective? *Int. J. Sustain. Devel. World Ecol.*, 16, 438-450.

Johnson, DT; Taconi, KA. 2007. The glycerin glut: Options for the value-added conversion of crude glycerol resulting from biodiesel production. *Environmental Progress*, 26(4), 338- 348.

Johnson, DW; Curtis, PS. 2001. Effects of forest management on soil C and N storage: Meta analysis. *Forest Ecology and Management*, 140(2-3), 227-238.

Johnson, JMF; Allmaras, RR; Reicosky, DC. 2006. Estimating source carbon from crop residues, roots and rhizodeposits using the national grain-yield database. *Agronomy Journal*, 98(3), 622-636.

Johnson, K; Scatena, FN; Pan, YD. 2010. Short- and long-term responses of total soil organic carbon to harvesting in a northern hardwood forest. *Forest Ecology and Management*, 259(7), 1262-1267.

Kahn, N; Warith, MA; Luk, G. 2007. A comparison of acute toxicity of biodiesel, biodiesel blends, and diesel on aquatic organisms. *Journal of the Air and Waste Management Association*, 57(3), 286-296.

Kalies, EL; Chambers, CL; Covington, WW. 2010. Wildlife responses to thinning and burning treatments in southwestern conifer forests: A meta-analysis. *Forest Ecology and Management*, 259, 333-342.

Kalogo, Y; Habibi, S; Maclean, HL; Joshi, SV. 2007. Environmental implications of municipal solid waste-derived ethanol. *Environmental Science and Technology*, 41(1), 35-41.

Kantor, LS; Lipton, K; Manchester, A; Oliveira, V. 1997. *Estimating and addressing America's food losses*. Food Review January-April 2-12. Available at: http://www.ers.usda.gov/Publications/FoodReview/Jan1997/Jan97a.pdf.

Karlen, DL; Lal, R; Follett, RF; Kimble, JM; Hatfield, JL; Miranowski, JM; Cambardella, CA; Manale, A; Anex, RP; Rice, CW. 2009. Crop residues: The rest of the story. *Environmental Science and Technology*, 43(21), 8011-8015.

Karraker, N; Welsh Jr., H. 2006. Long-term impacts of even-aged timber management on abundance and body condition of terrestrial amphibians in Northwestern California. *Biological Conservation*, 131, 132-140.

Kaufman, GA; Kaufman, DW. 2008. Effects of haying on small mammals in mixed grass prairie of central Kansas. *Transactions of the Kansas Academy of Science*, 111, 275-282.

Keeney, R; Hertel, TW. 2009. The indirect land use impacts of United States biofuel policies, The importance of acreage, yield, and bilateral trade responses. *American Journal of Agricultural Economics*, 91(4), 895-909.

Kenny, JF; Barber, NL; Hutson, SS; Linsey, KS; Lovelace, JK; Maupln: MA. 2009. Estimated use of water in the United States in 2005. *Circular*, 1344. U.S. Geological Survey.

Keoleian, GA; Volk, TA. 2005. Renewable energy from willow biomass crops, Life cycle energy, environmental and economic performance. *Critical Reviews in Plant Sciences*, 24(5-6), 385-406.

Keshwani, D; Cheng, J. 2009. Switchgrass for bioethanol and other value-added applications, A review. *Bioresource Technology*, 100(4), 1515-1523.

Keystone Alliance, 2009. Environmental Resource Indicators for Measuring Outcomes of On-Farm Agricultural Production in the United States. Available at: http://www.keystone.org/files/file/SPP/environment/field-to-market/Field-toMarket_Environmental-Indicator_First_Report_With_Appendices_01122009.pdf.

Khanina, L; Bobrovsky, M; Komarov, A; Mikhajilov, A. 2007. Modeling dynamics of forest ground vegetation diversity under different forest management regimes. *Forest Ecology and Management*, 248(1-2), 80-94.

Kim, H; Kim, S; Dale, BE. 2009. Biofuels, land use change, and greenhouse gas emissions: Some unexplored variables. *Environmental Science and Technology*, 43(3), 961-967.

Kim, S; Dale, BE. 2005. Life cycle assessment of various cropping systems utilized for producing biofuels, Bioethanol and biodiesel. *Biomass and Bioenergy*, 29(6), 426-439.

Kimble, J. n.d. Biofuels and emerging issues for emergency responders. U.S. Environmental Protection Agency. Available at: http://www.epa.gov/oem/docs/oil/fss/fss09/kimblebiofuels.pdf.

King, CW; Webber, ME. 2008. Water intensity of transportation. *Environmental Science and Technology*, 42, 7866-7872.

Kleter, GA; Bhula, R; Bodnaruk, K; Carazo, E; Felsot, AS; Harris, CA; Katayama, A; Kuiper, HA; Racke, KD; Rubln: B; Shevah, Y; Stephenson, GR; Tanaka, K; Unsworth, J; Wauchope, RD; Wong, SS. 2008. Altered pesticide use on transgenic crops and the associated general impact from an environmental perspective. *Pest Management Science*, 63(11), 1107-1115.

Klocke, NL; Watts, DG; Schneekloth, JP; Davidson, DR; Todd, RW; Parkhurst, AM. 1999. Nitrate leaching in irrigated corn and soybean in a semi-arid climate. *Transactions of the American Society of Agricultural Engineers*, 42(6), 1621-1630.

Knapp, AK; Seastedt, TR. 1986. Detritus accumulation limits productivity of tallgrass prairie. *BioScience*, 36, 662-668.

Knight, DH; Yavitt, JB; Joyce, GD. 1991. Water and nitrogen outflow from lodgepole pine forest after 2 levels of tree mortality. *Forest Ecology and Management*, 46(3-4), 215-225.

Knox, OGG; Constable, GA; Pyke, B; Gupta, VVSR. 2006. Environmental impact of conventional and Bt insecticidal cotton expressing one and two Cry genes in Australia. *Australian Journal of Agricultural Research*, 57(5), 501-509.

Koh, LP; Ghazoul, J. 2008. Biofuels, biodiversity, and people: Understanding the conflicts and finding opportunities. *Biological Conservation*, 141, 2450-2460.

Kosmala, A; Charvet, S; Roger, MC; Faessel, B. 1999. Impact assessment of a wastewater treatment plant effluent using instream invertebrates and the *Ceriodaphnia dubia* chronic toxicity test. *Water Research*, 33(1), 266-278.

Kreutzweiser, DP; Hazlett, PW; Gunn, JM. 2008. Logging impacts on the biogeochemistry of boreal forest soils and nutrient export to aquatic systems: A review. *Environmental Reviews*, 16, 157-179.

Krivanek, M; Pysek, P; Jarosik, V. 2006. Planting history and propagule pressure as predictors of invasion by woody species in a temperate region. *Conservation Biology*, 20, 1487-1498.

Lal, R. 1997. Residue management, conservation tillage and soil restoration for mitigating greenhouse effect by CO_2-enrichment. *Soil and Tillage Research*, 43(1-2), 81-107.

Lal, R. 2003. Global potential of soil carbon sequestration to mitigate the greenhouse effect. *Critical Reviews in Plant Sciences*, 22(2), 151-184.

Lal, R. 2004. Soil carbon sequestration impacts on global climate change and food security. *Science*, 304(5677), 1623-1627.

Lal, R. 2009. Soil quality impacts of residue removal for bioethanol production. *Soil and Tillage Research*, 102(2), 233-241.

Landis, A; Miller, S; Theis, T. 2007. Life cycle of the corn–soybean agroecosystem for biobased production. *Environmental Science and Technology*, 41(4), 1457-1464.

Landis, DA; Werling, BP. 2010. Arthropod and biofuel production systems in North America. *Insect Science*, 17, 220-236.

Landis, DA; Gardiner, MM; van der Werf, W; Swinton, SM. 2008. Increasing corn for biofuel production reduces biocontrol services in agricultural landscapes. Proceedings of the National Academy of Sciences, 105, 20552-20557. Available at: http://www.pnas.org/content/105/51/20552.full.pdf+html.

Lapuerta, M; Armas, O; Rodriguez-Fernandez, J. 2008. Effect of biodiesel fuels on diesel engine emissions. *Progress in Energy and Combustion Science*, 34(2), 198-223.

Laser, M; Larson, E; Dale, B; Wang, M; Greene, N; Lynd, LR. 2009. Comparative analysis of efficiency, environmental impact, and process economics for mature biomass refining scenarios. *Biofuels, Bioproducts, and Biorefining*, 3, 247-270.

Lavergne, S; Molofsky, J. 2004. Reed canarygrass (Phalaris arundinacea) as a biological model in the study of plant invasions. *Critical Reviews in Plant Sciences*, 23(5), 415-429.

Lee, D; Nair, R; Chen, A. 2009. Regulatory hurdles for transgenic biofuel crops. *Biofuels Bioproducts and Biorefining*, 3, 468-480.

Lee, LS; Carmosini, N; Sassman, SA; Dion, HM; Sepulveda, MS. 2007a. Agricultural contributions of antimicrobials and hormones on soil and water quality. *Advances in Agronomy*, (93), 1-68.

Lee, DK; Owens, VN; Doolittle, JJ. 2007b. Switchgrass and soil carbon sequestration response to ammonium nitrate, manure, and harvest frequency on conservation reserve program land. *Agronomy Journal*, 99(2), 462-468.

Lee, H; Sumner, DA. 2010. International trade patterns and policy for ethanol in the United States. *Handbook of Bioenergy Policy and Economics: Natural Resource Management and Policy*, 33(5), 327-345.

LevIn: LA; Ekau, W; Gooday, AJ; Jorissen, F; Middelburg, JJ; Naqvi, SWA; Neira, C; Rabalais, NN; Zhang, J. 2009. Effects of natural and human-induced hypoxia on coastal benthos. *Biogeosciences*, 6(10), 2063-2098.

LevIn: RB; EpsteIn: PR; Ford, TE; Harrington, W; Olson, E; Reichard, EG. 2002. U.S. drinking water challenges in the twenty-first century. *Environmental Health Perspectives*, 110(1), 43-52.

Lewandowski, I; Clifton-Brown, JC; Scurlock, JMO; Huisman, W. 2000. Miscanthus, European experience with a novel energy crop. *Biomass and Bioenergy*, 19, 209-227.

Lewandowski, I; Clifton-Brown, JC; Andersson, B; Basch, G; Christian, DG; Jorgensen, U; Jones, MB; Riche, AB; Schwarz, KU; Tayebi, K; Teixeira, F. 2003. Environment and harvest time affects the combustion qualities of Miscanthus genotypes. *Agronomy Journal*, 95, 1274-1280.

Liebig, MA; Schmer, MR; Vogel, KP; Mitchell, RB. 2008. Soil carbon storage by switchgrass grown for bioenergy. *Bioenergy Research*, 1(3-4), 215-222.

Liska, AJ; Yang, HS; Bremer, VR; KlopfensteIn: TJ; Walters, DT; Erickson, GE; Cassman, KG. 2008. Improvements in life cycle energy efficiency and greenhouse gas emissions of corn-ethanol. *Journal of Industrial Ecology*, 13(1), 58-74.

Liu, WD; Tollenaar, M; Stewart, G; Deen, W. 2004. Response of corn grain yield to spatial and temporal variability in emergence. *Crop Science*, 44(3), 847-854.

Loreau, M; Naeem, S; Inchausti, P (eds.). 2002. *Biodiversity and ecosystem functioning: Synthesis and perspectives*. Oxford University Press.

Luiro, J; Kukkola, M; Saarsalmi, A; Tamminen, P; Helmisaari, HS. 2010. Logging residue removal after thinning in boreal forests: Long-term impact on the nutrient status of Norway spruce and Scots pine needles. *Tree Physiology*, 30(1), 78-88.

Lundborg, A. 1997. Reducing nitrogen load: Whole tree harvesting, a literature review. *Ambio*, 26(6), 387-393.

MacDonald, JM; Ribaudo, MO; Livingston, MJ; Beckman, J; Huang, W. 2009. Manure use for fertilizer and energy: Report to congress. Administrative Publication No. AP-037. U.S. Department of Agriculture, Economic Research Service.

Mackay, DM; de Sieyes, NR; Einarson, MD; Feris, KP; Pappas, AA; Wood, IA; Jacobson, L; Justice, LG; Noske, MN; Scow, KM; Wilson, JT. 2006. Impact of ethanol on the natural attenuation of benzene, toluene, and o-xylene in a normally sulfate-reducing aquifer. *Environmental Science and Technology*, 40, 6123-6130.

Malcolm, SA; Aillery, M; Weinberg, M. 2009. Ethanol and a changing agricultural landscape. Economic Research Report 86. U.S. Department of Agriculture, Economic Research Service.

Malik, YS; Randall, GW; Goyal, SM. 2004. Fate of salmonella following application of swine manure to tile-drained clay loam soil. *Journal of Water and Health*, 2(2), 97-101.

McBroom, MW; Beasley, RS; Chang, MT; Ice, GG. 2008. Storm runoff and sediment losses from forest clearcutting and stand re-establishment with best management practices in East Texas, USA. *Hydrological Processes*, 22(10), 1509-1522.

McGuire, VL. 2009. Water-level changes in the High Plains aquifer, predevelopment to 2007, 2005-06, and 2006-07. U.S. Geological Survey Scientific Investigations Report 209 5019. Available at: http://pubs.usgs.gov/sir/2009/5019/.

McIsaac, GF; David, MB; Mitchell, CA. 2010. Miscanthus and switchgrass production in central Illinois: Impacts on hydrology and inorganic nitrogen leaching. *Journal of Environmental Quality*, 39, 1790-1799.

McLauchlan, K. 2006. The nature and longevity of agricultural impacts on soil carbon and nutrients: *A review. Ecosystems*, 9(8), 1364-1382.

McLaughlln: SB; Kszos, LA. 2005. Development of switchgrass (Panicum virgatum) as a bioenergy feedstock in the United States. *Biomass and Bioenergy*, 28(6), 515-535.

McLaughlln: JW; Phillips, SA. 2006. Soil carbon, nitrogen, and base cation cycling 17 years after whole-tree harvesting in a low-elevation red spruce (Picea rubens)–balsam fir (Abies balsainea) forested watershed in central Maine, USA. *Forest Ecology and Management*, 222(1-3), 234-253.

McLaughlln: SB; Walsh, ME. 1998. Evaluating environmental consequences of producing herbaceous crops for bioenergy. *Biomass and Bioenergy*, 14(4), 317-324.

Miguez, FE; Villamil, MB; Long, SP; Boller, GA. 2008 Meta-analysis of the effects of management factors on Miscanthus X giganteus growth and biomass production. *Agricultural and Forest Meteorology*, 148, 1280-1292.

Milbrandt, A. 2005. *A geographic perspective on the current biomass resource availability in the United States*. Technical Report NREL/TP-560-39181. U.S. Department of Energy.

Miller, EL; Beasley, RS, Lawson, ER. 1988. Forest harvest and site preparation effects on erosion and sedimentation in the Ouachita Mountains. *Journal of Environmental Quality*, 17, 219-225.

Mississippi River/Gulf of Mexico Watershed Nutrient Task Force. 2008. Gulf Hypoxia Action Plan 2008 for Reducing, Mitigating, and Controlling Hypoxia in the Northern Gulf of Mexico and Improving Water Quality in the Mississippi River Basin. Available at: http://water.epa.gov/type/watersheds/named/msbasin/actionplan.cfm#documents.

Moore, PA; Daniel, TC; Sharpley, AN; Wood, CW. 1995. Poultry manure management—environmentally sound options. *Journal of Soil and Water Conservation*, 50(3), 321-327.

Mosier, AR; Duxbury, JM; Freney, JR; Heinemeyer, O; Minami, K. 1996. Nitrous oxide emissions from agricultural fields, Assessment, measurement and mitigation. *Plant and Soil*, 181(1), 95-108.

Mubako, S; Lant, C. 2008. Water resource requirements of corn-based ethanol. *Water Resources Research*, 44(8), 1-5.

Mueller, S. 2010. 2008 National dry mill corn ethanol survey. *Biotechnology Letters*, 32, 1261 1264.

Muhs, J; Viamajala, S; Heydorn, B; Edwards, M; Hu, Q; Hobbs, R; Allen, M; Smith, DB; Fenk, T; Bayless, D; Cooksey, K; Kuritz, T; Crocker, M; Morton, S; Sears, J; Daggett, D; Hazlebeck, D; Hassenia, J. 2009. *Algae biofuels and carbon recycling: A summary of opportunities, challenges, and research needs*. Utah State University. Available at: www.utah.gov/ustar/documents/63.pdf.

Munoz, R; Guieysse, B. 2006. Algal-bacterial processes for the treatment of hazardous contaminants: A review. *Water Research*, 40, 2799-2815.

Murray, LD; Best, LB. 2003. Short-term bird response to harvesting switchgrass for biomass in Iowa. *Journal of Wildlife Management*, 67, 611-621.

Murray, LD; Best, LB; Jacobsen, TJ; Braster, ML. 2003. Potential effects on grassland birds of converting marginal cropland to switchgrass biomass production. *Biomass and Bioenergy*, 25, 167-175.

NACEPT (National Advisory Council for Environmental Policy and Technology). 2008.

NACEPT's Fourth Advice Letter to the Administrator on Biofuels. Available at: http://www.epa.gov/ocempage/nacept/reports/pdf/2008_1215_biofuels_letter.pdf.

NASS (United States Department of Agriculture, National Agricultural Statistics Service). 2006. Agricultural chemical usage 2005 field crops summary. Ag Ch 1 (06). Available at: http://usda.mannlib.cornell.edu/usda/nass/AgriChemUsFC//2000s/2006/AgriChemUsFC 05-17-2006.pdf.

NASS (United States Department of Agriculture, National Agricultural Statistics Service). 2007. Agricultural chemical usage 2006 field crops summary. Ag Ch 1 (07)a. Available at: http://usda.mannlib.cornell.edu/usda/nass/AgriChemUsFC//2000s/2007/AgriChem UsFC 05-16-2007_revision.pdf.

NASS (U.S. Department of Agriculture, National Agricultural Statistics Service). 2009a. 2007 census of agriculture. Available at: http://www.agcensus.usda.gov.

NASS (U.S. Department of Agriculture, National Agriculture Statistics Service). 2009b. 2008 farm and ranch irrigation survey. Available at: http://www.agcensus.usda.gov/ Publications/2007/Online_Highlights/Farm_and_Ranch_I rrigation_Survey/index.asp.

NASS (United States Department of Agriculture, National Agricultural Statistics Service). 2010a. Quick stats. Available at: http://www.nass.usda.gov/Data_and_Statistics/ Quick_Stats.

NASS (United States Department of Agriculture, National Agricultural Statistics Service). 2010b. County maps. Available at: http://www.nass.usda.gov/Charts_and_Maps/Crops_ County/index.asp.

NASS (United States Department of Agriculture, National Agricultural Statistics Service). 2010c. National statistics for corn—corn grain yield. Available at: http://www.nass.usda. gov/Statistics_by_Subject/result.php?AFBDFE1E-1AFC-35DE8A93-7FB72F0DA089& sector=CROPS&group=FIELD%20CROPS&comm=CORN.

National Invasives Species Council. 2009. Biofuels:Cultivating Energy, Not Invasive Species. Available at: http://www.doi.gov/NISC/home_documents/BiofuelWhitePaper.pdf

Nave, LE; Vance, ED; Swanston, CW; Curtis, PS. 2010. Harvest impacts on soil carbon storage in temperate forests. *Forest Ecology and Management*, 259(5), 857-866.

NCASI (National Council for Air and Stream Improvement, Inc.). 2009. Compendium of forestry best management practices for controlling nonpoint source pollution in North America. Technical Bulletin No. 0966.

Neary, DG; Ice, GG; Jackson, CR. 2009. Linkages between forest soils and water quality and quantity. *Forest Ecology and Management*, 258(10), 2269-2281.

Nelson, RG; Ascough, JC; Langemeier, MR. 2006. Environmental and economic analysis of switchgrass production for water quality improvement in northeast Kansas. *Journal of Environmental Management*, 79(4), 336-348.

Nickson, T. 2008. Planning environmental risk assessment for genetically modified crops: Problem formulation for stress-tolerant crops. *Plant Physiology*, 147, 494-502.

Niemelä, J; Koivula, M; Kotze, DJ. 2007. The effects of forestry on Carabid beetles (Coleoptera: Carabidae) in boreal forests. *Journal of Insect Conservation*, 11, 5-18.

NRC (National Research Council). 2000. *Genetically modified pest-protected plants: Science and regulation*. National Academies Press.

NRC (National Research Council). 2001. *Ecological monitoring of genetically modified crops*. A workshop summary. National Academies Press.

NRC (National Research Council). 2002. *Environmental effects of transgenic plants: The scope and adequacy of regulation*. National Academies Press.

NRC (National Research Council) 2008. *Water implications of biofuels production in the United States*. National Academies Press. Available at: http://www.nap.edu/catalog/12039.html.

NRC (National Research Council). 2010a. *The impact of genetically engineered crops on farm sustainability in the United States*. National Academies Press.

NRC (National Research Council). 2010b. *Energy for transportation*. Section 3. In: Hidden costs of energy: Unpriced consequences of energy production and use. National Academies Press. 154-197.

NRCS (United States Department of Agriculture, National Resources Conservation Service). 2009. NI_190_304, comprehensive nutrient management plan technical criteria. Available at: http://directives.sc.egov.usda.gov/viewerFS.aspx?hid=25686.

NRCS (United States Department of Agriculture, National Resources Conservation Service). 2010a. Assessment of the effects of conservation practices on cultivated cropland in the Upper Mississippi River Basin. Available at: http://www.nrcs.usda.gov/technical/NRI/ceap/umrb/index.html.

NRCS (United States Department of Agriculture, National Resources Conservation Service). 2010b. Mississippi River Basin Healthy Watersheds Initiative. Available at: http://www.nrcs.usda.gov/Internet/FSE_DOCUMENTS/nrcsdev11_023951.pdf

NREL (National Renewable Energy Laboratory). 2009a. *Biodiesel handling and use guide*. Fourth edition. Available at: http://www.nrel.gov/vehiclesandfuels/pdfs/43672.pdf.

NREL (National Renewable Energy Laboratory). 2009b. *Preliminary compatibility assessment of metallic dispenser materials for service in ethanol fuel blends*.

Nyakatawa, EZ; Mays, DA; Tolbert, VR; Green, TH; Bingham, L. 2006. Runoff, sediment, nitrogen, and phosphorus losses from agricultural land converted to sweetgum and switchgrass bioenergy feedstock production in north Alabama. *Biomass and Bioenergy*, 30(7), 655-664.

Ochoa-Acuna, HG; Bialkowski, W; Yale, G; Hahn, L. 2009. Toxicity of soybean rust fungicides to freshwater algae and Daphnia magna. *Ecotoxicology*, 18(4), 440-446.

Ogle, SM; Breidt, FJ; Paustian, K. 2005. Agricultural management impacts on soil organic carbon storage under moist and dry climatic conditions of temperate and tropical regions. *Biogeochemistry*, 72(1), 87-121.

OSTP (Office of Science and Technology Policy). 1986. Coordinated framework for regulation of biotechnology. *Federal Register*, 51, 23302.

Owen, MDK. 2005. Maize and soybeans—controllable volunteerism without ferality? In: Gressel, J (ed.). Crop ferality and volunteerism. CRC Press. 149-165.

Oyediran, IO; Hibbard, BE; Clark, TL. 2004. Prairie grasses as hosts of the western corn rootworm (Coleoptera: Chrysomelidae). *Environmental Entomology*, 33, 740-747.

Paerl, HW; Robln: D; Whitall, DR. 2002. Atmospheric deposition of nitrogen, Implications for nutrient over-enrichment of coastal waters. *Estuaries*, 25(4B), 677-693.

Page-Dumroese, DS; Jurgensen, M; Terry, T. 2010. Maintaining soil productivity during forest or biomass-to-energy thinning harvests in the Western United States. *Western Journal of Applied Forestry*, 25(1), 5-11.

Parrish, DJ; Fike, JH. 2005. The biology and agronomy of switchgrass for biofuels. *Critical Reviews in Plant Sciences*, 24, 423-459.

Pate, R, Hightower, M; Cameron, C; Einfeld, W. 2007. Overview of energy-water interdependencies and the emerging energy demands on water resources. Report SAND 2007-1349C. Sandia National Laboratories.

Paul, KI; Polglase, PJ; Nyakuengama, JG; Khanna, PK. 2002. Change in soil carbon following afforestation. *Forest Ecology and Management*, 168(1-3), 241-257.

Perkins, JH. 2009. Integrated pest management, biofuels, and a new green revolution: A case study of the American Midwest. In: Peshin: R; Dhawan, AK (eds.). Integrated pest management; Dissemination and impact. Springer Netherlands. 581-607.

Perlack, R; Wright, L; Turhollow, A; Graham, R; Stokes, B; Erbach, D. 2005. Biomass as feedstock for a bioenergy and bioproducts industry: The technical feasibility of a billion-ton annual supply. U.S. Department of Agriculture and U.S. Department of Energy. Oak Ridge National Laboratory. ORNL/TM-2005/66. Available at: http://feedstockreview. ornl.gov/pdf/billion_ton_vision.pdf.

Perrln: R; Vogel, K; Schmer, M; Mitchell, R. 2008. Farm-Scale Production Cost of Switchgrass for Biomass. *Bioenergy Research*, 1, 91-97.

Perry, CH; Brooks, KN; Grigal, DF; Isebrands, JG; Tolbert, VR. 1998. A comparison of nutrient export from short-rotation hybrid poplar plantations and natural forest stands. In: Expanding Bioenergy Partnerships; Proceedings of the 8th National Bioenergy Conference. Great Lakes Regional Biomass Energy Program, Chicago, IL. 1252 1261.

Perry, CH; Miller, RC; Brooks, KN. 2001. Impacts of short-rotation hybrid poplar plantations on regional water yields. *Forest Ecology and Management*, 143, 143-151.

Perttu, KL. 1995. Ecological, biological balances and conservation. *Biomass and Bioenergy*, 9, 107-116.

Pheloung, PC; Williams, PA; Halloy, SR. 1999. A weed risk assessment model for use as a biosecurity tool evaluating plant introductions. *Journal of Environmental Management*, 57, 239-251.

Phillips, MJ; Swift Jr., LW; Blinn, CR. 2000. *Best management practices for riparian areas*. In: Verry, ES; Hornbeck, JW; Dolloff, CA (eds). Riparian management in forests of the continental Eastern United States. CRC Press. 273-286.

Pimentel, D. 2006. Soil erosion: A food and environmental threat. *Environment, Development and Sustainability*, 8(1), 119-137.

Pimentel, D; Markleln: A; Toth, MA; Karpoff, MN; Paul, GS; McCormack, R; Kyriazis, J; Krueger, T. 2009. Food versus biofuels: Environmental and economic costs. *Human Ecology*, 37, 1-12.

Poff, NL; Zimmerman, JKH. 2010. Ecological responses to altered flow regimes: A literature review to inform the science and management of environmental flows. *Freshwater Biology*, 55(1), 194-205.

Pollard, SJT; Kemp, RV; Crawford, M; Duarte-Davidson, R; IrwIn: JG; Yearsley, R. 2004. Characterizing environmental harm: Developments in an approach to strategic risk assessment and risk management. *Risk Analysis*, 24, 1551-1560.

Pollock, MS; Clarke, LMJ; Dube, MG. 2007. The effects of hypoxia on fishes, from ecological relevance to physiological effects. *Environmental Reviews*, 15, 1-14.

Powers, RF; Scott, DA; Sanchez, FG; Voldseth, RA; Page-Dumroese, D; Elioff, JD; Stone, DM. 2005. The North American long-term soil productivity experiment, Findings from the first decade of research. *Forest Ecology and Management*, 220(1-3), 31-50.

Powers, SE. 2005. *Quantifying cradle-to-farm gate life-cycle impacts associated with fertilizer used for corn, soybean, and stover production*. NREL/TP-510-27500. National Renewable Energy Laboratory. Available at: http://www1.eere.energy.gov/biomass/pdfs/37500.pdf.

Powers, SE. 2007. Nutrient loads to surface water from row crop production. *International Journal of Life Cycle Assessment*, 12(6), 399-407.

Powers, SE; Hunt, CS; Heermann, SE; Corseuil, HX; Rice, D; Alvarez, PJJ. 2001. The transport and fate of ethanol and BTEX in groundwater contaminated by gasohol. *Critical Reviews in Environmental Science and Technology*, 31(1), 79-123.

Powlson, DS; Riche, AB; Shield, I. 2005. Biofuels and other approaches for decreasing fossil fuel emissions from agriculture. *Annals of Applied Biology*, 146 (2), 193-201.

Pradhan, A; Shrestha, DS; Van Gerpen, J; Duffield, J. 2008. The energy balance of soybean oil biodiesel production: a review of past studies. *Transactions of the American Society of Agricultural and Biological Engineers*, 51, 185-194.

Prasifka, JR; Bradshaw, JD; Meagher, RL; Nagoshi, RN; Steffey, KL; Gray, ME. 2009. Development and feeding of fall armyworm on Miscanthus x giganteus and switchgrass. *Journal of Economic Entomology*, 102, 2154-2159.

Propheter, JL; Staggenborg, SA; Wu, X; Wang, D. 2010. Performance of annual and perennial biofuel crops: Yield during the first two years. *Agronomy Journal*, 102(2), 806-814.

Puppan, D. 2002. Environmental evaluation of biofuels. *Periodica Polytechnica Ser. Soc. Man. Sci.*, 10(1), 95-116.

Pysek, P; Krivanek, M; Vojtech, J. 2009. Planting intensity, residence time, and species traits determine invasion success of alien woody species. *Ecology*, 90(10), 2734–2744.

Rabalais, NN; Turner, RE; Díaz, RJ; Justić, D. 2009. Global change and eutrophication of coastal waters. *ICES Journal of Marine Science*, 66(7), 1528-1537.

Ragauskas, AJ; Williams, CK; Davison, BH; Britovsek, G; Cairney, J; Eckert, CA; Frederick, WJ Jr; Hallett, JP; Leak DJ; Liotta, CL; Mielenz, JR; Murphy, R; Templer, R; Tschaplinski, T. 2006. The path forward for biofuels and biomaterials. *Science*, 311, 484- 489.

Raghu, S; Anderson, RC; Daehler, CC; Davis, AS; Wiedenmann, RN; Simberloff, D; Mack, RN. 2006. Adding biofuels to the invasive species fire? *Science*, 313, 1742.

Randall, GW; Huggins, DR; Russelle, MP; Fuchs, DJ; Nelson, WW; Anderson, JL. 1997. Nitrate losses through subsurface tile drainage in conservation reserve program, alfalfa, and row crop systems. *Journal of Environmental Quality*, 26(5), 1240-1247.

Ranney, JW; Mann, LK. 1994. Environmental considerations in energy crop production. *Biomass Bioenergy*, 6(3), 211.

Raybould, A. 2007. Ecological versus ecotoxicological methods for assessing the environmental risks of transgenic crops. *Plant Science*, 173, 589-602.

Regassa, T; Koelsch, R; Erickson, G. 2008. Impact of feeding distillers grains on nutrient planning for beef cattle systems. University of Nebraska-Lincoln Extension—RP190. Available at: http://www.ianrpubs.unl.edu/epublic/live/rp190/build/rp190.pdf.

Reichard, SH; Hamilton, CW. 1997. Predicting invasions of woody plants introduced into North America. *Conservation Biology*, 11,193-203.

Reilly, TE; Dennehy, KF; Alley, WM; Cunningham, WL. 2008. Ground-water availability in the United States. *Circular*, 1323. U.S. Geological Survey.

Reiss, J; Bridle, JR; Montoya, JM; Woodward, G. 2009. Emerging horizons in biodiversity and ecosystem functioning research. *Trends in Ecology and Evolution*, 24(9), 505-514.

Reiter, M; Heffner, JT; Beech, S; Turner, T; Bilby, RE. 2009. Temporal and spatial turbidity patterns over 30 years in a managed forest of western Washington. *Journal of the American Water Resources Association*, 45(3), 793-808.

Reukema, D. L. 1975. Guidelines for precommercial thinning of Douglas-fir. Gen. Tech. Rep. PNW-GTR030, U.S. Department of Agriculture, Forest Service, Pacific Northwest Research Station, Portland, OR. 10.

RFA (Renewable Fuels Association). 2010. *2010 ethanol industry outlook: Climate of opportunity*. Available at: http://ethanolrfa.3cdn.net/32b7ed69bd366321cb_r1m626lb0.pdf.

Rice, ME. 2003. Transgenic rootworm corn: Assessing potential agronomic, economic, and environmental benefits. Plant Health Progress doi: 10.1094/PHP-2004-0301-01-RV. Available at: http://www.plantmanagementnetwork.org/pub/php/review/2004/rootworm/.

Richardson, CW; Bucks, DA; Sadler, EJ. 2008. The Conservation Effects Assessment Project benchmark watersheds: Synthesis of preliminary findings. *Journal of Soil and Water Conservation*, 63(6), 590-604.

Richardson, DM. 1998. Forestry trees as invasive aliens. *Conservation Biology*, 12(1), 18-26.

Richter, DD; Markewitz, D; Trumbore, SE; Wells, CG. 1999. Rapid accumulation and turnover of soil carbon in a re-establishing forest. *Nature*, 400(6739), 56-58.

Richter, GM; Riche, AB; Dailey, AG; Gezan, SA; Powlson, DS. 2008. Is UK biofuel supply from Miscanthus water-limited? *Soil Use and Management*, 24, 235-245.

Rittmann, BE. 2008. Opportunities for renewable bioenergy using microorganisms. *Biotechnology and Bioengineering*, 100, 203-212.

Rockwood, DL; Rudie, AW; Ralph, SA; Zhu, JY; Winandy, JE. 2008. Energy product options for Eucalyptus species grown as short rotation woody crops. *International Journal of Molecular Sciences*, 9, 1361-1378.

Romeis, J; Bartsch, D; Bigler, F; Candolfi, MP; Gielkens, MMC; Hartley, SE; Hellmich, RL; Huesing, JE; Jepson, PC; Layton, R; Quemada, H; Raybould, A; Rose, RI; Schiemann, J; Sears, MK; Shelton, AM; Sweet, J; Vaituzis, Z; Wolt, JD. 2008. Assessment of risk of insect-resistant transgenic crops to nontarget arthropods. *Nature Biotechnology*, 26, 203-208.

Rose, LW; Das, MK; Taliaferro, CM. 2008. Estimation of genetic variability and heritability for biofuel feedstock yield in several populations of switchgrass. *Annals of Applied Biology*, 152(1), 11-17.

Roth, AM; Sample, DW; Ribic, CA; Paine, L; Undersander, DJ; Bartelt, GA. 2005. Grassland bird response to harvesting switchgrass as a biomass energy crop. *Biomass and Bioenergy*, 28, 490-498.

Ruiz-Aguilar, GML; Fernandez-Sanchez, JM; Kane, SR; Kim, D; Alvarez, PJJ. 2002. Effect of ethanol and methyl-tert-butyl ether on monoaromatic hydrocarbon biodegradation: Response variability for aquifer materials under various electron-accepting conditions. *Environmental Toxicology and Chemistry*, 21(12), 2631-2639.

Rupert, MG. 2008. Decadal-scale changes in nitrate in ground water of the United States, 1988 2004. *Journal of Environmental Quality*, 37, S-240-S-248.

SAB (United States Environmental Protection Agency, Science Advisory Board). 2007. Hypoxia in the northern Gulf of Mexico, an update by the EPA Science Advisory Board. EPA SAB-08-003.

Sage, R; Cunningham, M; Haughton, AJ; Mallott, MD; Bohan, DA; Riche, A; Karp, A. 2010. The environmental impacts of biomass crops, Use by birds of miscanthus in summer and winter in southwestern England. *IBIS*, 152, 487-499.

Samson, F; Knopf, F. 1994. Prairie Conservation in North-America. *Bioscience*, 44(6), 418-421.

Sanchez, FG; Scott, DA; Ludovici, KH. 2006. Negligible effects of severe organic matter removal and soil compaction on loblolly pine growth over 10 years. *Forest Ecology and Management*, 227(1-2), 145-154.

Sarath, G; Mitchell, RB; Sattler, SE; Funnell, D; Pedersen, JF; Graybosch, RA; Vogel, KP. 2008. Opportunities and roadblocks in utilizing forages and small grains for liquid fuels. *Journal of Industrial Microbiology and Biotechnology*, 35(5), 343-354.

Sartori, F; Lal, R; Ebinger, MH; Parrish, DJ. 2006. Potential soil carbon sequestration and CO_2 offset by dedicated energy crops in the USA. *Critical Reviews in Plant Sciences*, 25(5), 441-472.

Scheller, RM; Mladenoff, DJ. 2002. Species diversity, composition, and spatial patterning of understory plants in old-growth and managed northern hardwood forests. *Ecological Applications*, 12, 1329-1343.

Schiffman, PM; Johnson, WC. 1989. Phytomass and detrital carbon storage during forest regrowth in the southeastern United-States Piedmont. *Canadian Journal of Forest Research* [Revue Canadienne de Recherche Forestiere] 19(1), 69-78.

Schilling, KE; Jha, MK; Zhang, Y-K; Gassman, PW; Wolter, CF. 2008. Impact of land use and land cover change on the water balance of a large agricultural watershed, Historical effects and future directions. *Water Resources Research*, 44(W00A09), 1-12.

Schlossberg, SR; King, DI. 2009. Postlogging Succession and Habitat Usage of Shrubland Birds. *Journal of Wildlife Management*, 73, 226-231.

Schmer, MR; Vogel, KP; Mitchell, RB; PerrIn: RK. 2008. Net energy of cellulosic ethanol from switchgrass. *Proceedings of the National Academy of Sciences*, 105(2), 464-469.

Schneckenberger, K; Kuzyakov, Y. 2007. Carbon sequestration under Miscanthus in sandy and loamy soils estimated by natural C-13 abundance. *Journal of Plant Nutrition and Soil Science* [Zeitschrift fur Pflanzenernahrung und Bodenkunde], 170(4), 538-542.

Schwer, CB; Clausen, JC. 1989. Vegetative filter treatment of diary milkhouse wastewater. *Journal of Environmental Quality*, 18, 446-451.

Searchinger, T; Heimlich, R; Houghton, RA; Dong, FX; Elobeid, A; Fabiosa, J; Tokgoz, S; Hayes, D; Yu, TH. 2008. Use of US croplands for biofuels increases greenhouse gases through emissions from land use change. *Science*, 319(5867), 1238-1240.

Secchi, S; Gassman, PW; Williams, JR; Babcock, BA. 2009. Corn-based ethanol production and environmental quality: A case of Iowa and the conservation reserve program. *Environmental Management*, 44(4), 732-744.

SECEX (Brazilian Ministry of Development, Industry and Trade). n.d. Table: Destino das exportações brasileiras de álcool etílico. Available at: http://www.desenvolvimento.gov.br/sitio/interna/interna.php?area=2&menu=999.

Semere, T; Slater, FM. 2007a. Ground flora, small mammal and bird species diversity in miscanthus (Miscanthus x giganteus) and reed canary-grass (Phalaris arundinacea) fields. *Biomass and Bioenergy*, 31(1), 20-29.

Semere, T; Slater, FM. 2007b. Invertebrate populations in miscanthus (Miscanthus x giganteus) and reed canary-grass (Phalaris arundinacea) fields. *Biomass and Bioenergy*, 31(1), 30-39.

Sexton, S; Zilberman, D; Rajagopal, D; Hochman, G. 2009. The role of biotechnology in a sustainable biofuel future. *AgBioForum*, 12(1), 130-140.

Shah, T; Burke, J; Villholth, K; Angelica, M; Custodio, E; Daibes, F; Hoogesteger, J; Giordano, M; Girman, J; van der Gun, J; Kendy, E; Kijne, J; Llamas, R; Masiyandima, M; Margat, J; MarIn: L; Peck, J; Rozelle, S; Sharma, B; Vincent, L; Wang, J. 2007. Groundwater: A global assessment of scale and significance. Section 10. In: Molden, D. (ed.). Water for food, water for life: A comprehensive assessment of water management in agriculture. Earthscan London, UK and International Water Management Institute Colombo, Sri Lanka. 395-423.

Shapouri, H; Duffield, JA; Wang, M. 2002. *The Energy Balance of Corn Ethanol: An Update*. U.S. Department of Agriculture, Office of the Chief Economist: Washington, DC. Available at: http://www.transportation.anl.gov/pdfs/AF/265.pdf.

Shapouri, H; Gallagher, PW; Nefstead, W; Schwartz, R; Noe, S; Conway, R. 2010. *2008 Energy Balance for the Corn-Ethanol Industry*. USDA Agricultural Economic Report Number 846. Available at: http://www.usda.gov/oce/reports/energy/2008Ethanol_June_final.pdf.

Sheehan, J; Camobreco, V; Duffield, J; Graboski, M; Shapouri, H. 1998a. *Life cycle inventory of biodiesel and petroleum diesel for use in an urban bus*. NREL/SR-580-24089. National Renewable Energy Laboratory.

Sheehan, J; Camboreco, V; Duffield, J; Graboski, M; Shapouri, H. 1998b. *An overview of biodiesel and petroleum diesel life cycles*. National Renewable Energy Laboratory.

Sheehan, J; Aden, A; Paustian, K; Killian, K; Brenner, J; Walsh, M; Nelson, R. 2004. Energy and environmental aspects of using corn stover for fuel ethanol. *Journal of Industrial Ecology*, 7(3-4), 117-146.

Sheehan, J; Dunahay, T; Benemann, J; Roessler, P. 1998. *A Look Back at the U.S. Department of Energy's Aquatic Species Program: Biodiesel from Algae*. Available at: http://www.nrel.gov/docs/legosti/fy98/24190.pdf.

Shepard, JP. 2006. Water quality protection in bioenergy production: The US system of forestry Best Management Practices. *Biomass and Bioenergy*, 30(4), 378-384.

Shi, AZ; Koh, LP; Tan, HTW. 2009. The biofuel potential of municipal solid waste. *Global Change Biology Bioenergy*, 1, 317-320.

Shinners, KJ; Boettcher, GC; Hoffman, DS; Munk, JT; Muck, RE; Weimer, PJ. 2009. Single-pass harvest of corn grain and stover, Performance of three harvester configurations. *Transactions of the ASABE*, 52(1), 51-60.

Shipitalo, MJ; Edwards, WM. 1998. Runoff and erosion control with conservation tillage and reduced-input practices on cropped watersheds. *Soil and Tillage Research*, 46(1-2), 1-12.

Simpson, TW; Sharpley, AN; Howarth, RW; Paerl, HW; MankIn: KR. 2008. The new gold rush, Fueling ethanol production while protecting water quality. *Journal of Environmental Quality*, 37(2), 318-324.

Simpson, TW; Martinelli, LA; Sharpley, AN; Howarth, RW. 2009. *Impact of ethanol production on nutrient cycles and water quality: The United States and Brazil as case studies.* In: Howarth, RW; Bringezu, S (eds). Biofuels: Environmental consequences and interactions with changing land use. Proceeding of the Scientific Committee on Problems of the Environment (SCOPE) International Biofuels Project Rapid Assessment. 153-167. Available at: http://cip.cornell.edu/DPubS?service=UI&version=1.0&verb=Display&handle=scope.

Six, J; Ogle, SM; Breidt, FJ; Conant, RT; Mosier, AR; Paustian, K. 2004. The potential to mitigate global warming with no-tillage management is only realized when practised in the long term. *Global Change Biology*, 10(2), 155-160.

Smith, VH; Sturm, BSM; deNoyelles, FJ; Billings, SA. 2009. The ecology of algal biodiesel production. *Trends in Ecology and Evolution*, 25(5), 302-309.

South Dakota State University. 2009. The cost of wet corn at harvest. Extension Extra 5056. Available at: http://agbiopubs.sdstate.edu/articles/ExEx5056.pdf.

Spencer, JL; Raghu, S. 2009. Refuge or reservoir? The potential impacts of the biofuel crop Miscanthus x giganteus on a major pest of maize. *PLoS One*, 4(12), e8336.

Stanton, B; Eaton, J; Johnson, J; Rice, D; Schuette, B; Moser, B. 2002. Hybrid poplar in the Pacific Northwest—the effects of market-driven management. *Journal of Forestry*, 100(4), 28-33.

Stewart, JR; Toma, Y; Fernandez, FG; Nishiwaki, A; Yamada, T; Bollero, GA. 2009. The ecology and agronomy of Miscanthus sinensis, a species important to bioenergy crop development, in its native range in Japan: A review. *Global Change Biology Bioenergy*, 1(2), 126-153.

Sticklen, MB. 2007. Feedstock crop genetic engineering for alcohol fuels. *Crop Science*, 47, 2238-2248.

Sticklen, MB. 2009. Expediting the biofuels agenda via genetic manipulations of cellulosic bioenergy crops. *Biofuels, Bioproducts and Biorefining*, 3, 448-455.

Stohlgren, TJ; Binkley, D; Chong, GW; Kalkhan, MA; Schell, LD; Bull, KA; Otsuki, Y; Newman, G; BashkIn: M; Son, Y. 1999. Exotic plant species invade hot spots of native plant diversity. *Ecological Monographs*, 69 (1), 25-46.

Stout, BM; Benfield, EF; Webster, JR. 1993. Effects of forest disturbance on shredder production in Southern Appalachian headwater streams. *Freshwater Biology*, 29(1), 56-69.

Suter, GW. 2007. Ecological risk assessment. CRC Press, Taylor & Francis Group Boca Raton. Taylor, RL; Maxwell, BD; Boik, RJ. 2006. Indirect effects of herbicides on bird food resources and beneficial arthropods. *Agriculture, Ecosystems and Environment*, 116(3-4), 157-164.

Teasdale, JR; Brandsaeter, LO; Calegari, A; Skora Neto, F. 2007. *Cover crops and weed management*. Non-chemical weed management. Available at: http://www.ars.usda.gov/sp2UserFiles/Place/12650400/Non-ChemWeedMgmt.pdf.

Taheripour, F; Hertel, TW; Tyner, WE; Beckman, JF; Birur, DK. 2010. Biofuels and their by-products, Global economic and environmental implications. *Biomass and Bioenergy*, 34(3), 278-289.

Tesfamariam, T; Bott, S; Cakmak, I; Römheld, V; Neumann, G. 2009. Glyphosate in the rhizosphere—role of waiting times and different glyphosate binding forms in soils for phytotoxicity to non-target plants. *European Journal of Agronomy*, 31(2), 126-132.

Thiffault, E; Pare, D; Belanger, N; Munson, A; Marquis, F. 2006. Harvesting intensity at clear- felling in the boreal forest, Impact on soil and foliar nutrient status. *Soil Science Society of America Journal*, 70(2), 691-701.

Thomas, SC; Halpern, CB; Falk, DA; Liguori, DA; AustIn: KA. 1999. Plant diversity in managed forests, Understory responses to thinning and fertilization. *Ecological Applications*, 9(3), 864-879.

Thompson, TR; Boal, CW; Lucia, D. 2009. Grassland bird associations with introduced and native grass conservation reserve program fields in the southern high plains. *Western North American Naturalist*, 69, 481-490.

Thomson, AM; Izarrualde, RC; West, TO; Parrish, DJ; Tyler, DD; Williams, JR. 2009. Simulating Potential Switchgrass Production in the United States. PNNL-19072.

Thornton, KW; Holbrook, SP; Stolte, KL; Landy, RB. 2000. Effects of forest management practices on Mid-Atlantic streams. *Environmental Monitoring and Assessment*, 63, 31-41.

Tilman, D; Hill, J; Lehman, C. 2006. Carbon-negative biofuels from low-impact high-diversity grassland biomass. *Science*, 314 (5805), 1598-1600.

Tilman, D; Socolow, R; Foley, JA; Hill, J; Larson, E; Lynd, L; Pacala, S; Reilly, J; Searchinger, T; Somerville, C; Williams, R. 2009. Beneficial biofuels—the food, energy, and environment trilemma. *Science*, 325, 270-271.

Timmons, D; Allen, G; Damery, D. 2008. Biomass energy crops: Massachusetts' potential. http://www.mass.gov/Eoeea/docs/doer/renewables/biomass/bio-ma-potential-crop.pdf.

Titus, BD; Roberts, BA; Deering, KW. 1997. Soil solution concentrations on three white birch sites in central Newfoundland following different harvesting intensities. *Biomass and Bioenergy*, 13(4-5), 313-330.

Troendle, CA; MacDonald, LH; Luce, CH; Larsen, IJ. 2010. *Fuel management and water yield*. In: Elliot, WJ; Miller, IS; AudIn: L (eds.). Cumulative watershed effects of fuel management in the western United States. Gen. Tech. Rep. RMRS-GTR-231. U.S. Department of Agriculture, Forest Service, Rocky Mountain Research Station. 124- 148. Available at: http://www.treesearch.fs.fed.us/pubs/34312.

Unc, A; Goss, MJ. 2004. Transport of bacteria from manure and protection of water resources. *Applied Soil Ecology*, 25(1), 1-18.

UNEP/GRID-Arendal. 2009. Kick the habit: A UN guide to climate neutrality. Available at: http://maps.grida.no/go/graphic/biofuel-production-map.

Updegraff, K; Baughman, MJ; Taff, SJ. 2004. Environmental benefits of cropland conversion to hybrid poplar: economic and policy considerations. *Biomass and Bioenergy*, 27, 411-428.

USDA (United States Department of Agriculture). 1991. Agricultural resources: Inputs situation and outlook report AR-21. Economic Research Service.

USDA (United States Department of Agriculture). 2010a. USDA long-term agricultural projection tables. Table 18. U.S. corn. Available at: http://usda.mannlib.cornell.edu/ MannUsda/ viewStaticPage.do?url=http://usda.mannlib.cornell.edu/usda/ers/94005/./ 2010/index.html.

USDA (United States Department of Agriculture). 2010b. USDA long-term agricultural projection tables. Table 23. Soybeans and products. Available at: http://usda.mannlib. cornell.edu/MannUsda/viewStaticPage.do?url=http://usda.mannlib.cornell.edu/usda/ers/9 4005/./2010/index.html.

USDA (United States Department of Agriculture). 2010c. USDA long-term agricultural projection tables. Table 17. Acreage for major field crops and Conservation Reserve Program (CRP) assumptions, long-term projections. Available at: http://usda.mannlib. cornell.edu/MannUsda/viewStaticPage.do?url=http://usda.mannlib.cornell.edu/usda/ers/9 4005/./2010/index.html.

USDA (United States Department of Agriculture). 2010d. *Sugar and sweeteners outlook.* Effects of global sugar markets on U.S. ethanol. Economic Research Service. Available at: http://www.ers.usda.gov/publications/sss/2010/Nov/SSSM267.pdf.

USDA (United States Department of Agriculture). 2010e. *Biomass crop assistance program: Final rule*. Available at: http://www.regulations.gov/#!documentDetail;D=CCC-2010 0001-0242.

USDA (United States Department of Agriculture) n.d. Global Agricultural Trade System (GATS) online. Foreign Agriculture Service. Available at: http://www.fas.usda.gov/ gats/ExpressQuery1.aspx.

USDA (United States Department of Agriculture, Forest Service). 2005. *A strategic assessment of forest biomass and fuel reduction treatments in western states*. Rocky Mountain Research Station, Fort Collins, CO, 17.

U.S. DOE (United States Department of Energy). 2006. *Energy demands on water resources: Report to Congress on the interdependency of energy and water*. Available at: http:// www.sandia.gov/energy-water/docs/121-RptToCongress-EWwEIAcomments-FINAL. pdf.

U.S. DOE (United States Department of Energy). 2010. *National algal biofuels technology roadmap*. Available at: http://www1.eere.energy.gov/biomass/pdfs/algal_biofuels_ roadmap.pdf.

U.S. EIA (United States Energy Information Administration). 2007. *Methodology for allocating municipal solid waste to biogenic and non-biogenic energy*. Available at: http://www.eia.doe.gov/cneaf/solar.renewables/page/mswaste/msw.pdf.

U.S. EIA (United States Energy Information Administration). 2009. *Short-term energy outlook supplement: Biodiesel supply and consumption in the short-term energy outlook*. Available at: http://www.eia.doe.gov/emeu/steo/pub/special/2009_sp_01.pdf.

U.S. EIA (United States Energy Information Administration). 2010. Biodiesel monthly report: 2009 edition. Available at: http://www.eia.doe.gov/cneaf/solar.renewables/page/ biodiesel/biodiesel.html.

U.S. EIA (United States Energy Information Administration). n.d.[a]. Biodiesel production (thousand barrels per day). International energy statistics. Available at: http://tonto.eia.doe.gov/ cfapps/ipdbproject/iedindex3.cfm?tid=79&pid=81&aid=1&cid= &syid=2004&eyid=2008&unit=TBPD.

U.S. EIA (United States Energy Information Administration). n.d.[b]. Total biofuels production (thousand barrels per day). *International energy statistics*. Available at: http://tonto.eia.gov/cfapps/ipdbproject/IEDIndex3.cfm?tid=79&pid=79&aid=1.

U.S. EIA (United States Energy Information Administration). n.d.[c]. U.S. imports of fuel ethanol (thousands of barrels per day). *Petroleum Navigator*. Available at: http://www.eia.doe.gov/dnav/pet/hist/LeafHandler.ashx?n=PET&s=MFEIMUS1&f=A.

U.S. EPA (United States Environmental Protection Agency). 1999. 5. Livestock manure management. In: U.S. methane emissions 1990-2020: Inventories, projections, and opportunities for reductions. Available at: http://epa.gov/methane/reports/05-manure.pdf.

U.S. EPA (United States Environmental Protection Agency). 2000. *National water quality inventory 2000 report*. Available at: http://www.epa.gov/305b/2000report.

U.S. EPA (United States Environmental Protection Agency). 2001. Control of air pollution from new motor vehicles: Heavy-duty engine and vehicle standards and highway diesel fuel sulfur control requirements. 40 CFR Parts 69, 80, and 86. Available at: http://www.epa.gov/otaq/highway-diesel/regs/2007-heavy-duty-highway.htm.

U.S. EPA (United States Environmental Protection Agency). 2002. *A comprehensive analysis of biodiesel impacts on exhaust emissions*. EPA-420-P-02-001. Available at: http://www.epa.gov/otaq/models/analysis/biodsl/p02001.pdf.

U.S. EPA (United States Environmental Protection Agency). 2006a. Framework for developing suspended and bedded sediments (SABS) water quality criteria. EPA-822-R-06-001. Available at: http://cfpub.epa.gov/ncea/cfm/recordisplay.cfm?deid=164423.

U.S. EPA (United States Environmental Protection Agency). 2006b. *Air quality criteria for ozone and related photochemical oxidants* (2006 final). EPA-600-R-05-004aF-cF.

U.S. EPA (United States Environmental Protection Agency). 2006c. *Wadeable Streams Assessment: A collaborative survey of the nation's streams*. EPA-841-B-06-002. Available at: http://www.epa.gov/owow/streamsurvey.

U.S. EPA (United States Environmental Protection Agency). 2006d. *Water quality assessment and total maximum daily loads information*. Available at: http://www.epa.gov/waters/ir/index.html.

U.S. EPA (United States Environmental Protection Agency). 2007. *Environmental law applicable to construction and operation of ethanol plants*. EPA-907-B-07-001. Available at: http://www.epa.gov/region07/priorities/agriculture/pdf/ethanol_plants_manual.pdf.

U.S. EPA (United States Environmental Protection Agency). 2008a. *Environmental law applicable to construction and operation of biodiesel production facilities*. EPA-907-B 08-001. Available at: http://www.epa.gov/Region7/priorities/agriculture/pdf/biodiesel_manual.pdf.

U.S. EPA (United States Environmental Protection Agency). 2008b. National coastal condition report—III. EPA-842-R-08-002. Available at: http://www.epa.gov/nccr.

U.S. EPA (United States Environmental Protection Agency). 2008c. *Applicability of effluent guidelines and categorical pretreatment standards to biodiesel manufacturing*. Available at: http://www.epa.gov/npdes/pubs/memo_biodieselpretreatment_aug08.pdf.

U.S. EPA (United States Environmental Protection Agency). 2009a. National water quality inventory: Report to Congress: 2004 reporting cycle. EPA-841-R-08-001. Available at: http://www.epa.gov/305b.

U.S. EPA (United States Environmental Protection Agency). 2009b. *Biofuels compendium.* Available at: http://www.epa.gov/oust/altfuels/bfcompend.htm.

U.S. EPA (United States Environmental Protection Agency). 2009c. *Biofuels compendium—ethanol—equipment compatibility.* Available at: http://www.epa.gov/oust/altfuels/ethcompat.htm.

U.S. EPA (United States Environmental Protection Agency). 2009d. Report to Congress on public health, air quality, and water resource impacts of fuel additive substitutes for MTBE, as required by Section 1505 of the Energy Policy Act 2005.

U.S. EPA (United States Environmental Protection Agency). 2010a. *Renewable fuel standard program (RFS2) regulatory impact analysis.* EPA-420-R-10-006. Available at: http://www.epa.gov/otaq/renewablefuels/420r10006.pdf.

U.S. EPA (United States Environmental Protection Agency). 2010b. *Water quality assessment and total maximum daily loads information* (ATTAINS). Available at: http://www.epa.gov/waters/ir/index.html.

U.S. EPA (United States Environmental Protection Agency). 2010c. *National causes of impairment.* In: National summary of state information. Available at: http://iaspub.epa.gov/waters10/attains_nation_cy.control#causes.

U.S. EPA (United States Environmental Protection Agency). 2010d. Regulation of fuels and fuel additives: Changes to renewable fuel standard program. EPA-HQ-OAR-2005-0161.

U.S. EPA (United States Environmental Protection Agency). 2010e. *Definition of solid waste for RCRA Subtitle C hazardous waste.* Available at: http://www.epa.gov/osw/hazard/dsw/.

U.S. EPA (United States Environmental Protection Agency). 2010f. Partial grant and partial denial of Clean Air Act waiver Application submitted by Growth Energy to increase the allowable ethanol content of gasoline to 15 percent: Decision of the Administrator. Available at: http://www.epa.gov/otaq/regs/fuels/additive/e15#waiver.

U.S. EPA (United States Environmental Protection Agency). 2010g. Background on ecological watershed monitoring program. Available at http://www.epa.gov/oppsrrd1/reregistration/atrazine/atrazine_update.htm#ewmp.

U.S. ITC (United States International Trade Commission). 2010. Interactive Tariff and Trade DataWeb. U.S. imports for consumption. Report: HTS-2207: Ethyl alcohol, undenatured of an alcoholic strength by volume of 80% vol. or higher; ethyl alcohol and other spirits, denatured, of any strength first unit of quantity by first unit of quantity.

Vadas, PA; Barnett, KH; Undersander, DJ. 2008. Economics and energy of ethanol production from alfalfa, corn, and switchgrass in the Upper Midwest, USA. *Bioenergy Research*, 1, 44-55.

Van Buskirk, J; Willi, Y. 2004. Enhancement of farmland biodiversity within set-aside land. *Conservation Biology*, 18(4), 987-994.

Van der Steen, NP; Nakiboneka, P; Mangalika, L; Ferrer, AVM; Gijzen, HJ. 2003. Effect of duckweed cover on greenhouse gas emissions and odor release from waste stabilization ponds. *Water Science and Technology*, 48(2), 341-348.

VanLoocke, A; Bernacchi, C; Twine, T. 2010. The Impacts of Miscanthus x giganteus production on the Midwest US hydrologic cycle. Global Change *Biology Bioenergy*, 2(4), 180-191.

Varvel, GE; Vogel, KP; Mitchell, RB; Follett, RF; Kimble, JM. 2008. Comparison of corn and switchgrass on marginal soils for bioenergy. *Biomass and Bioenergy*, 32(1), 18-21.

Vogel, KP; Masters, RA. 1998. Developing switchgrass into a biomass fuel crop for the midwestern USA. Paper presented at BioEnergy '98: Expanding Bioenergy Partnerships, Madison, WisconsIn: 4–8 October 1998.

Volk, TA; Abrahamson, LP; Nowak, CA; Smart, LB; Tharakan, PJ; White, EH. 2006. The development of short-rotation willow in the northeastern United States for bioenergy and bioproducts, agroforestry and phytoremediation. *Biomass and Bioenergy*, 30(8-9), 715-727.

Von Blottnitz, H; Curran, MA. 2007. A review of assessments conducted on bio-ethanol as a transportation fuel from a net energy, greenhouse gas, and environmental life cycle perspective. *Journal of Cleaner Production*, 15, 607-619.

Waddell, KL. 2002. Sampling coarse woody debris for multiple attributes in extensive resource inventories. *Ecological Indicators*, 1, 139-153.

Walsh, ME; de la Torre Ugarte, DG; Shapouri, H; Slinsky, SP. 2003. Bioenergy crop production in the United States, Potential quantities, land use changes, and economic impacts on the agricultural sector. *Environmental and Resource Economics*, 24(4), 313-333.

Walter, WD; Vercauteren, KC; Gilsdorf, JM; Hygnstrom, SE. 2009. Crop, native vegetation, and biofuels, Response of white-tailed deer to changing management priorities. *Journal of Wildlife Management*, 73, 339-344.

Wang, D; Lebauer, DS; Dietze, MC. 2010. A quantitative review comparing the yield of switchgrass in monocultures and mixtures in relation to climate and management factors. *Global Change Biology Bioenergy*, 2(1), 16-25.

Wang, L; Robertson, DM; Garrison, PJ. 2007a. Linkages between nutrients and assemblages of macroinvertebrates and fish in wadeable streams. *Environmental Management*, 39, 194- 212. Available at , http://www.springerlink.com/content/ 4k441384336k1x28/ fulltext.pdf.

Wang, M; Wu, M; Huo, H. 2007b. Life-cycle energy and greenhouse gas emission impacts of different corn ethanol plant types. *Environmental Research Letters*, 2, 1-13.

Warburton, DB; Klimstra, WD. 1984. Wildlife use of no-till and conventionally tilled corn fields. *Journal of Soil and Water Conservation*, 39(5), 327-330.

Weed, DAJ; Kanwar, RS. 1996. Nitrate and water present in and flowing from root-zone soil. *Journal of Environmental Quality*, 25, 709-719.

Weibe, K; Gollehon, N. 2006. *Agricultural resources and environmental indicators*, 2006 edition. Economic Information Bulletin No. (EIB-16). U.S. Department of Agriculture, Economic Research Service. Available at: http://www.ers.usda.gov/publications/arei/ eib16/.

Weigelt, A; Weisser, WW; Buchmann, N; Scherer-Lorenzen, M. 2009. Biodiversity for multifunctional grasslands, Equal productivity in high-diversity low-input and low-diversity high-input systems. *Biogeosciences*, 6, 1695-1706.

West, TO; Post, WM. 2002. Soil organic carbon sequestration rates by tillage and crop rotation, A global data analysis. *Soil Science Society of America Journal*, 66(6), 1930-1946.

Whalen, J; Cissel, B. 2009. Soil insect management in field corn. University of Delaware Cooperative Extension.

Whicker, JJ; Pinder, JE; Breshears, DD. 2008. Thinning semiarid forests amplifies wind erosion comparably to wildfire, Implications for restoration and soil stability. *Journal of Arid Environments*, 72(4), 494-508.

White, EM. 2010. Woody biomass for bioenergy and biofuels in the United States—a briefing paper. Gen. Tech. Rep. PNW-GTR-825. U.S. Department of Agriculture.

Wilhelm, WW; Johnson, JME; Karlen, DL; Lightle, DT. 2007. Corn stover to sustain soil organic carbon further constrains biomass supply. *Agronomy Journal*, 99(6), 1665-1667.

Wilkinson, M; Tepfer, M. 2009. Fitness and beyond: Preparing for the arrival of GM crops with ecologically important novel characters. *Environmental Biosafety Research*, 8, 1-14.

Williams, PR; Inman, DD; Aden, A; Heath, GA. 2009. Environmental and sustainability factors associated with next-generation biofuels in the U.S.: What do we really know? *Environmental Science and Technology*, 43, 4763-4775.

Wiltsee, G. 1998. Urban waste grease resource assessment. NREL/SR-570-26141. National Renewable Energy Laboratory.

Winebrake, J; Wang, M; He, D. 2001. Toxic emissions from mobile sources: A total fuel-cycle analysis for conventional and alternative fuel vehicles. *Journal of the Air and Waste Management Association*, 51, 1073-1086.

Wolt, JD. 2009. Advancing environmental risk assessment for transgenic biofeedstock crops. *Biotechnology for Biofuels*., 2, 27.

Wright, L; Turhollow, A. 2010. Switchgrass selection as a “model” bioenergy crop, A history of the process. *Biomass and Bioenergy*, 34, 851-868.

Wright, LL. 1994. Production technology status of woody and herbaceous crops. *Biomass and Bioenergy*, 6(3), 191-210.

Wu, M. 2008. Analysis of the efficiency of the U.S. ethanol industry. Center for Transportation Research, Argonne National Laboratory. Report No. 10229. U.S. Department of Energy.

Wu, M; Wang, M. 2006. Energy and emission benefits of alternative transportation liquid fuel derived from switchgrass: A fuel life-cycle assessment. *Biotechnology Progress*; 22(4), 1012-1024.

Wu, M; Mintz, M; Wang, M; Arora, S. 2009. Water consumption in the production of ethanol and petroleum gasoline. *Environmental Management*, 44, 981-997.

Wullschleger, SD; Davis, EB; Borsuk, ME; Gunderson, CA; Lynd, LR. 2010. Biomass production in switchgrass across the United States: Database description and determinants of yield. *Agronomy Journal*, 102(4), 1158-1168.

Yacobucci, B. 2005. CRS report to Congress; Ethanol imports and the Caribbean Basin Initiative. Congressional Research Service. Order Code RS21930. Available at: http://opencrs.com/document/RS21930/2005-01-06/.

Yang, JD; Worley, E; Wang, MY; Lahner, B; Salt, DE; Saha, M; Udvardi, M. 2009. Natural variation for nutrient use and remobilization efficiencies in switchgrass. Bioenergy Research 2(4), 257-266.

Yanowitz, J; McCormick, R. 2009. Effect of E85 on tailpipe emissions from light-duty vehicles. *Journal of the Air and Waste Management Association*, 59, 172-182.

Zah, R; Ruddy, TF. 2009. International trade in biofuels: An introduction to the special issue. *Journal of Cleaner Production*, 17, S1-S3.

Zaimes, GN; Schultz, RC; Isenhart, TM. 2004. Stream bank erosion adjacent to riparian forest buffers, row-crop fields, and continuously-grazed pastures along bear creek in central Iowa. *Journal of Soil and Water Conservation*, 59(1), 19-27.

Zavaleta, ES; Pasari, JR; Hulvey, KB; Tilman, GD. 2010. Sustaining multiple ecosystem functions in grassland communities requires higher biodiversity. *Proceedings of the National Academy of Sciences of the United States of America*, 107, 1443-1446.

Zhu, Y; Fox, RH. 2003. Corn-soybean rotation effects on nitrate leaching. *Agronomy Journal*, 95, 1028-1033.

End Notes

[1] The 2005 Energy Policy Act amended the Clean Air Act and established the first national renewable fuel standards. The statute specifies the total volume of renewable fuel that is to be used based on the volume of gasoline sold in the U.S. each year, with the total volume of renewable fuel increasing over time to 7.5 billion gallons in 2012.

[2] To be considered "renewable," fuels produced by biorefineries constructed after EISA's enactment on December 19, 2007, must generally achieve at least a 20 percent reduction in life cycle greenhouse gas emissions compared to petroleum fuels.

[3] EISA 2007 amended Section 211(o) of the Clean Air Act to include the definitions and requirements of RFS2.

[4] Pesticides include antimicrobials, fungicides, herbicides, insecticides, and rodenticides.

[5] Transportation fuel includes fuels used in motor vehicles, motor vehicle engines, non-road vehicles, or non-road engines (except for ocean-going vessels).

[6] EISA defines "additional renewable fuel" as "fuel produced from renewable biomass that is used to replace or reduce fossil fuels used in heating oil or jet fuel." Though RFS2 does not specify a volume standard for this fuel category, it does allow renewable fuel blended into heating oil or jet fuel to count toward achieving the standard for total renewable fuel. More information about "additional renewable fuel" can be found in Section II.b.e of the final RFS2 rule, available at http://www.epa.gov/otaq/fuels/renewablefuels/regulations

[7] EISA requires that all renewable fuel be made from feedstocks that meet the new definition of renewable biomass, which includes certain land use restrictions. For full details, see Section 3.1.

[8] Although EISA specified a 2010 cellulosic biofuel requirement of 100 million gallons/year, as shown in Table 2-1, EPA determined that this level was not achievable for 2010. The U.S. EIA projected 5 million gallons/year of cellulosic production for 2010 (6.5 million gallons ethanol equivalent), and EPA accepted this as the 2010 standard. While this is lower than the level specified in EISA, no change to the advanced biofuel and total renewable fuel standards was warranted due to the inclusion of an energy-based equivalence value for biodiesel and renewable diesel.

[9] Percentages calculated as bushels consumed for ethanol fuel as a function of total consumption of grain, including exports.

[10] Percentage of soybeans allocated to biodiesel calculated by dividing soybean oil allocated to production of biodiesel by annual crushing yields and expressed as a function of total disposition.

[11] Numbers in text were calculated by summing miles/acres reported by each state in their 305(b) assessments as impaired by "nutrients"; "ammonia, un-ionized"; "nitrogen, total"; "nutrient/eutrophication"; "phosphorus, total"; "ammonia, total"; "nitrogen, nitrate"; and "ammonia."

[12] SPARROW (SPAtially Referenced Regressions On Watershed) is a watershed model developed by USGS relating water quality measurements at monitoring stations to other watershed attributes. The model estimates nitrogen and phosphorus entering a stream per acre of land, and evaluates the contributions of nutrient sources and watershed properties that control nutrient transport.

[13] The Soil and Water Assessment Tool (SWAT) is a public domain model jointly developed by USDA Agricultural Research Service and Texas A&M University System. SWAT is a river basin–scale model to simulate the quality and quantity of surface and ground water and predict the environmental impact of land management practices on different soil patterns and land use patterns.

[14] See also http://cfpub.epa.gov/npdes/stormwater/menuofbmps/index.cfm?action=factsheet_results&view=specific &bmp=82.

[15] Numbers in text were calculated by summing miles/acres reported by each state in their 305(b) assessments as impaired by "sedimentation/siltation" or "turbidity."

[16] *No-till* refers to the absence of soil tillage to establish a seed bed, meaning the farmer plants the crop directly into the previous year's crop residue. In *strip-till*, only the portion of the soil that is to contain the seed row is disturbed. In *ridge-till*, plants grow on hills that are the product of cultivation of the previous crop and are not

tilled out after harvest. In *mulch-till*, plant residues are conserved but a field cultivator or disks are used to till before planting to partially incorporate the residue into the soil.

[17] Defined as any tillage practice that leaves less than 15 percent of crop residues on the soil surface after planting.

[18] It should be noted that the removal of crop residues by percent mass is not the same as by percent soil coverage. All the percentages from the studies discussed here are by percent mass, unless otherwise noted.

[19] http://bioenergy

[20] See http://plantsfor a full explanation of the assessment and approved management practices.

[21] See http://plantsfor a full explanation of the assessment and approved management practices.

[22] The term "microalgae" refers to photosynthetic and heterotrophic organisms too small to be easily seen with the naked eye—distinguished from macroalgae, otherwise known as seaweed. Macroalgae is generally not grown as an energy crop. In this report the terms "algae" and "microalgae" are used interchangeably.

[23] Specifically, a partnership between the Solid Waste Authority of Central Ohio and Quasar Energy Group to produce ethanol from municipal solid waste (see http://www.quasarenergygroup.com/pages/home.html), as well as the "Deploying Renewable Energy—Transforming Waste to Value" grant program (see http://www.biomassintel.com/ohio-10-million-available-waste-to-energy-grant-program/).

[24] Sources include the Renewable Fuels Association's Ethanol Biorefinery Locations (updated October 22, 2009) and *Ethanol Producer Magazine*'s producing plant list (last modified on October 22, 2009), in addition to information gathered from producer websites and follow-up correspondence.

[25] FAPRI-CARD is a worldwide agricultural sector economic model. For the RIA, the model was run by the Center for Agricultural and Rural Development at Iowa State University on behalf of EPA.

In: Biofuels and the Environment
Editor: Tobias M. Fischer

ISBN: 978-1-62948-092-3

Chapter 2

RENEWABLE FUEL STANDARD (RFS): OVERVIEW AND ISSUES*

Randy Schnepf and Brent D. Yacobucci

SUMMARY

Federal policy has played a key role in the emergence of the U.S. biofuels industry. Policy measures include minimum renewable fuel usage requirements, blending and production tax credits, an import tariff, loans and loan guarantees, and research grants. One of the more prominent forms of federal policy support is the Renewable Fuel Standard (RFS)—whereby a minimum volume of biofuels is to be used in the national transportation fuel supply each year. This report describes the general nature of the RFS mandate and its implementation, and outlines some emerging issues related to the continued growth of U.S. biofuels production needed to fulfill the expanding RFS mandate, as well as the emergence of potential unintended consequences of this rapid expansion.

Congress first established the RFS with the enactment of the Energy Policy Act of 2005 (EPAct, P.L. 109-58). This initial RFS (referred to as RFS1) mandated that a minimum of 4 billion gallons be used in 2006, rising to 7.5 billion gallons by 2012. Two years later, the Energy Independence and Security Act of 2007 (EISA, P.L. 110-140) greatly expanded the biofuel mandate volumes and extended the date through 2022. The expanded RFS (referred to as RFS2) required the annual use of 9 billion gallons of biofuels in 2008, rising to 36 billion gallons in 2022, with at least 16 billion gallons from cellulosic biofuels, and a cap of 15 billion gallons for corn-starch ethanol.

In addition to the expanded volumes and extended date, RFS2 has three important distinctions from RFS1. First, the total renewable fuel requirement is divided into four separate, but nested categories—total renewable fuels, advanced biofuels, biomass-based diesel, and cellulosic biofuels—each with its own volume requirement. Second, biofuels qualifying under each category must achieve certain minimum thresholds of lifecycle greenhouse gas (GHG) emission reductions, with certain exceptions applicable to existing facilities. Third, all renewable fuel must be made from feedstocks that meet an amended definition of renewable biomass, including certain land use restrictions.

---

* This is an edited, reformatted and augmented version of Congressional Research Service, Publication No. R40155, dated March 14, 2013.

The Environmental Protection Agency (EPA) is responsible for establishing and implementing regulations to ensure that the nation's transportation fuel supply contains the mandated biofuels volumes. EPA's initial regulations for administering RFS1 (issued in April 2007) established detailed compliance standards for fuel suppliers, a tracking system based on renewable identification numbers (RINs) with credit verification and trading, special treatment of small refineries, and general waiver provisions. EPA rules for administering RFS2 (issued in February 2010) built upon the earlier RFS1 regulations and include specific deadlines for announcing annual standards, as well as greater specificity on potential waiver requests and RIN oversight.

Over the long term, the RFS is likely to play a dominant role in the development of the U.S. biofuels sector, but with considerable uncertainty regarding potential spillover effects in other markets and on other important policy goals. Emerging resource constraints related to the rapid expansion of U.S. corn ethanol production have provoked questions about its long-run sustainability and the possibility of unintended consequences in other markets as well as on the environment. Questions also exist about the ability of the U.S. biofuels industry to meet the expanding mandate for biofuels from non-corn sources such as cellulosic biomass materials, whose production capacity has been slow to develop, or biomass-based biodiesel, which remains expensive to produce owing to the relatively high prices of its feedstocks. Finally, considerable uncertainty remains regarding the development of the infrastructure capacity (e.g., trucks, pipelines, pumps, etc.) needed to deliver the expanding biofuels mandate to consumers.

INTRODUCTION

Increasing dependence on foreign sources of crude oil, concerns over global climate change, and the desire to promote domestic rural economies have raised interest in renewable biofuels as an alternative to petroleum in the U.S. transportation sector. In response to this interest, U.S. policymakers have enacted an increasing variety of policies, at both the state and federal levels, to directly support U.S. biofuels production and use (although some of these policies have expired).[1] Policy measures have included blending and production tax credits to lower the cost of biofuels to end users, an import tariff to protect domestic ethanol from cheaper foreign-produced ethanol, research grants to stimulate the development of new biofuels technologies, loans and loan guarantees to facilitate the development of biofuels production and distribution infrastructure, and, perhaps most important, minimum usage requirements to guarantee a market for biofuels irrespective of their cost.[2] As a result of expanding policy support, biofuels (primarily corn-based ethanol and biodiesel) production has grown significantly (up over 600%) since the early 2000s. However, despite the rapid growth, U.S. biofuels consumption remains small as a component of U.S. motor fuels, comprising about 5.7% of total transportation fuel consumption (on a gasoline-equivalent basis) in 2012.[3]

Initially, the most significant federal programs for supporting biofuels were tax credits for the production or blending of ethanol and biodiesel into the nation's fuel supply. However, under the Renewable Fuel Standard (RFS)—first established in 2005, then greatly expanded in 2007 (as described below)—Congress mandated biofuels use. In the long term, the expanded RFS usage mandate is likely to prove more significant than tax incentives in promoting the use of these fuels.

This report focuses specifically on the RFS. It describes the general nature of the biofuels RFS and its implementation, and outlines some of the emerging issues related to the sustainability of the continued growth in U.S. biofuels production needed to fulfill the expanding RFS mandate, as well as the emergence of potential unintended consequences of this rapid expansion. This report does not address the broader public policy issue of how best to support U.S. energy policy.

THE RENEWABLE FUEL STANDARD (RFS)

Congress first established a Renewable Fuel Standard (RFS)—a mandatory minimum volume of biofuels to be used in the national transportation fuel supply—in 2005 with the enactment of the Energy Policy Act of 2005 (EPAct, P.L. 109-58). The initial RFS (referred to as RFS1) mandated that a minimum of 4 billion gallons of renewable fuel be used in the nation's gasoline supply in 2006, and that this minimum usage volume rise to 7.5 billion gallons by 2012 (**Table 1**). Two years later, the Energy Independence and Security Act of 2007 (EISA, P.L. 110-140) superseded and greatly expanded the biofuels mandate to 36 billion gallons by 2022. In addition to gasoline, the expanded RFS (referred to as RFS2) applies to most transportation fuel used in the United States—including diesel fuel intended for use in highway motor vehicles, non-road, locomotive, and marine diesel (MVNRLM).[4]

RFS2 directly supports U.S. biofuels production by providing a mandatory market for qualifying biofuels—fuel blenders must incorporate minimum volumes of biofuels in their annual transportation fuel sales irrespective of market prices. By guaranteeing a market for biofuels, RFS2 substantially reduces the risk associated with biofuels production, thus providing an indirect subsidy for capital investment in the construction of biofuels plants. As such, the expanding RFS is expected to continue to stimulate growth of the biofuels industry.

EISA was passed on December 19, 2007, and EPA issued its final rule to implement and administer the RFS2 on February 3, 2010.[5] The new rule builds upon the earlier rule for RFS1. However, there are four major distinctions between RFS1 and RFS2:

- First and foremost, RFS2 increases the mandated usage volumes and extends the time frame over which the volumes ramp up through at least 2022 (**Table 1**).
- Second, RFS2 subdivides the total renewable fuel requirement into four separate but nested categories—total renewable fuels, advanced biofuels, biomass-based diesel, and cellulosic biofuels—each with its own volume requirement or standard (described below).
- Third, biofuels qualifying under each nested RFS2 category must achieve certain minimum thresholds of lifecycle greenhouse gas (GHG) emission performance, with exceptions applicable to facilities existing or under construction when EISA was enacted (**Table 2**).[6]
- Fourth, under RFS2 all renewable fuel must be made from feedstocks that meet a revised definition of renewable biomass, including certain land use restrictions.[7]

The RFS is administered by the Environmental Protection Agency (EPA).[8] EPA issued its final rule for administering RFS1 in April 2007.[9] This rule established detailed compliance standards for fuel suppliers, a tracking system based on renewable identification numbers (RINs) with credit verification and trading, provisions for treatment of small refineries, and general waiver provisions. EPA rules for administering RFS2 (issued in February 2010) built upon the earlier RFS1 regulations and include specific deadlines for announcing annual standards, as well as greater specificity on potential waiver requests and RIN oversight.

Table 1. EISA 2007 Expansion of the Renewable Fuel Standard (in billions of gallons)

		RFS2 biofuel mandate					
				Portion to be from advanced biofuels			
Year	RFS1 biofuel Mandate in EPAct of 2005	Total renewable fuels	Cap on corn starchderived ethanol	Total noncorn starch	Cellulosic	Biomassbased diesel	Other[a]
2006	4.0	—	—	—	—	—	—
2007	4.7	—	—	—	—	—	—
2008	5.4	9.00	9.0	0.00	0.00	0.00	0.00
2009	6.1	11.10	10.5	0.60	0.00	0.00	0.10
2010	6.8	12.95	12.0	0.95	0.0065[b]	1.15[c]	0.29
2011	7.4	13.95	12.6	1.35	0.006[d]	0.80	0.54
2012	7.5	15.20	13.2	2.00	0.00[e]	1.00	1.00
2013	7.6 (est.)	16.55	13.8	2.75	0.014[f]	1.28 [f]	1.46
2014	7.7 (est.)	18.15	14.4	3.75	1.75	[g]	1.00
2015	7.8 (est.)	20.50	15.0	5.50	3.00	[g]	1.50
2016	7.9 (est.)	22.25	15.0	7.25	4.25	[g]	2.00
2017	8.1 (est.)	24.00	15.0	9.00	5.50	[g]	2.50
2018	8.2 (est.)	26.00	15.0	11.00	7.00	[g]	3.00
2019	8.3 (est.)	28.00	15.0	13.00	8.50	[g]	3.50
2020	8.4 (est.)	30.00	15.0	15.00	10.50	[g]	3.50
2021	8.5 (est.)	33.00	15.0	18.00	13.50	[g]	3.50
2022	8.6 (est.)	36.00	15.0	21.00	16.00	[g]	4.00
2023	—	[h]	[h]	[h]	[h]	[h]	[h]

Source: RFS1 is from EPAct (P.L. 109-58), Section 1501; RFS2 is from EISA (P.L. 110-140), Section 202.

[a] "Other" advanced biofuels is a residual category left over after the ethanol-equivalent gallons of cellulosic and biodiesel biofuels are subtracted from the "Total" advanced biofuels mandate.

[b] The initial EISA cellulosic biofuels mandate for 2010 was for 100 million gallons. On February 3, 2010, EPA revised this mandate downward to 6.5 million ethanol-equivalent gallons.

[c] The biomass-based diesel mandate for 2010 combines the original EISA mandate of 0.65 billion gallons (bgals) with the 2009 mandate of 0.5 bgals.

[d] The initial RFS for cellulosic biofuels for 2011 was 250 million gallons. In November 2010 EPA revised this mandate downward to 6.0 million ethanol-equivalent gallons.

[e] The initial RFS for cellulosic biofuels for 2012 was 500 million gallons. In December 2011 EPA revised this mandate downward to 10.45 million ethanol-equivalent gallons. In January 2013, the U.S. Court of Appeals for D.C. vacated EPA's initial cellulosic mandate for 2012 and remanded EPA to replace it with a revised mandate. On February 28, 2013, EPA dropped the 2012 RFS for cellulosic biofuels to zero.

[f] The initial 2013 cellulosic RFS was 1 bgals. In January 2013, EPA revised this mandate to 14 million ethanol-equivalent gals. The 2013 biodiesel mandate was revised upwards from 1 bgals to 1.28 bgals actual volume.

[g] To be determined by EPA through a future rulemaking, but no less than 1.0 billion gallons.

[h] To be determined by EPA through a future rulemaking.

Four Biofuel Categories

RFS2 includes four biofuel categories, each with a specific volume mandate and lifecycle GHG emission reduction threshold (as compared to the lifecycle GHG emissions of the 2005 baseline average gasoline or diesel fuel that it replaces). Each is subject to biomass feedstock criteria.

- **Total renewable fuels**. The mandate grows from nearly 13 billion gallons (bgals) in 2010 to 36 bgals in 2022. Biofuels must reduce lifecycle GHG emissions by at least 20% relative to conventional fuels to qualify as a renewable fuel. Most biofuels, including corn-starch ethanol from new facilities, qualify for this mandate. However, the volume of corn-starch ethanol included in the RFS is capped at 13.8 bgals in 2013, but grows to 15 bgals by 2015 and is fixed thereafter.
- **Advanced biofuels**.[10] The mandate grows from nearly 1 bgals in 2010 to 21 bgals in 2022. Advanced biofuels must reduce lifecycle GHG emissions by 50% to qualify. A subcomponent of the total renewable fuels mandate, this category includes biofuels produced from non-corn feedstocks—corn-starch ethanol is expressly excluded from this category. Potential feedstock sources include grains such as sorghum and wheat. Imported Brazilian sugarcane ethanol, as well as biomass-based biodiesel and biofuels from cellulosic materials (including non-starch parts of the corn plant such as the stalk and cob) also qualify. The total advanced biofuel mandate for 2013 is 2.75 bgals (ethanol equivalent).
- **Cellulosic and agricultural waste-based biofuel**. The mandate grows from 100 million gallons in 2010 to 16 bgals in 2022 (subsequently, RFS mandates were lowered for 2010, 2011, and 2012, and are proposed to be lowered in 2013—see discussion under "Waivers to Annual Biofuel Standards"). Cellulosic biofuels must reduce lifecycle GHG emissions by at least 60% to qualify. Cellulosic biofuels are renewable fuels derived from cellulose, hemicellulose, or lignin. This includes cellulosic biomass ethanol as well as any biomass-to-liquid fuel such as cellulosic gasoline or diesel.
- **Biomass-based biodiesel (BBD)**. The mandate grows from 0.5 bgals in 2009 to 1 bgals in 2012.[11] Any diesel fuel made from biomass feedstocks (including algae) qualifies, including biodiesel (mono-alkyl esters) and non-ester renewable diesel (e.g., cellulosic diesel).[12] The lifecycle GHG emissions reduction threshold is 50%. EPA established the 2013 mandate at 1.28 bgals (actual volume).

Usage Volume Requirements

RFS2 is essentially a biofuels mandate with limits on corn-starch ethanol inclusion and carve-outs for higher-performing biofuels (**Figure 1**)—as measured by reductions in lifecycle GHG emissions. The cap on the volume of ethanol derived from corn starch that can be counted under the RFS is intended to encourage the use of non-corn-based biofuels. As a result, corn-starch ethanol blended in excess of its annual cap is not credited toward the annual total RFS.

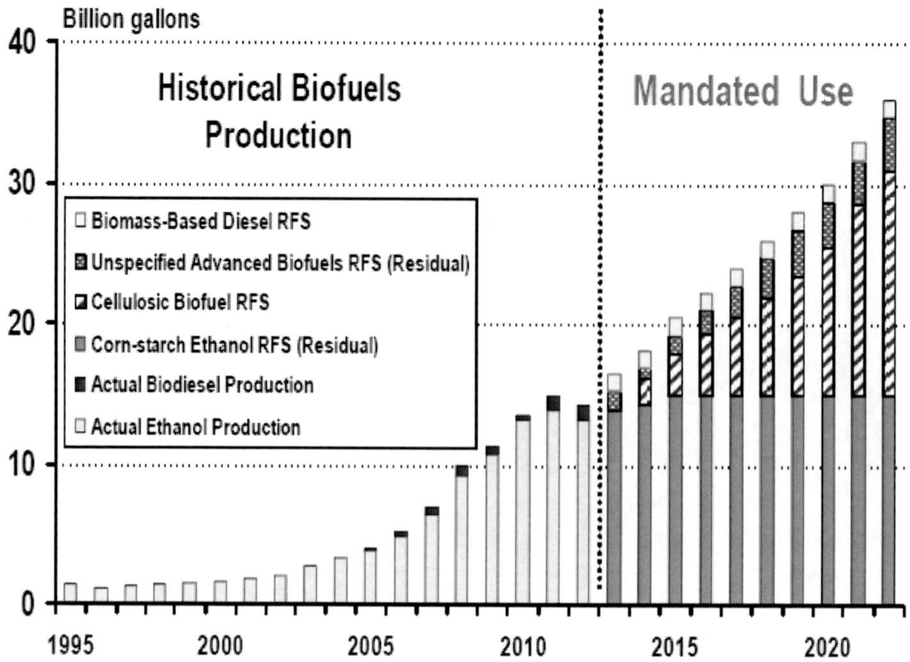

Source: Actual ethanol and biodiesel production data for 1995-2012 are from the Energy Information Agency (EIA), Department of Energy; the RFS2 mandates by category for 2013-2022 are from EISA (P.L. 110-140).

Figure 1. Renewable Fuels Standard (RFS2) vs. U.S. Biofuel Production Since 1995.

Nested Categories

Because of the nested nature of the biofuel categories, any renewable fuel that meets the requirement for cellulosic biofuels or biomass-based diesel (BBD) is also valid for meeting the advanced biofuels requirement. Thus, if any combination of cellulosic biofuels or BBD were to exceed their individual mandates, the surplus volume would count against the advanced biofuels mandate, thereby reducing the potential need for imported sugar-cane ethanol or other fuels to meet the unspecified portion of the advanced biofuels mandate (which grows to 21 bgals by 2022).

Similarly, any renewable fuel that meets the requirement for advanced biofuels is also valid for meeting the overall total renewable fuel requirement (which grows to 36 bgals by 2022). As a result, any combination of cellulosic biofuels, BBD, or imported sugar-cane ethanol that exceeds the advanced biofuel mandate would reduce the potential need for corn-starch ethanol to meet the overall mandate.

Required Reduction in Lifecycle Greenhouse Gas Emissions

In addition to volume mandates, EISA specified that the lifecycle GHG emissions of a qualifying renewable fuel must be less than the lifecycle GHG emissions of the 2005 baseline average gasoline or diesel fuel that it replaces.[13] EISA established lifecycle GHG emission thresholds for each of the RFS2 biofuels categories (**Table 2**). With respect to the GHG emissions assessments, EISA specifically directed EPA to evaluate the aggregate quantity of GHG emissions (including direct emissions and significant indirect emissions, such as significant emissions from land use changes) related to the full lifecycle, including all stages of fuel and feedstock production, distribution, and use by the ultimate consumer.

Table 2. EISA-Mandated Reductions in Lifecycle GHG Emissions by Biofuel Category (percent reduction from 2005 baseline for gasoline or diesel fuel)

Biofuels category	Threshold reduction
Renewable fuela	20%
Advanced biofuels	50%
Biomass-based diesel	50%
Cellulosic biofuel	60%

Source: "Regulatory Announcement: EPA Finalizes Regulations for the National Renewable Fuel Standard Program for 2010," EPA-420-F-10-007, Office of Transportation and Air Quality, EPA, February 3, 2010.

[a] The 20% criteria applies to renewable fuel from facilities that commenced construction after December 19, 2007, the date EISA was signed into law.

Fuel Pathways (including ILUC) Meeting Lifecycle GHG Thresholds

EPA's initial measurement of lifecycle GHG reductions for various biofuels pathways under RFS1 was viewed as contentious by some due to the explicit requirement to incorporate so-called "indirect land use changes" (ILUC) in the GHG emissions assessment.[14] ILUC refers to the idea that diversion of an acre of traditional field cropland in the United States to grow a biofuels feedstock crop might result (due to market price effects) in that same acre reappearing at another location and potentially on virgin soils, such as the Amazon rainforest. Such a transfer—when included in the lifecycle GHG calculation of a particular biofuel—could result in an estimated net increase in GHG emissions.

Several environmental and academic groups argued that, as a result of ILUC costs, corn ethanol should not be permissible under the RFS2. Biofuels proponents argued that ILUC was too vague a concept to be measurable in a meaningful way, and that it alone should not determine the fate of the U.S. biofuels industry. As a result, EPA reconsidered all of the evidence (including ILUC) and made relevant adjustments to its analytical tools. The resultant changes were announced as part of its final RFS2 rule of February 3, 2010.[15]

In addition, EPA has pointed out that other pathways are likely to be similar enough to existing qualifying pathways (**Table 3**) that they can be extended the same GHG reduction compliance determinations.[16] However, EPA stated that, although the announced determinations for the qualifying fuel pathways (**Table 3**) are final for the time being, its lifecycle methodology remains subject to new developments in the state of scientific knowledge, and that future reassessments may alter the current status of these fuel pathways.

EPA says that it will be able to make determinations on several other potential biomass crops and their fuel pathways in the future.[17] For example, in a February 2013 rule that qualified several new fuel pathways for cellulosic biofuel production, EPA stated that it hoped to provide opportunities to increase the volume of advanced, low-GHG renewable fuels.[18] For other biofuel pathways not yet modeled, EPA encourages parties to use a petition process to request EPA to examine additional pathways.

Table 3. Qualifying Fuel Pathways for Lifecycle GHG Emissions by Biofuel Category

Renewable Fuel—20% GHG Reduction
• Biofuel from the capacity of plants or production facilities that either existed or commenced construction prior to December 19, 2007. • Ethanol produced from corn starch at a new natural gas-fired facility (or expanded capacity from an existing facility) using advanced efficient technologies. • Biobutanol from corn starch. • Ethanol made from grain sorghum at dry mill facilities that use natural gas for process energy. • Note: biodiesel and renewable diesel produced from palm oil do not meet the minimum 20% lifecycle GHG reduction threshold needed to qualify as renewable fuel under the RFS program.
Advanced biofuels—50% GHG Reduction
• Ethanol produced from sugarcane (as in Brazil). • Naphtha and liquefied petroleum gas (LPG) from camelina oil. • Ethanol made from grain sorghum at dry mill facilities that use specified forms of biogas for both process energy and most electricity production.
Biomass-based diesel (including jet fuel and heating oil)—50% GHG Reduction
• Biodiesel and non-ester renewable diesel from soy oil, non-food grade corn oil, camelina oil, algal oils, waste oils, fats, and greases. • Biodiesel produced using esterification (a new process method) from soybean oil, oil from annual cover crops, algal oil, biogenic waste oils, fats, and greases, non-food grade corn oil, Canola or rapeseed oil, and camelina oil. • Non-ester renewable diesel based on electricity or natural gas for process energy and feedstocks of soybean oil, oil from annual cover crops, algal oil, biogenic waste oils, fats, and greases, or the non-cellulosic portions of food wastes (i.e., non-food grade corn oil, and camelina oil). • Biodiesel produced using a glycerolysis production process element combined with the traditional transesterification process from free fatty acids (FFA).
Cellulosic biofuels (either cellulosic ethanol or cellulosic diesel)—60% GHG Reduction
• Cellulosic biofuels based on perennial grasses including switchgrass, miscanthus, energy cane, giant reed, and napier grass. • Cellulosic biofuels produced from crop residue (e.g., corn stover, wheat straw, rice straw, and citrus residue), forest material (including slash, pre-commercial thinnings, and solid tree residue remaining from forest product production), secondary annual cover crops planted on existing crop land (e.g., winter cover crops), cellulosic components of separated food and yard waste (including biogenic waste from food processing), and separated municipal solid waste using certain processes identified below. • The following processes—all utilizing natural gas, biogas, and/or biomass as the only process energy sources— qualify as cellulosic biofuel: thermochemical pyrolysis; thermochemical gasification; biochemical direct fermentation; biochemical fermentation with catalytic upgrading; and any other process that uses biogas and/or biomass as the only process energy sources.

Source: EPA announcements of various rules and determinations as posted at "Renewable Fuels: Regulations and Standards," at http://www.epa.gov/otaq/fuels/renewablefuels/regulations.htm.

Notes: This table is not an official, inclusive listing of EPA certified biofuel pathways. For information on potential new biofuel feedstock or production process pathways currently under review, as well as completed pathways, see EPA, "Guidance on New Fuel Pathway Approval Process" at http://www.epa.gov/otaq/fuels/ renewablefuels/compliancehelp/rfs2-lca-pathways.htm.

Grandfathered Plants

Fuel from the capacity of facilities that either existed or commenced construction prior to December 19, 2007 (the date of enactment of EISA), is exempt from the 20% lifecycle GHG threshold requirement. The exemption is extended to ethanol facilities that commenced construction on or before December 31, 2009, provided that those facilities use natural gas, biofuels, or a combination thereof as processing fuel. However, any new expansion of production capacity at existing facilities must be designed to achieve the 20% GHG reduction threshold if the facility wants to generate RINs for that volume.

Feedstock Requirements

EISA changed the definition of renewable fuel to require that it be made from feedstocks that qualify under an amended definition of "renewable biomass."[19] As such, EISA limits not only the types of feedstocks that can be used to make renewable fuel, but also the land that these renewable fuel feedstocks may come from. Specifically excluded under the EISA definition are virgin agricultural land cleared or cultivated after December 19, 2007, as well as tree crops, tree residues, and other biomass materials obtained from federal lands. These restrictions are applicable to both domestic and foreign feedstock and biofuels producers.

Existing agricultural land includes three land categories—cropland, pastureland, and Conservation Reserve Program (CRP) land. Rangeland is excluded. Fallow land is defined as idled cropland and is therefore included within the definition of agricultural land.

EPA determined that fuels produced from five categories of feedstocks (primarily targeted for cellulosic biofuels) were expected to have less or no indirect land use change and thereby qualify as renewable biomass:[20]

- crop residues such as corn stover, wheat straw, rice straw, citrus residue;
- forest material including eligible forest thinnings and solid residue remaining from forest product production;
- secondary annual crops planted on existing cropland, such as winter cover crops;
- separated food and yard waste, including biogenic waste from food processing; and
- perennial grasses, including switchgrass and miscanthus.

Implementation of the RFS

The EPA is responsible for revising and implementing regulations to ensure that the national transportation fuel supply sold in the United States during a given year contains the mandated volume of renewable fuel in accordance with the four nested volume mandates of the RFS2.[21] To accomplish this task, EPA first calculates annual percentage standards for the four biofuel categories of RFS2. The percentage standards apply to refiners, blenders, and importers of gasoline and diesel fuels and are used to determine each individual company's renewable volume obligation (RVO). To facilitate meeting the requirements, while taking into consideration regional differences in biofuels production and availability, EPA established a system of tradable RINs. Percentage standards, RVOs, and RINs are described in this section.

Determining Annual Percentage Standards

In order to ensure that the requisite volumes of biofuels are used each year, EPA first estimates the total volume of transportation fuel that is expected to be used in the United

States during the upcoming year. EPA relies on projections from the Department of Energy's Energy Information Agency (EIA) for this estimate.[22] The percentage obligation (or standard) is computed as the total amount of renewable fuels mandated to be used in a given year expressed as a percentage of expected total U.S. transportation fuel use (**Table 4**). This ratio is adjusted to account for the small refinery exemptions. A separate ratio is calculated for each of the four biofuel categories. Under EISA, EPA is required to set the biofuel standards on a final basis by November 30 for the following year, based in part on information provided by the Energy Information Agency (EIA) of the Department of Energy (DOE). In order to accommodate this deadline, EPA announced that it intended to issue a notice of proposed rulemaking (NPRM) by summer of the preceding year, and on a final basis by November 30 of the preceding year.[23] These announcements are to include the cellulosic biofuel waiver credit price (see section on "Cellulosic Biofuel Waiver Credits") and the status of the aggregate compliance approach to land-use restrictions under the definition of renewable biomass for both the United States and Canada.

Table 4. Proposed RFS Standards for 2013

RFS Category	Percentage Ratio (%)	Volume of Renewable Fuel (billion gallons)
Cellulosic biofuels	0.008%	0.014
Biomass-based diesel	1.12%	1.28
Advanced biofuels	1.60%	2.75
Total renewable fuel	9.63%	16.55

Source: EPA, "Regulation of Fuels and Fuel Additives: 2013 Renewable Fuel Standards; Proposed Rule," *Federal Register,* vol. 78, no. 26, February 7, 2013, pp. 9282-9306.

Notes: All volumes are given in ethanol-equivalent gallons except for biomass-based diesel, which is given in terms of physical volume.

Determining an Individual Company's Obligation

The RFS mandates (by biofuel category) are ultimately enforced on retail fuel blenders and exporters (not on biofuels producers or importers). Companies that supply gasoline or diesel transportation fuel for the retail market are obligated to include a quantity of biofuels equal to a percentage of their total annual fuel sales—referred to as a renewable volume obligation (RVO). The RVO is obtained by applying the EPA-announced standards for each of the four biofuel categories to the firm's annual fuel sales to compute the mandated biofuels volume. At the end of the year, each supplier must have enough RINs to show that it has met its share of each of the four mandated standards. Failure to acquire sufficient RINs to meet a party's RVO (see section "Renewable Identification Numbers (RINs)" for details) is subject to civil penalties of up to $32,500 per day, plus the amount of any economic benefit or savings resulting from the violation.[24]

Equivalence Values

The equivalence value (EV) of a renewable fuel represents the number of gallons that can be claimed for compliance purposes for every physical gallon of renewable fuel. Under RFS1, the EV was based on the energy content of each renewable fuel relative to ethanol. As a result, the EV for ethanol was 1.0; butanol was 1.3; biodiesel (mono-alkyl ester) was 1.5; and non-ester renewable diesel was 1.7.

Cellulosic biofuel was granted a 2.5-to-1 credit.

In general, these equivalence values were continued in the RFS2, with one key exception. For purposes of meeting the biomass-based biodiesel standard, each gallon of BBD will count as 1.0; however, for purposes of meeting the advanced biofuel standard, cellulosic standard and/or the total renewable biofuel standard, each gallon of BBD will count as 1.5 or 1.7 (depending on the type of fuel) in order to reflect its higher energy content. Under the RFS2, the 2.5-to-1 bonus for cellulosic biofuel was eliminated.

Waivers to Annual Biofuel Standards

EISA requires that EPA evaluate and make an appropriate market determination for setting the cellulosic standard each year. As part of this process, EPA announced that it will issue a notice of proposed rulemaking by summer and a final rule by November 30 of each year to set the renewable fuel standard for each ensuing year.[25] Pursuant to this task, the EPA Administrator has the authority to waive the RFS requirements, in whole or in part, if, in her determination, there is inadequate domestic supply to meet the mandate, or if "implementation of the requirement would severely harm the economy or environment of a State, a region, or the United States."[26] Further, under certain conditions, the EPA administrator may waive (in whole or in part) the specific carve-outs for cellulosic biofuel and biomass-based diesel fuel.

For example, in each of the years 2010 through 2013 EPA has waived or proposed waiving most of the original RFS mandates for cellulosic biofuels, as follows:

- In February 2010, EPA lowered the 2010 RFS for cellulosic biofuels to 6.5 million gallons (mgals), on an ethanol-equivalent basis, down from its original 100 mgals scheduled by EISA.[27]
- In November 2010, EPA lowered the 2011 RFS for cellulosic biofuels to 6 mgals (ethanol equivalent), down from its original 250 mgals.[28]
- In December 2011, EPA lowered the 2012 RFS for cellulosic biofuels to 8.65 mgals (ethanol equivalent), down from its original 500 mgals.[29]
- In January 2013, EPA proposed to lower the 2013 RFS for cellulosic biofuels to 14 mgals (ethanol equivalent), down from its original 1 billion gallons.[30]

EPA cited a lack of current and expected production capacity, driven largely by a lack of investment in commercial-scale refineries—for example, only a limited number of cellulosic biofuel RINs were registered in 2012 (20,069 gallons), while no commercial production was reported in 2010 and 2011.[31] The downward revisions for 2010 through 2013 suggest that the actual cellulosic biofuels standard for future years, although explicitly listed in **Table 1**, is uncertain.

In addition to waivers associated with the annual RFS review process, EPA may respond to waiver requests resulting from unusual circumstances. For example, in 2008 the governor of Texas requested a waiver of the RFS because of high grain prices; however, that waiver request was denied because EPA determined that the RFS requirements alone did not "severely harm the economy ... of a State, a region, or the United States," a standard required by the statute. A similar waiver was requested in 2012 by the governors of Arkansas and North Carolina. In both cases, the petition was ultimately denied.[32]

Cellulosic Biofuel Waiver Credits

If EPA reduces the required volume of cellulosic biofuel according to the waiver provisions in EISA, EPA will offer a number of credits to obligated parties no greater than the reduced cellulosic biofuel standard. These waiver credits are not allowed to be traded or banked for future use, and are only allowed to be used to meet the cellulosic biofuel standard for the year that they are offered. Since the cellulosic standard was lowered for each year from 2010 through 2013, cellulosic waiver credits were made available to obligated parties at announced prices per gallon—$1.56 in 2010; $1.13 in 2011; $0.78 in 2012; and a proposed $0.42 (or the amount by which $3.00 per gallon—adjusted for inflation—exceeds the average wholesale price of a gallon of gasoline in the United States) in 2013.

Unachieved Cellulosic Biofuels Mandates

After three successive years (2010-2012) where, first, EPA lowered the cellulosic biofuels mandate and, then, cellulosic biofuels production failed to achieve the lowered mandates, the American Petroleum Institute (API), in 2012, challenged the obligation under the RFS to use cellulosic biofuels that do not exist in sufficient amounts in commercial markets or pay a fee. API petitioned the U.S. Court of Appeals, D.C., charging that EPA exceeded its authority by setting unachievable standards in an effort to promote cellulosic biofuel development. On January 25, 2013, the appeals court agreed with API's charge, ruling that the EPA's cellulosic biofuels mandate for 2012 was vacated and that EPA must replace it with a revised mandate. On February 27, 2013, EPA announced that the 2012 cellulosic biofuel standard was vacated (dropped to zero).[33] Then, on March 13, 2013, EPA also announced that it was voluntarily, retroactively lowering the 2011 RFS to zero.[34]

Renewable Identification Numbers (RINs)

A RIN is a unique 38-character number that is issued (in accordance with EPA guidelines) by the biofuel producer or importer at the point of biofuel production or the port of importation.[35] Each qualifying gallon of renewable fuel has its own unique RIN. RINs are generally assigned by batches of renewable fuel production as follows:

RIN	= KYYYYCCCCFFFFFBBBBBRRDSSSSSSSSEEEEEEEE Where
K	= code distinguishing RINs still assigned to a gallon from RINs already detached
YYYY	= the calendar year of production or import
CCCC	= the company ID
FFFFF	= the company plant or facility ID
BBBBB	= the batch number
RR	= the biofuel equivalence value (described below)
D	= the renewable fuel category
SSSSSSSS	= the start number for this batch of biofuel
EEEEEEEE	= the end number for this batch of biofuel

Under the RFS2 RIN formulation, Code D has been redefined to identify which of the four RFS categories—total, advanced, cellulosic, or biodiesel—the biofuel satisfies (**Table 5**).

Together, SSSSSSSS and EEEEEEEE identify the RIN block that demarcates the number of gallons of renewable fuel the batch represents in the context of compliance with the RFS—that is, RIN gallons. The RIN-gallon total equals the product of the liquid volume of renewable fuel times its equivalence value. For example, since biodiesel has an equivalence value of 1.5 when being used as an advanced biofuel, 1,000 gallons of biodiesel would equal 1,500 RIN gallons of advanced biofuels. If the RIN block start for that batch was 1 (i.e., SSSSSSSS = 00000001), then the end value (EEEEEEEE) would be 00001500, and the RR code would be RR = 15).

Table 5. RFS D Code Definitions

D value	RFS1	RFS2
1	Cellulosic biomass ethanol	na
2	Any other renewable fuel	na
3	na	Cellulosic biofuel
4	na	Biomass-based diesel
5	na	Advanced biofuel
6	na	Renewable fuel
7	na	Cellulosic diesel

Source: EPA, 40 C.F.R. Part 80, "Regulation of Fuels and Fuel Additives: Changes to Renewable Fuel Standard Program, Final Rule," Feb. 3, 2010.

Notes: na = not applicable.

Any party that owns RINs at any point during the year (including domestic and foreign producers, refiners, exporters, and importers of renewable fuels) must register with the EPA and follow RIN record-keeping and reporting guidelines. RINs can only be generated if it can be established that the feedstock from which the fuel was made meets EISA's definitions of renewable biomass, including land restrictions. The feedstock affirmation and record-keeping requirements apply to RINs generated by both domestic renewable fuel producers and RIN-generating foreign renewable fuel producers or importers. After a RIN is created by a biofuel producer or importer, it must be reported to the EPA (usually on a quarterly basis). When biofuels change ownership (e.g., are sold by a producer to a blender), the RINs are also transferred. When a renewable fuel is blended for retail sale or at the port of embarkation for export, the RIN is separated from the fuel and maybe used for compliance or trade. The Code K status of the RIN is changed at separation.

Small Refinery Exemption

A permanent exemption is available to any parties who produce or import less than 10,000 gallons of renewable fuel in a year—they are not required to generate RINs for that volume, and are not required to register with the EPA if they do not take ownership of RINs generated by other parties. Under EISA, this exemption is temporarily extended (for up to three years beginning with the calendar year in which the refinery produces its first gallon of renewable fuel) to renewable fuel producers who produce less than 125,000 gallons per year from new production facilities. This exemption is intended to allow pilot and demonstration plants to focus on developing the technology and obtaining financing during their early stages rather than complying with RFS2 regulations.

Flexibility in Administering the RIN Requirements

RINs generated during the current year may be used to satisfy either the current year's or the following year's RVO. A RIN would not be viable for any year's RVO beyond the immediately successive year; thus giving it essentially up to a two-year lifespan. For any individual company, up to 20% of the current year's RVO may be met by RINs from the previous calendar year.

In addition to compliance demonstration, RINs can be used for credit trading. When a fuel supplier has blended or sold a quantity of biofuel, the RINs are detached from the biofuels. If a supplier has already met its mandated share and has supplied surplus biofuels for a particular biofuel category, it can sell the extra RINs to another supplier (who has failed to meet its mandate for that same biofuel standard) or it can hold onto the RINs for future use (either to satisfy the succeeding year's requirement or for sale in the succeeding year). Since biofuels supply and demand can vary over time and across regions, a market has developed for RINs.

The marketability of RINs allows fuel suppliers who have not bought enough biofuels to fulfill their RFS requirement for each of the four RFS categories by purchasing the biofuels-specific RINs instead. As a result, RINs have value as a replacement for the actual purchase of biofuels. Because four separate biofuel mandates must be met, the RIN value may vary across the individual biofuel categories.[36] Since the RFS biofuels categories are nested, the price of RINs for specific sub-mandates (e.g., cellulosic biofuels or biodiesel) must be equal to or greater than the price of RINs for advanced biofuels which, in turn is equal to or greater than the RIN value for total renewable biofuels. Thus, RIN values may vary across RFS categories as well as geographically with variations in specific biofuels supply and demand conditions.

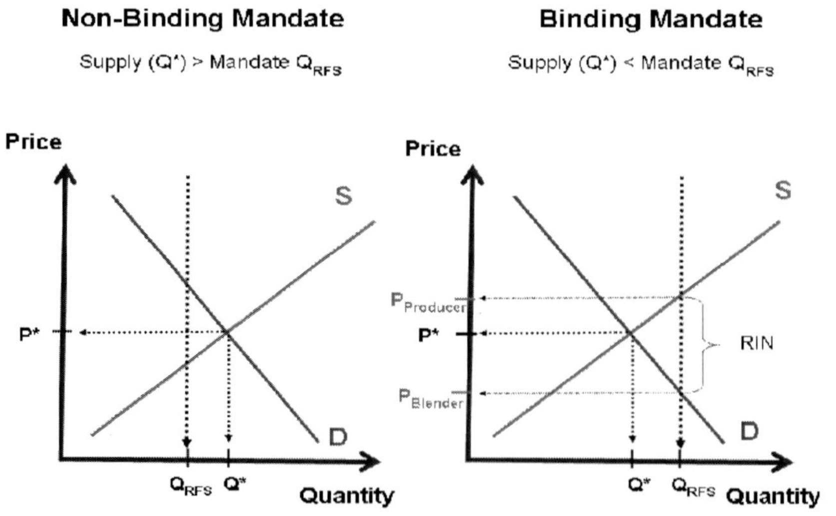

Source: "Renewable Identification Numbers are the Tracking Instrument and Bellwether of U.S. Biofuel Mandates," by Wyatt Thompson, Seth Meyer, and Pat Westhoff, *EuroChoices* 8(3), 2009.

Note: Supply equals domestic production and imports; demand equals both blenders and exporters demand.

Figure 2. How a Mandate May or May Not Affect RIN Values.

Differences in RIN values also reflect the degree to which the mandate associated with a specific RIN biofuel category is binding on the market equilibrium.[37] For example, if the supply of a specific biofuel—including both domestically produced as well as imported—available to the market exceeds the RFS mandate (see left-hand side of **Figure 2** where $Q^* > Q_{RFS}$), then the RIN's "core" value (i.e., its price minus transaction costs and speculative component) would be zero at the mandated level (Q_{RFS}).

In contrast, if the mandated biofuel usage level exceeds what is offered by the market (see right-hand side of **Figure 2** where $Q_{RFS} > Q^*$), the biofuels mandate is binding because it forces biofuels blenders to use more biofuels than they would without the mandate. The price of the biofuel purchased by the blender has to rise to $P_{producer}$ to solicit the extra production from the biofuels producers, while the biofuels price must fall to $P_{blender}$ to encourage greater blender purchases. The RIN's core value would be equal to the gap between these two prices, $P_{producer}$ minus $P_{blender}$. However, the blender must pay the full price of $P_{producer}$, which includes both $P_{blender}$ plus the RIN's core value, to acquire the mandated Q_{RFS}.

A RIN also may have speculative value, even when in surplus, if an investor were to anticipate a shortage in the near future (i.e., within the period for which a RIN is valid) and seek to acquire RINs cheaply in advance of the shortage. To date, the overall biofuels mandates have not been binding and until recently general renewable fuel RIN values generally have been small. It is expected that, if the RFS becomes binding, fuel suppliers will pass the added cost of biofuels acquisition (i.e., the RIN value), on to motor fuel consumers in the form of higher fuel prices.[38]

Because RINs have value, they are not immune to fraudulent activity. In late 2011 and early 2012, EPA issued notices of violations (NOVs) to three companies (Clean Green Fuels, LLC, Absolute Fuels, LLC, and Green Diesel, LLC) that the agency alleges fraudulently generated a combined 140 million biodiesel RINs in 2010 and 2011. Subsequently, individuals representing two of these companies have also faced criminal prosecution.[39]

EPA Analysis of RFS Impacts

As part of its final rule determination, EPA included an analysis of the market and environmental impact of the increased use of renewable fuels under the RFS2 standards.[40] The analytical results include the following.

- **Reduced dependence on foreign sources of crude oil.** By 2022, the mandated 36 bgals of renewable fuel will displace about 13.6 bgals of petroleum-based gasoline and diesel fuel, representing about 7% of expected annual U.S. transportation fuel consumption.
- **Reduced price of domestic transportation fuels.** By 2022, the increased use of renewable fuels is expected to decrease gasoline costs by $0.024 per gallon and diesel costs by $0.121 per gallon, producing a combined annual savings of nearly $12 billion.
- **Reduced GHG emissions.** When fully implemented in 2022, the expanded use of biofuels under the RFS is expected to reduce annual GHG emissions by 138 million metric tons—equivalent to taking about 27 million vehicles off the road.

- **Increased U.S. farm income.** By 2022, the expanded market for agricultural products such as corn and soybeans resulting from biofuels production is expected to increase annual net farm income by $13 billion.
- **Decreased corn and soybean exports.** The expanded use of corn starch and soybean oil for biofuels is expected to reduce corn exports by 8% and soybean exports by 14% by 2022.
- **Increased cost of food in the United States.** The increased demand for U.S. agricultural products is expected to raise the overall commodity price structure, leading to an annual increase in the cost of food per capita of about $10 by 2022, or over $3 billion.
- **Increased emissions of certain air contaminants, but decreased emissions of others.** Contaminants expected to increase include hydrocarbons, nitrogen oxides (NOx), acetaldehyde, and ethanol; those expected to decrease include carbon monoxide (CO) and benzene. The effects are expected to vary widely across regions, but in the net, increases in population-weighted annual average ambient PM and ozone concentrations are anticipated to lead to up to 245 cases of adult premature mortality.

RFS as Public Policy

Proponents' Viewpoint

Supporters of an RFS claim it serves several public policy interests[41] in that it:

- reduces the risk of investing in renewable biofuels by guaranteeing biofuels demand for a projected period (such risk would otherwise keep significant investment capital on the sidelines);
- enhances U.S. energy security via the production of liquid fuel from a renewable domestic source resulting in decreased reliance on imported fossil fuels (the United States currently imports over half of its petroleum,[42] two-thirds of which is consumed by the transportation sector);
- provides an additional source of demand—renewable biofuels—for U.S. agricultural output that has significant agricultural and rural economic benefits via increased farm and rural incomes and substantial rural employment opportunities;[43]
- underwrites the environmental benefits of renewable biofuels over fossil fuels (most biofuels are non-toxic, biodegradable, and produced from renewable feedstocks); and
- responds to climate change concerns because agricultural-based biofuels emit substantially lower volumes of direct greenhouse gases (GHGs) than fossil fuels when produced, harvested, and processed under certain circumstances.

Critics' Viewpoints

Critics of an RFS, particularly of the EISA expansion of the original RFS, have taken issue with many specific aspects of biofuels production and use, including the following:

- By picking the "winner," policymakers may exclude or retard the development of other, potentially preferable alternative energy sources.[44] Critics contend that biofuels are given an advantage via billions of dollars of annual subsidies that distort investment markets by redirecting venture capital and other investment dollars away from competing alternative energy sources. Instead, these critics have argued for a more "technology-neutral" policy such as a carbon tax, a capand-trade system of carbon credits, or a floor price on imported petroleum.
- Continued large federal incentives for corn-starch ethanol production are no longer necessary since the sector is no longer in its "economic infancy" and would have been profitable in most months since 2006 without federal subsidies.[45]
- The expanded mandate could have substantial unintended consequences in other areas of policy importance, including energy/petroleum security, pollutant and greenhouse gas emissions, agricultural commodity and food markets, land use patterns, soil and water quality, conservation, the ability of the gasoline-marketing infrastructure and auto fleet to accommodate higher ethanol concentrations in gasoline, the likelihood of modifications in engine design, and other considerations.
- Taxpayers are being asked to finance continued biofuels subsidies in support of existing and future biofuels infrastructure that have the potential to affect future federal budgetary choices.

The Increasing Cost of Biofuels Policy

A 2007 survey of federal and state government subsidies in support of ethanol production reported that total annual federal support fell somewhere in the range of \$5.4 billion to \$6.6 billion per year—nearly \$1 per gallon of biofuel produced in the United States.[46] In 2011 (prior to the expiration of various tax credits on December 31, 2011), federal subsidies were estimated at over \$7.8 billion, including nearly \$7.5 billion in tax credits.[47] In 2012, following the expiration of the corn ethanol tax credit, a preliminary estimate of federal biofuel subsidies (including Title IX farm bill energy programs)[48] is approximately \$1.3 billion, of which slightly more than \$1 billion is attributable to the biodiesel tax credit of \$1.00 per gallon.

Historically, the major direct federal costs associated with the implementation of the RFS have been the federal tax credits available to the various biofuels that are blended to meet the RFS mandate.[49] Most of these tax credits expired at the end of 2011, while the cellulosic biofuels production tax credit expired at the end of 2012. On January 2, 2013, both the cellulosic biofuels production tax credit of \$1.01 per gallon and the biodiesel and renewable diesel fuel mixtures tax credit of \$1.00 per gallon were extended through 2013 by the American Taxpayers Relief Act of 2012 (ATRA: P.L. 112-240). ATRA also retroactively applied the extension to biodiesel and renewable diesel fuel mixtures produced in 2012.

POTENTIAL ISSUES WITH THE EXPANDED RFS

Most U.S. biofuel production is ethanol produced from corn starch. As a result, as the U.S. ethanol industry has grown over the years, so too has its usage share of the annual corn crop. In 2000, national ethanol production was using about 6% of U.S. corn supplies; by 2012 it was expected to use about 40%.[50]

Under the expanded RFS, the 2015 corn ethanol cap of 15 billion gallons—coupled with the existing U.S. ethanol production capacity of nearly 14 billion gallons—suggests that corn ethanol will continue to use about 40% of the volume of U.S. corn production for several years to come, depending on yield and area developments.[51] This significant shift towards greater corn use for biofuels has meant higher prices for other corn users, including both the livestock and export sectors (**Figure 3**).

The biofuels-driven expansion in feedstocks production (especially corn for grain and stover) has heightened competition for available cropland between biofuels feedstocks and other field crops, and has resulted in more intense agricultural activity on U.S. cropland to meet growing demand for food, feed, and fuel resources. This has consequences for several important agricultural markets, including

- grains—because corn competes with other grains for land;
- livestock—because animal feed costs increase with the price of corn;
- agricultural inputs—because corn is more input-intensive (in terms of fertilizers and pesticides) than other major field crops; and
- land—because the value of cropland, as well as total harvested acreage, increases with commodity prices and returns per acre.

In addition to agricultural effects, an increase in corn-based ethanol production could potentially have other market effects, including effects on:

- energy markets—because natural gas is a key input in both corn and ethanol production (although the recent growth in U.S. production of natural gas under new fracking methods has largely diminished this aspect);
- water quality—because expanding corn-based ethanol production likely involves heavier use of farm chemicals with increased potential for run-off or leaching;
- water resource availability—because water plays a crucial role in all stages of biofuels production, from cultivation of feedstocks through their conversion into biofuels, yet there remain many uncertainties about national and regional effects of increased biofuels production on water resources;[52]
- soil fertility—because several potential biofuels activities (including intensive year-over-year corn production, diversion of corn stover to cellulosic biofuels production and away from field retention as a soil amendment under low-till cultivation, and the expansion of biofuels feedstock cultivation on marginal land) could result in diminished soil fertility and/or increased erosion;
- wildlife habitat—because expanding biofuels feedstock production on marginal lands traditionally left fallow under a conserving practice could compete with wildlife and fowl habitat; and

- federal budget exposure—because applying the federal cellulosic biofuels production tax credit of $1.01 per gallon to the RFS requirements produces an annual budget liability of $16 billion by 2022 (assuming that the tax credit is extended and that the cellulosic mandate is met).

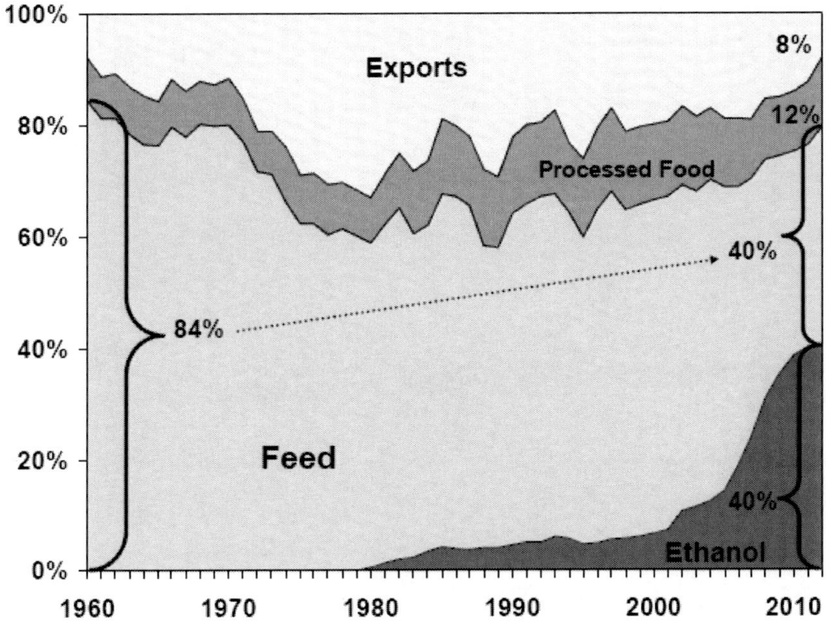

Source: USDA, Production, Supply, and Demand (PSD) database, February 8, 2013.

Note: Data do not account for use of ethanol by-products as animal feed; about 30% (by weight) of corn used for biofuels is left over from the production process and is useful as a relatively high-protein animal feed.

Figure 3. Ethanol Uses an Increasing Share of U.S. Corn Production, Particularly Since 2005, While Feed Use Has Fallen Sharply; (annual U.S. corn disappearance categories, as a percent of total use, excluding ending stocks).

Overview of Long-Run Corn Ethanol Supply Issues

The ability of the U.S. corn industry to continue to expand production and satisfy the steady growth in demand depends, first and foremost, on continued productivity gains. U.S. corn yields have shown strong, steady growth since the late 1940s, with some acceleration occurring since the mid-1990s as bio-engineered advances in seed technology have heightened drought and pest resistance in corn plants (**Figure 4**). However, since achieving a record 164.7 bushels/acre in 2009, U.S. corn yields have experienced three consecutive years of decline (related to unfavorable precipitation and temperature conditions at critical points in the growing season), to the point where the 2012 national corn yield estimate of 122.3 bushels/acre is the lowest yield since 1995.

Corn Prices

Expanding U.S. corn production has only partially offset the rapid growth in demand following the rapid expansion of the U.S. ethanol industry that has occurred since 2005. As a result, corn prices have trended steadily upward in direct relation to the added growth in demand from the ethanol sector (**Figure 5**). USDA projects corn prices to remain in the \$4 to \$5 per bushel range through 2020, compared with an average farm price of \$2.15 per bushel during the 10-year period from 1997 to 2006.[53]

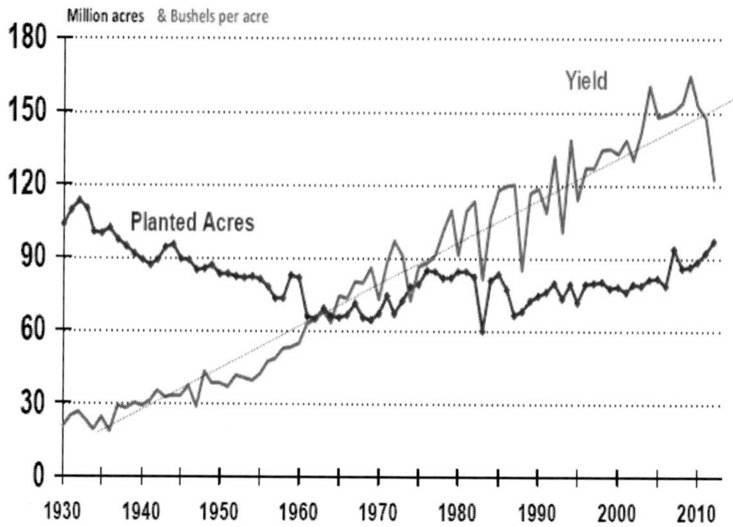

Source: USDA, PSD database, as of February 8, 2013.

Figure 4. U.S. Annual Corn Planted Acres and Yield.

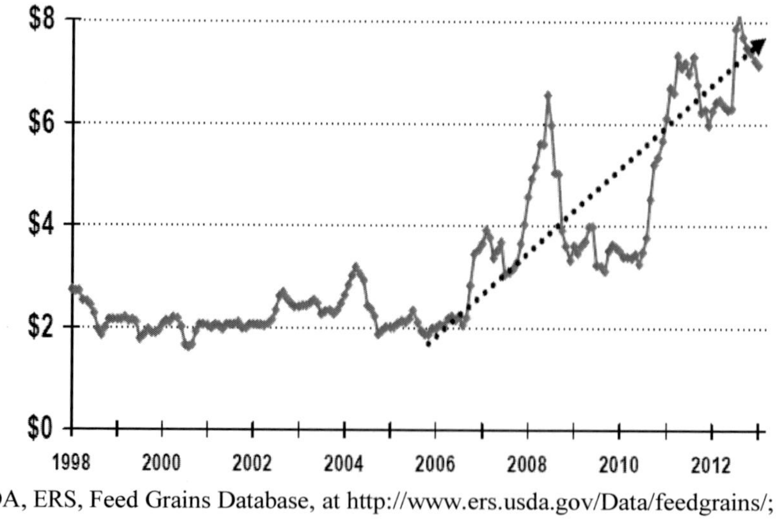

Source: USDA, ERS, Feed Grains Database, at http://www.ers.usda.gov/Data/feedgrains/; as of Feb. 26, 2013.

Figure 5. Monthly U.S. Corn Prices Have Trended Upward Since Late 2005; (Central Illinois cash price for no. 2, yellow corn).

Corn Yields

It is likely that upward-trending farm prices (**Figure 5**) will encourage continued research investments to move corn yields steadily higher in the future. However, even slight differences in the long-run growth rate portend large impacts in the price outlook. Some economists think that yield increases will slow in coming decades because of land degradation and the impact of climate change. Others suggest that dramatic developments in bio-engineering and seed technology will push corn yields sharply higher. A prime example of the differences in U.S. corn yield outlooks is the contrast between USDA, whose economists project U.S. corn yields to reach about 240 bushels per acre by 2050, and the scientists of the biotech seed company Monsanto, who predict that corn yields will be much higher—as much as 300 bushels per acre—by 2030.[54]

According to USDA, achieving "300-bushel corn" by 2030 would require an extraordinary deviation (a tripling) from both projected and accelerated corn yield trends, and would be historically unprecedented.[55]

Corn Area

U.S. cropland planted to corn has increased in recent years from the 1983 low of 60.3 million acres to as high as 97.2 million acres in 2012. Prospects for further expansion in crop area are far less certain, as corn is an energy-intensive crop that prefers deep, fertile soils and timely precipitation. Within the prime corn-growing regions of the Corn Belt, per-acre returns for corn easily dwarf other field crops that vie for the same acreage. Recent seed developments have allowed corn production to expand dramatically into the central and northern Plains states. However, the risk of investing up front in high operating costs to be offset at harvest by strong returns increases as production moves into less traditional regions, such as the northern Plains, the Delta, and the Southeast. In addition, planting successive corn crops—referred to as corn-on-corn cultivation—rather than in rotation with soybeans, wheat, or fallow generally produces a yield drag on successive corn crops that can lower yields anywhere from 5% to 15%, depending on soil, climate, and cultivation practices.

Corn-Soybean Rotation

The most likely source of new corn acreage will come from shifts in crop rotation from soybeans to corn.[56] However, crop intensification also has its limits. Corn (of the grass family) is traditionally planted in an annual rotation with soybeans (a broad-leaf legume) that offers important agronomic benefits including pest and disease control, as well as enhanced soil fertility.[57] When farmers shift away from this rotation, corn yields tend to suffer. Planting cornon-corn in two consecutive years usually results in a 5% to 15% yield decline in the second year.[58]

As a result, the corn-to-soybean price ratio would have to tilt fairly strongly in favor of corn for corn-on-corn production to be profitable. Given the limitations on corn area expansion and rotational intensification, it is likely that the sustainable long-run corn planted area is probably in the range of 90 million to 95 million acres. If this is the case, then it would mean that future growth in U.S. corn production will be increasingly dependent on yield growth.

Overview of Non-Corn-Starch-Ethanol RFS Issues

EISA defines "advanced biofuels" very broadly as biofuels other than corn-starch ethanol. As such, advanced biofuels would include imported Brazilian sugar-cane ethanol, as well as homegrown biodiesel. However, the principal focus of advanced biofuels is on biofuels based on cellulosic biomass. Under the RFS2, advanced biofuels use is mandated to reach a minimum of 21 billion gallons by 2022, of which at least 16 billion gallons must be some type of cellulosic biofuel. However, many obstacles must first be overcome before commercially competitive cellulosic biofuels production occurs.[59]

In the near term, it is likely that corn stover[60] will be a primary biomass of choice for cellulosic biofuels production. This is because many ethanol plants already exist in corn production zones and an extension of those plants to include cellulosic biofuels production from stover would offer some scale economies. However, stover-to-biofuel conversion has its own set of potential environmental trade-offs, paramount of which is the dilemma of sacrificing soil fertility gains by harvesting the stover rather than returning it to the soil under no- or minimum-tillage practices.

Cellulosic Biofuels Production Uncertainties

There are substantial uncertainties regarding both the costs of producing cellulosic feedstocks and the costs of producing biofuels from those feedstocks. Dedicated perennial crops are often slow to establish, and it can take several years before a marketable crop is produced. Crops heavy in cellulose tend to be bulky and represent significant problems in terms of harvesting, transporting, and storing. New harvesting machinery would need to be developed to guarantee an economic supply of cellulosic feedstocks.[61] Seasonality issues involving the operation of a biofuels plant year-round based on a four- or five-month harvest period of biomass suggest that storage is likely to matter a great deal. In addition, most marginal lands (i.e., the low-cost biomass production zones) are located far from major urban markets, making it difficult to reconcile plant location with the cost of fuel distribution.

Following the EPA's substantial revisions to the first three years of cellulosic biofuels mandates (2010-2012) and a proposed revision for the fourth year (2013), there has been considerable uncertainty surrounding current cellulosic biofuel conversion technologies and the cost of the conversion process (including physical, chemical, enzymatic, and microbial treatment and conversion of the biomass feedstocks into motor fuel). These uncertainties, plus the financial crisis of 2008 and the ensuing recession and credit crunch, severely curtailed new investment in the biofuels sector.[62] However, it appears that 2013 will experience the first substantial commercial production levels of cellulosic biofuels. EPA has proposed a cellulosic biofuels RFS of 14 million gallons under the expectation that two plants will begin commercial production during the first quarter of 2013, with at least two more plants expected to follow by the fourth quarter of 2013.[63] In addition, industry reports suggest that new cellulosic biofuels plants are either in the planning stages or under construction in as many as 20 states and Canadian provinces.[64]

Unintended Policy Outcomes of the "Advanced Biofuels" Mandate

Because the advanced biofuels mandate in the RFS is a fixed mandate, irrespective of prices, the above uncertainties about the production of cellulosic ethanol could have significant implications for fuel supply and fuel prices. If cellulosic ethanol production is unable to advance rapidly enough to meet the RFS mandate for non-corn-starch ethanol, then other unexpected biofuels sources may be forced to step in and fill the void:

- production of domestic sorghum-starch ethanol may expand across the prairie states and in other regions less suitable for corn production;
- costly domestic sugar-beet ethanol or biodiesel production may be undertaken to fill the mandate; or
- imports of Brazilian sugar-cane ethanol could expand.

Energy Supply Issues

Biofuels are not primary energy sources. Energy is first stored in biological material (through photosynthesis), and then must be converted into a more useful, portable fuel. This conversion requires energy. The amount and types of energy used to produce biofuels (e.g., coal versus natural gas), and the feedstocks for biofuels production (e.g., corn versus cellulosic biomass), are critical in determining a biofuels net energy balance and the environmental benefits of a biofuel.

Energy Balance

To analyze the net energy consumption of ethanol, the entire fuel cycle must be considered. The fuel cycle consists of all inputs and processes involved in the development, delivery, and final use of the fuel. For corn-based ethanol, these inputs include the energy needed to produce fertilizers, operate farm equipment, transport corn, convert corn to ethanol, and distribute the final product.

Estimates of the net energy output/input ratio for corn-starch ethanol production are dated. The most recent study by USDA estimated an energy output/input ratio of 2.3 based on a 2005 survey of corn growers and 2008 data for ethanol plants (and assuming the then-most-advanced technology for corn and ethanol production).[65] A 2.3 ratio implies that the energy contained in a gallon of corn ethanol was 130% higher than the amount of energy needed to produce and distribute it. Ethanol industry sources argue that technological innovation will continue to improve corn ethanol's energy balance. However, other analyses have resulted in a significantly lower output/input ratio.

If feedstocks other than corn are used to produce biofuels, it is expected that lower nitrogen fertilizer use would greatly improve the energy balance. Further, if biomass were used to provide process energy at the biofuels refinery (rather than coal or natural gas), the energy savings would be even greater.[66] Some estimates are that cellulosic ethanol could have an energy balance of 8.0 or more.[67] Similarly high energy balances have been calculated for sugar-cane ethanol and certain types of biodiesel.

Natural Gas Demand

As biofuels production increases, the total energy needed to process biomass into liquid fuel can be expected to increase. The resultant increase in energy demand will likely support higher energy prices. The two principal processing fuels used in the United States are natural gas and coal. Other fuels include electricity and biomass.

The United States has been a net importer of natural gas since the early 1980s, although the vast majority of this gas comes via pipeline from Canada. However, recent technological breakthroughs in accessing gas shale have the potential to alter long-run U.S. natural gas supplies. As a result, processing energy has diminished as an issue for biofuels production.

Energy Security[68]

Ethanol displaces gasoline, and the benefits to energy security from ethanol, while relatively small, are still potentially important. However, biofuels' potential to play a larger role in energy security is questionable. Roughly 40% of the U.S. corn crop was used for ethanol in 2012, and the resultant ethanol only accounts for about 7% of gasoline consumption on an energy-equivalent basis.[69] There is considerable uncertainty regarding how quickly or how much U.S. corn production can expand. If the entire 2012 U.S. corn crop of 10.8 billion bushels were used as ethanol feedstock, the resultant 30 billion gallons of ethanol (20.6 billion gasoline-equivalent gallons, or GEG) would represent about 16% of estimated national gasoline use of approximately 132 billion gallons.[70] In contrast, the import share of U.S. liquid fuel consumption (crude oil and other petroleum products) was estimated at 66% in 2011.[71]

An expanded RFS would certainly displace petroleum consumption, but the overall effect on life-cycle fossil fuel consumption is questionable, especially if there is a large reliance on corn-based ethanol. Under the EISA RFS mandate, by 2022 biofuels will still represent about 20% of gasoline energy transportation fuel demand and 2.4% of diesel transportation fuel demand.[72]

The specific definition of "advanced biofuels" also affects the overall energy security picture for biofuels. For example, an expanded RFS provides an incentive to increase imports of sugar-cane ethanol, especially from Brazil. The expanded RFS also provides an incentive for imports of biodiesel and other renewable diesel substitutes from tropical countries. This would represent a "diversification" of fuel sources, not the "domestication" that some claim is true energy security.

Energy Prices

The effects of the expanded RFS on energy prices are uncertain. If wholesale biofuels prices are higher than gasoline prices (after all economic incentives are taken into account), then mandating higher and higher levels of biofuels would likely lead to higher gasoline pump prices. However, if petroleum prices—and thus gasoline prices—are high, the use of some biofuels might help to mitigate high gasoline prices.

Current production costs are thought to be so high for some biofuels, especially cellulosic biofuels and biodiesel from algae, that significant technological advances—or significant increases in petroleum prices—would be necessary to make them competitive with gasoline.[73] Without cost reductions, mandating large amounts of these fuels would likely raise fuel prices. If a price were placed on greenhouse gas emissions—perhaps through the enactment of a carbon tax—then the economics could shift in favor of these fuels despite their high production costs, as they have lower fuel-cycle and life-cycle greenhouse gas emissions (see below).

Ethanol Infrastructure and Distribution Issues

In addition to the above concerns about feedstock supply for ethanol production, there also are issues involving ethanol distribution and infrastructure. Expanding ethanol production likely will strain the existing supply infrastructure. Further, expansion of ethanol use beyond the current 10% blend will require investment in entirely new infrastructure that would be necessary to handle an increasing percentage of ethanol in gasoline, or retrofitting and recertification of existing equipment. On the other hand, if drop-in fuels (i.e., petroleum-like biofuels such as biobutanol or biomass-based diesel substitutes) are produced in large quantities, some of these infrastructure issues may be mitigated, since these fuels can be used in existing infrastructure.

Distribution Issues

Unlike petroleum products, ethanol and ethanol-blended gasoline cannot be shipped in existing U.S. pipeline infrastructure. Ethanol-blended gasoline tends to separate in pipelines due to the presence of water in the lines.[74] Further, ethanol is corrosive and may damage existing pipelines and storage tanks. Also, corn ethanol must be moved from rural areas in the Midwest to more populated areas, which are often located along the coasts. This shipment is in the opposite direction of existing pipeline transportation, which moves gasoline from refiners along the coast to other coastal cities and into the interior of the country. While some studies have concluded that shipping ethanol or ethanol-blended gasoline via pipeline could be feasible, no major U.S. pipeline has made the investments to allow such shipments.[75]

The current distribution system for ethanol is dependent on rail cars, tanker trucks, and barges. These deliver ethanol to fuel terminals where it is blended with gasoline before shipment via tanker truck to gasoline retailers. However, these transport modes lead to prices higher than for pipeline transport, and the supply of current shipping options (especially rail cars) is limited. Because of these distribution issues, some pipeline operators are seeking ways to make their systems compatible with ethanol or ethanol-blended gasoline. These modifications could include coating the interior of pipelines with epoxy or some other, corrosion-resistant material. Another potential strategy could be to replace all susceptible pipeline components with newer, hardier components. However, even if such modifications are technically possible, they likely will be expensive, and could further increase ethanol transportation costs.

As non-corn biofuels play a larger role, some of the supply infrastructure concerns may be alleviated. Cellulosic biofuels potentially can be produced from a variety of feedstocks that are more widely distributed throughout the country, unlike current dependency on a single crop (corn) from one region of the country. For example, municipal solid waste is ubiquitous across the United States, and could serve as a ready feedstock for biofuels production if the technology were developed to convert it economically to fuel. Further, increased imports of biofuels from other countries could allow for greater use of biofuels, especially along the coasts.

The Blend Wall and Higher-Level Ethanol Blends

More than half of all U.S. gasoline contains some ethanol (mostly blended at the 10% level or lower). A key benefit of gasoline-ethanol blends up to 10% ethanol is that they are compatible with existing vehicles and infrastructure (fuel tanks, retail pumps, etc.). All automakers that produce cars and light trucks for the U.S. market warranty their vehicles to run on gasoline with up to 10% ethanol (E10). As a result, this 10% blend has represented an upper bound (sometimes referred to as the "blend wall") to the amount of ethanol that can be introduced into the gasoline pool.[76] If most or all gasoline in the country contained 10% ethanol, this would allow only for roughly 13 billion gallons, far less than the RFS mandates for 2013 onward.

For ethanol consumption to exceed the so-called "blend wall" and meet the RFS mandates, increased consumption at higher blending ratios is needed. For example, raising the blending limit from 10% to a higher ratio such as 15% or 20% would immediately expand the "blend wall" to somewhere in the range of 20 billion to 27 billion gallons. The U.S. ethanol industry is a strong proponent of raising the blending ratio.

In response to industry concerns regarding the impending blend wall, the EPA, after substantial vehicle testing, issued a partial waiver for gasoline that contains up to a 15% ethanol blend (E15) for use in model year 2001 or newer light-duty motor vehicles (i.e., passenger cars, light-duty trucks, and sport utility vehicles), but announced that no waiver would be granted for E15 use in model year 2000 and older light-duty motor vehicles, as well as in any motorcycles, heavy duty vehicles, or non-road engines.[77] According to the Renewable Fuel Association (RFA), the approval of E15 use in model year 2001 and newer passenger vehicles covered 62% of passenger vehicles on U.S. roads at the end of 2010.[78]

In addition to the EPA waiver decision, fuel producers needed to register the new fuel blends and submit health effects testing to EPA. Further, numerous other changes have to occur before large numbers of gasoline stations will begin selling E15, including many approvals by states and potentially significant infrastructure changes (pumps, storage tanks, etc.). As a result, the vehicle limitation to newer models, coupled with infrastructure issues, are likely to limit rapid expansion of blending rates. Moreover, a group of engine and equipment manufacturers challenged the partial waiver in court, arguing that EPA failed to estimate the likelihood of misfueling (using E15 in equipment denied a waiver), and the economic and environmental consequences of that misfueling.[79] In response to these concerns, EPA requires E15 suppliers to submit to the agency misfueling mitigation plans (MMP).[80] Concerns over a preliminary MMP that required a four-gallon minimum purchase from some pumps supplying both E15 and E10 led to a new MMP that EPA approved in February 2013 that eliminates the four-gallon requirement as long as a fuel station has at least one dedicated E10 (or lower) pump to fuel older passenger cars and light trucks as well as non-road vehicles/engines.[81]

The blend wall problem is made more acute by substantial revisions in EIA's projections of U.S. transportation fuel consumption rates since the RFS was first passed into law in 2007 (**Figure 6**). At that time, EIA estimated that U.S. transportation consumers were using about 145 billion gallons of gasoline (including ethanol) per year, but that consumption would grow strongly to 176 billion gallons of gasoline by 2022—as a result, RFS mandated biofuels would represent about 19% of annual gasoline consumption. By 2013, EIA had substantially lowered its fuel consumption outlook—partly due to sustained high petroleum prices, the prolonged effects of the 2008 financial crises on consumer incomes, and significantly higher fuel economy standards on new vehicles. Instead of growth, EIA projects gasoline consumption to fall to about 120 billion gallons by 2022, thus causing the RFS mandate's share of the gasoline transportation fuel market to grow to nearly 20% of annual consumption (in gasoline-equivalent gallons).[82]

As of March 2012, RFA listed only 13 stations in three states offering E15 for retail sale.[83] Thus concerns about a rapid expansion of E15 by its opponents may be overstated, at least as of the first quarter of 2013.

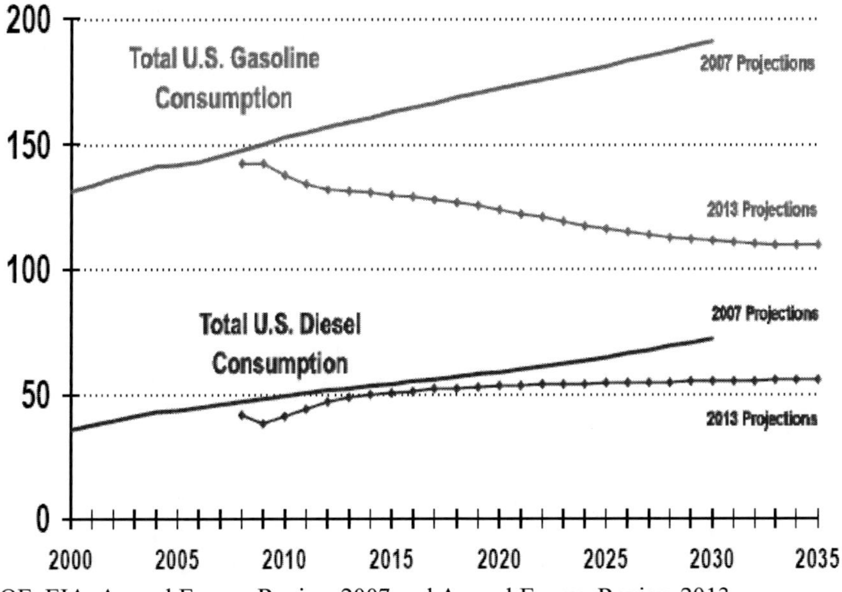

Source: DOE, EIA, Annual Energy Review 2007 and Annual Energy Review 2013.

Figure 6. EIA Long-Term Projections of U.S. National Transportation Fuel Use; Billion gallons.

Two additional options to resolving this bottleneck exist but appear to be long-run alternatives. First, increased use of ethanol in flex-fuel vehicles (FFVs) at ethanol-to-gasoline blend ratios as high as 85% (referred to as E85) is a possibility. However, increased E85 use involves substantial infrastructure development, particularly in the number of designated storage tanks and E85 retail pumps, as well as a rapid expansion of the FFV fleet to absorb larger volumes of ethanol. Infrastructure expansion will require significant investments, especially at the retail level. Installation of a new E85 pump and underground tank can cost as much as $100,000 to $200,000.[84] However, if existing equipment can be used with little modification, the cost could be less than $10,000.

A second alternative is to expand use of processing technologies at the biofuel plant to produce biofuels in a "drop-in" form (e.g., butanol) that can be used by existing petroleum-based distribution and storage infrastructure and the current fleet of U.S. vehicles. However, more infrastructure-friendly biofuels generally require more processing than ethanol and are therefore more expensive to produce.

Vehicle Infrastructure Issues

As was stated above, if a large portion of any increased RFS is met using ethanol, then the United States likely does not have the vehicles to consume the fuel. The 10% blend wall on ethanol in gasoline for conventional vehicles still poses a significant barrier to expanding ethanol consumption beyond 14 billion gallons per year.[85] To allow more ethanol use, vehicles will need to be certified and warranted for higher-level ethanol blends, or the number of ethanol FFVs will need to increase. Turnover of the U.S. automobile fleet has slowed during the recession, making it more difficult to integrate FFVs into the fleet.

Ethanol RINs and the Blend Wall

The price of renewable fuel RINs have increased dramatically in the first quarter of 2013. Spot prices for ethanol RINs averaged between $0.07 and $0.08 per gallon in the first weeks of January. However, in the first week of March 2013, ethanol RINs averaged roughly $0.76 per gallon—a nine-fold increase.[86] It is unclear what is driving the increase, but concerns over the blend wall and a reduction in output from U.S. ethanol producers in the first quarter have increased concerns that the RFS mandates will be binding in 2013 or 2014 and that there will be a scarcity of RINs.[87]

CONCLUSION

There is continuing interest in expanding the U.S. biofuels industry as a strategy for promoting energy security and achieving environmental goals. However, it is possible that increased biofuel production may place desired policy objectives in conflict with one another. There are limits to the amount of biofuels that can be produced from current feedstocks and questions about the net energy and environmental benefits they might provide. Further, rapid expansion of biofuels production may have many unintended and undesirable consequences for agricultural commodity costs, fossil energy use, and environmental degradation. Owing to these concerns, alternative strategies for energy conservation and alternative energy production are widely seen as warranting consideration.

End Notes

[1] For more information, see CRS Report R41282, Agriculture-Based Biofuels: Overview and Emerging Issues.

[2] For more information on incentives (both tax and non-tax) for biofuels, see CRS Report R42566, Alternative Fuel and Advanced Vehicle Technology Incentives: A Summary of Federal Programs.

[3] In gasoline-equivalent shares with 7.3% for ethanol and 2.1% for biodiesel. CRS estimates based on data from Energy Information Agency (EIA), Department of Energy (DOE).

[4] Heating oil, jet fuel, and fuels for ocean-going vessels are excluded from RFS2's national transportation fuel supply; however, renewable fuels used for these purposes may count towards the RFS2 mandates. EPA, 40 C.F.R. Part 80, "Regulation of Fuels and Fuel Additives: Changes to Renewable Fuel Standard Program, Final Rule," Feb. 3, 2010.

[5] Ibid.

[6] CRS Report R40460, Calculation of Lifecycle Greenhouse Gas Emissions for the Renewable Fuel Standard (RFS).

[7] CRS Report R40529, Biomass: Comparison of Definitions in Legislation Through the 112th Congress.

[8] For more information, see the section "Implementation of the RFS" later in this report.

[9] "Renewable Fuels: Regulations & Standards," EPA's online chronicle of RFS rule making, available at http://www.epa.gov/otaq/renewablefuels/regulations.htm.

[10] The term "advanced biofuels" comes from legislation in the 110th Congress, and is defined in Section 201 of the Energy Independence and Security Act of 2007 (EISA). EISA defines "advanced biofuels" as biofuels other than ethanol derived from corn starch (kernels) having 50% lower lifecycle greenhouse gas emissions relative

to gasoline. In some cases, the definition of "advanced biofuels" includes mature technologies and fuels that are currently produced in large amounts. For example, the EISA definition of "advanced biofuels" potentially includes ethanol from sugar cane, despite the fact that Brazilian sugar growers have been producing fuel ethanol for decades.

[11] As part of its February 3, 2010, final rule, EPA announced a revision in the BBD standard for 2010 to 1.15 bgals. This revision represents a summation of the 2009 standard of 0.5 bgals with the 2010 standard of 0.65 bgals. The RFS1 regulatory system, which was in effect during 2009 and which was based on national gasoline supply, did not provide any mechanism for implementing the 2009 BBD standard. As a result, it was integrated into the 2010 standard. Qualifying RINs accumulated during 2009 were acceptable in compliance.

[12] A diesel fuel product produced from cellulosic feedstocks that meets the 60% GHG threshold can qualify as either cellulosic biofuel or BBD.

[13] CRS Report R40460, Calculation of Lifecycle Greenhouse Gas Emissions for the Renewable Fuel Standard (RFS).

[14] EISA (P.L. 110-140), Title II, Sec. 201 Definitions, "(H) Lifecycle Greenhouse Gas Emissions."

[15] For more information on EPA's determination of lifecycle GHG emissions see CRS Report R40460, Calculation of Lifecycle Greenhouse Gas Emissions for the Renewable Fuel Standard (RFS).

[16] See "Section V. Lifecycle Analysis of Greenhouse Gas Emissions," Preamble, EPA RFS2 Final Rule, February 3, 2010, at http://epa.gov/otaq/renewablefuels/rfs2-preamble.pdf.

[17] For information on adding a potential new biofuel feedstock or production process pathway, see EPA, "Guidance on New Fuel Pathway Approval Process" at http://www.epa.gov/otaq/fuels/renewablefuels/compliance help/rfs2-lcapathways.htm.

[18] EPA, "EPA Issues Final Rule Additional Qualifying Renewable Fuel Pathways Under the Final RFS2 Program," EPA-420-F-13-014, February 2013.

[19] CRS Report R40529, Biomass: Comparison of Definitions in Legislation Through the 112th Congress.

[20] From various EPA announcements on "Renewable Fuels Regulations and Standards," at http://www.epa.gov/otaq /fuels/renewablefuels/regulations.htm.

[21] EPA, 40 C.F.R. Part 80, "Regulation of Fuels and Fuel Additives: Changes to Renewable Fuel Standard Program, Final Rule," Feb. 3, 2010. EPA's official "Renewable Fuel Standard (RFS)" website, with links to all official documents, is available at http://www.epa.gov/otaq/fuels/renewablefuels/.

[22] The data are taken from EIA's October issue of its monthly Short-Term Energy Outlook Report, "Table 4a. U.S. Crude Oil and Liquid Fuels Supply, Consumption, and Inventories," and "Table 8. U.S. Renewable Energy Consumption," available at http://www.eia.gov/forecasts/steo/.

[23] EPA, 40 C.F.R. Part 80, Final Rule, Feb. 3, 2010, p. 14675.

[24] U.S. Code, "Regulation of fuels," 42 USC 7545.

[25] However, EPA has missed this statutory deadline in multiple years. For 2013, EPA did not even propose a standard until February 2013, with a public comment period running through late March.

[26] For more information, see CRS Report RS22870, Waiver Authority Under the Renewable Fuel Standard (RFS).

[27] The 2010 RFS was revised as part of a final rulemaking implementing the RFS as expanded by EISA, available at http://www.epa.gov/otaq/renewablefuels/420f10007.pdf.

[28] EPA finalized the 2011 requirements in November 2010. EPA, "Regulation of Fuels and Fuel Additives: 2011 Renewable Fuel Standards; Final Rule," 75 Federal Register 76790-76830, Dec. 9, 2010.

[29] EPA finalized the 2012 requirements in December 2011. EPA, "Regulation of Fuels and Fuel Additives: 2012 Renewable Fuel Standards; Final Rule," 77 Federal Register 1320-1358, Jan. 9, 2012.

[30] EPA Proposes the 2013 requirements in February 2013. EPA, "EPA Proposes 2013 Renewable Fuel Standards," EPA-420-F-13-007, January 2013.

[31] This is the number of cellulosic RINs generated under the RFS2 program in 2012 as reported by EPA (as of March 1, 2013) on its "RFS2 EMTS Informational Data" online reporting system at http://www.epa.gov/otaq/fuels/rfsdata/.

[32] See, "EPA Decision on Texas Request for Waiver of Portion of Renewable Fuel Standard (RFS)," EPA 420-F-08- 029, August 2008; at http://www.epa.gov/otaq/renewablefuels/420f08029.htm.

[33] EPA, "Update—2012 Cellulosic Biofuel Standard Mandate Issued," EnviroFlash, mailto:enviroflash@ epa.govFebruary 27, 2013. As part of the news release, EPA announced that since the 2012 mandate was zero, no compliance was necessary and any parties who had already submitted payment for 2012 cellulosic biofuel waiver credits would be issued refunds.

[34] Amanda Peterka, "EPA to File Motion Taking Back 2011 Cellulosic Decision," Greenwire, E&E Publishing, LLC, March 13, 2013.

[35] See CRS Report R42824, Analysis of Renewable Identification Numbers (RINs) in the Renewable Fuel Standard (RFS). Other sources include Robert Wisner, "Renewable Identification Numbers (RINs) and Government Biofuels Blending Mandates," AgMRC Renewable Energy Newsletter, Agricultural Marketing Research Center, Iowa State University, April 2009; or Wyatt Thompson, Seth Meyer, and Pat Westhoff, "Renewable Identification Numbers are the Tracking Instrument and Bellwether of U.S. Biofuel Mandates," EuroChoices 8(3), 2009, pp. 43-50.

[36] For example, this was the case in late 2010. The biodiesel production tax credit of $1.00 per gallon had expired at the end of 2009 and subsequent biodiesel production dropped in 2010, potentially below that needed to meet the combined 2009/2010 mandate. As a result, biomass-based diesel RINs were trading at dramatically higher levels than one year previously. By the second week in December 2010, biomass-based diesel RINs were trading at $0.85 to $0.90 as compared to $0.01 to $0.02 the same week in 2009. However, with the enactment of an extension of the tax credit, biomass-based diesel RIN prices dropped by more than 30% in one week. "RIN Quotes," The Ethanol Monitor, vol. 6, no. 48 (December 20, 2010), p. 12.

[37] This discussion is based on "Renewable Identification Numbers are the Tracking Instrument and Bellwether of U.S. Biofuel Mandates," by Wyatt Thompson, Seth Meyer, and Pat Westhoff, EuroChoices 8(3), 2009.

[38] Ibid., p. 46.

[39] For more information, see CRS Report R42824, Analysis of Renewable Identification Numbers (RINs) in the Renewable Fuel Standard (RFS).

[40] EPA, "Renewable Fuel Standard Program (RFS2) Regulatory Impact Analysis," Assessment and Standards Division, Office of Transportation and Air Quality, EPA-420-R-10-006, February 2010.

[41] See Renewable Fuels Assoc. (RFA), Position Papers at http://www.ethanolrfa.org/pages/position-papers.

[42] DOE, EIA, Annual Energy Review 2013, Table A1, "Total Energy Supply and Disposition Summary," Washington, December 5, 2012, at http://www.eia.gov/forecasts/aeo/er/pdf/tbla1.pdf.

[43] For example, see John M. Urbanchuk (Technical Director, Environmental Economics), Contribution of the Ethanol Industry to the Economy of the United States, white paper prepared for Renewable Fuels Assoc., January 31, 2013.

[44] For example, see Bruce A. Babcock, "High Crop Prices, Ethanol Mandates, and the Public Good: Do They Coexist?" Iowa Ag Review, Vol. 13, No. 2, Spring 2007; and Robert Hahn and Caroline Cecot, "The Benefits and Costs of Ethanol," Working Paper 07-17, AEI-Brookings Joint Center for Regulatory Studies, November 2007.

[45] Chris Hurt, Wally Tyner, and Otto Doering, Department of Agricultural Economics, Purdue University, Economics of Ethanol, December 2006, West Lafayette, IN.

[46] Ronald Steenblik, Biofuels—At What Cost? Government Support for Ethanol and Biodiesel in the United States, Global Subsidies Initiative of the International Institute for Sustainable Development, Geneva, Switzerland, September 2007, p. 37; available at http://www.globalsubsidies.org.

[47] CRS projection based on available data various USDA and DOE sources.

[48] See CRS Report R41985, Renewable Energy Programs and the Farm Bill: Status and Issues.

[49] See CRS Report R42566, Alternative Fuel and Advanced Vehicle Technology Incentives: A Summary of Federal Programs.

[50] World Agr. Outlook Board (WAOB), USDA, World Agr. Supply and Demand Estimates (WASDE), Feb. 8, 2013.

[51] USDA Agricultural Projections to 2022, OCE-2013-1, Office of the Chief Economist, February 2013.

[52] "Many Uncertainties Remain about National and Regional Effects of Increased Biofuel Production on Water Resources," GAO-10-116, U.S. Government Accountability Office, November 2009.

[53] USDA Agricultural Projections to 2022, OCE-2013-1, Office of the Chief Economist, February 2013.

[54] Philip Brasher, "2050 Corn Harvest Will Affect Food, Fuel Policies," Des Moines Register, November 15, 2009.

[55] Paul W. Heisey, "Science, Technology, and Prospects for Growth in U.S. Corn Yields," Amber Waves, vol. 7, no. 4, Economic Research Service, USDA, December 2009.

[56] Chad E. Hart, "Feeding the Ethanol Boom: Where Will the Corn Come From?" Iowa Ag Review, vol. 12, no. 4 (Fall 2006), pp. 4-5.

[57] Bruce A. Babcock and David A. Hennessy, "Getting More Corn Acres from the Corn Belt," Iowa Ag Review, vol. 12, no. 4 (Fall 2006), pp. 6-7.

[58] Michael Duffy and David Correll, "The Economics of Corn on Corn," Integrated Crop Management, IC-498 (1), February 12, 2007.

[59] See CRS Report R41106, Meeting the Renewable Fuel Standard (RFS) Mandate for Cellulosic Biofuels: Questions and Answers, and CRS Report R41460, Cellulosic Ethanol: Feedstocks, Conversion Technologies, Economics, and Policy Options.

[60] Stover is the above-soil part of the corn plant excluding the kernels.

[61] To economically supply field residues to biofuels producers, farm equipment manufacturers likely would need to develop one-pass harvesters that could collect and separate crops and crop residues at the same time.

[62] CRS Report R41106, Meeting the Renewable Fuel Standard (RFS) Mandate for Cellulosic Biofuels: Questions and Answers.

[63] EPA, "Regulation of Fuels and Fuel Additives: 2013 Renewable Fuel Standards; Proposed Rule," Federal Register, Vol. 78, No. 26, February 7, 2013, pp. 9282-9306.

[64] For example see, Advanced Ethanol Council (AEC), "Cellulosic Plants: Industry Progress Report 2012/13," December 19, 2012; and "Visible Progress in Commercialization of Advanced Biofuels, Biobased Products and Renewable Chemicals," Biotechnology Industry Organization (BIO), June 15, 2012.

[65] H. Shapouri, Paul W. Gallagher, Ward Nefstead, Rosalie Schwartz, Stacey Noe, and Roger Conway, 2008 Energy Balance for the Corn-Ethanol Industry, AER No. 846, Office of the Chief Economist, USDA, June 2010; hereinafter referred to as Shapouri et al. (2010).

[66] "Ethanol Energy Balance," Alternative Fuels & Advanced Vehicles Data Center, Dept. of Energy, available at http://www.afdc.energy.gov/afdc/ethanol/balance.html.

[67] David Andress, Ethanol Energy Balances, November 2002.

[68] A key question in evaluating the energy security benefits or costs of an expanded RFS is "what is the definition of energy security." For many policymakers, "energy security" and "energy independence" (i.e., producing all energy within our borders) are synonymous. For others, "energy security" means guaranteeing that we have reliable supplies of energy regardless of their origin. For this section, the former definition is used.

[69] By volume, ethanol accounted for nearly 10% of gasoline consumption in the United States in 2012, but a gallon of ethanol yields only about 68% of the energy of a gallon of gasoline.

[70] This estimate is based on DOE, EIA, Annual Energy Review 2013, Table A1, "Total Energy Supply and Disposition Summary," Washington, December 5, 2012,, and USDA's February 8, 2013, WASDE Report, using comparable conversion rates.

[71] DOE, EIA, Annual Energy Review 2013, Table A1, "Total Energy Supply and Disposition Summary," Washington, December 5, 2012, at http://www.eia.gov/forecasts/aeo/er/pdf/tbla1.pdf.

[72] Calculated by CRS based on EIA's 2013 Energy Outlook and the assumption that the BBD RFS remains fixed at 1.28 bgals through 2022.

[73] Plant-level cost-of-production (COP) data is proprietary and not readily available to the general public. News reports vary widely in their estimates of per-gallon COP for cellulosic biofuels. Furthermore, since commercial production of cellulosic biofuels is as yet minimal, most COP estimates are based on laboratory data which may not be representative of commercial scale costs.

[74] John Whims, "Pipeline Considerations for Ethanol," AgMRC, Sparks Companies, Inc., August 2002.

[75] For a DOE study of the economic feasibility of a hypothetical ethanol pipeline linking large East Coast demand centers with a stable supply from the Midwest, see "Report to Congress: Dedicated Ethanol Pipeline Feasibility Study," March 2010, at http://www1.eere.energy.gov/biomass/pdfs/report_to_congress_ethanol_pipeline.pdf.

[76] See CRS Report R40445, Intermediate-Level Blends of Ethanol in Gasoline, and the Ethanol "Blend Wall."

[77] For details, see EPA online information site, "E15 (a blend of gasoline and ethanol)," at http://www.epa.gov/otaq/regs/fuels/additive/e15/.

[78] "E15 Decision Opens Blend to 2 Out of 3 Vehicles; More Work Yet to be Done," RFA news release, Jan. 21, 2011.

[79] Outdoor Power Equipment Institute (OPEI), Fact Sheet: E-15 Partial Waiver Legal Challenge, December 17, 2010. The case is Alliance of Automobile Manufacturers et. al v. Environmental Protection Agency. In August 2012 the D.C. Circuit held that the petitioners lacked standing because no engine or equipment owners had been injured in fact because little or no E15 had been introduced into commerce.

[80] See EPA online information site, "E15: Misfueling Mitigation Plans," at http://www.epa.gov/otaq/regs/fuels/additive/e15/e15-mmp.htm.

[81] Ibid.

[82] Data is from EIA/DOE's 2013 Annual Energy Outlook. EIA also projects the U.S. national biodiesel transportation fuel market to show slow but steady growth (at about 1% per year) from about 47 bgals in 2012

to nearly 54 bgals by 2022. As a result, RFS BBD's share of the biodiesel transportation fuel market is projected to remain steady at about 2.5% through 2022.

[83] See http://www.chooseethanol.com/pages/e15-station-locations/. Accessed March 11, 2013.

[84] David Sedgwick, Automotive News, January 29, 2007. p. 112.

[85] Note that 15 billion gallons is the corn starch ethanol limit for the expanded RFS in the EISA.

[86] "Ethanol & Gasoline Component Spot Market Prices," OPIS Ethanol and Biodiesel Information Service, January 14, 2013 and March 11, 2013.

[87] "Soaring RIN Values Unleash New Interest in Evolving Market," OPIS Ethanol and Biodiesel Information Service, vol. 10, no. 10 (March 11, 2013), pp. 3-4.

INDEX

#

A

B

C

D

E

F

G

H

I

J

K

L

M

N

O

P

Q

R

T